Economic and Social Implications of Information and Communication Technologies

Yilmaz Bayar
Bandirma Onyedi Eylul University, Turkey

Lina Karabetyan
Independent Researcher, Turkey

IGI Global
PUBLISHER of TIMELY KNOWLEDGE

A volume in the Advances in Wireless Technologies and Telecommunication (AWTT) Book Series

Published in the United States of America by
IGI Global
Information Science Reference (an imprint of IGI Global)
701 E. Chocolate Avenue
Hershey PA, USA 17033
Tel: 717-533-8845
Fax: 717-533-8661
E-mail: cust@igi-global.com
Web site: http://www.igi-global.com

Library of Congress Cataloging-in-Publication Data

Names: Bayar, Yilmaz, 1977- editor. | Karabetyan, Lina, 1983- editor.
Title: Economic and social implications of information and communication technologies / Yilmaz Bayar, and Lina Karabetyan, editor.
Description: Hershey, PA : Information Science Reference, [2023] | Includes bibliographical references and index. | Summary: "The publication will explore the economic and social impacts of ICT development and penetration from a multidisciplinary perspective"-- Provided by publisher.
Identifiers: LCCN 2022039752 (print) | LCCN 2022039753 (ebook) | ISBN 9781668466209 (hardcover) | ISBN 9781668466247 (paperback) | ISBN 9781668466216 (ebook)
Subjects: LCSH: Information technology--Economic aspects. | Information technology--Social aspects. | Knowledge management--Economic aspects. | Knowledge management--Social aspects. | Technological innovations--Economic aspects. | Technological innovations--Social aspects.
Classification: LCC HD30.2 .E25 2022 (print) | LCC HD30.2 (ebook) | DDC 303.48/33--dc23/eng/20220907
LC record available at https://lccn.loc.gov/2022039752
LC ebook record available at https://lccn.loc.gov/2022039753

This book is published in the IGI Global book series Advances in Wireless Technologies and Telecommunication (AWTT) (ISSN: 2327-3305; eISSN: 2327-3313)

British Cataloguing in Publication Data
A Cataloguing in Publication record for this book is available from the British Library.

Advances in Wireless Technologies and Telecommunication (AWTT) Book Series

Xiaoge Xu
University of Nottingham Ningbo China, China

ISSN:2327-3305
EISSN:2327-3313

Mission

The wireless computing industry is constantly evolving, redesigning the ways in which individuals share information. Wireless technology and telecommunication remain one of the most important technologies in business organizations. The utilization of these technologies has enhanced business efficiency by enabling dynamic resources in all aspects of society.

The **Advances in Wireless Technologies and Telecommunication Book Series** aims to provide researchers and academic communities with quality research on the concepts and developments in the wireless technology fields. Developers, engineers, students, research strategists, and IT managers will find this series useful to gain insight into next generation wireless technologies and telecommunication.

Coverage

- Mobile Web Services
- Mobile Technology
- Virtual Network Operations
- Mobile Communications
- Wireless Sensor Networks
- Network Management
- Radio Communication
- Wireless Technologies
- Wireless Broadband
- Broadcasting

IGI Global is currently accepting manuscripts for publication within this series. To submit a proposal for a volume in this series, please contact our Acquisition Editors at Acquisitions@igi-global.com or visit: http://www.igi-global.com/publish/.

The Advances in Wireless Technologies and Telecommunication (AWTT) Book Series (ISSN 2327-3305) is published by IGI Global, 701 E. Chocolate Avenue, Hershey, PA 17033-1240, USA, www.igi-global.com. This series is composed of titles available for purchase individually; each title is edited to be contextually exclusive from any other title within the series. For pricing and ordering information please visit http://www.igi-global.com/book-series/advances-wireless-technologies-telecommunication/73684. Postmaster: Send all address changes to above address.

Titles in this Series

For a list of additional titles in this series, please visit: www.igi-global.com/book-series/advances-wireless-technologies-telecommunication/73684

Modelling and Simulation of Fast-Moving Ad-Hoc Networks (FANETs and VANETs)
T.S. Pradeep Kumar (Vellore Institute of Technology, India) and M. Alamelu (Kumaraguru College of Technology, India)
Information Science Reference • © 2023 • 251pp • H/C (ISBN: 9781668436103) • US $240.00

Role of 6G Wireless Networks in AI and Blockchain-Based Applications
Malaya Dutta Borah (NIT Silchar, Cachar, Assam, India) Steven A. Wright (Georgia State University, USA) Pushpa Singh (Delhi Technical Campus, India) and Ganesh Chandra Deka (RDSDE,DGT, Ministry of Skill Development and Entrepreneurship, Govt of India, India)
Information Science Reference • © 2023 • 320pp • H/C (ISBN: 9781668453766) • US $250.00

Challenges and Risks Involved in Deploying 6G and NextGen Networks
A.M. Viswa Bharathy (GITAM University, Bengaluru, India) and Basim Alhadidi (Al-Balqa Applied University, Jordan)
Information Science Reference • © 2022 • 258pp • H/C (ISBN: 9781668438046) • US $250.00

Achieving Full Realization and Mitigating the Challenges of the Internet of Things
Marcel Ohanga Odhiambo (Mangosuthu University of Technology, South Africa) and Weston Mwashita (Vaal University of Technology, South Africa)
Engineering Science Reference • © 2022 • 263pp • H/C (ISBN: 9781799893127) • US $240.00

Handbook of Research on Design, Deployment, Automation, and Testing Strategies for 6G Mobile Core Network
D. Satish Kumar (Nehru Institute of Engineering and Technology , India) G. Prabhakar (Thiagarajar College of Engineering, India) and R. Anand (Nehru Institute of Engineering and Technology, India)
Engineering Science Reference • © 2022 • 490pp • H/C (ISBN: 9781799896364) • US $360.00

Implementing Data Analytics and Architectures for Next Generation Wireless Communications
Chintan Bhatt (Charotar University of Science and Technology, India) Neeraj Kumar (Thapar University, India) Ali Kashif Bashir (Manchester Metropolitan University, UK) and Mamoun Alazab (Charles Darwin University, Australia)
Information Science Reference • © 2022 • 227pp • H/C (ISBN: 9781799869887) • US $240.00

Handbook of Research on Advances in Data Analytics and Complex Communication Networks
P. Venkata Krishna (Sri Padmavati Mahila University, India)
Engineering Science Reference • © 2022 • 275pp • H/C (ISBN: 9781799876854) • US $380.00

IGI Global
PUBLISHER of TIMELY KNOWLEDGE

701 East Chocolate Avenue, Hershey, PA 17033, USA
Tel: 717-533-8845 x100 • Fax: 717-533-8661
E-Mail: cust@igi-global.com • www.igi-global.com

Editorial Advisory Board

Table of Contents

Preface ... xv

Chapter 1
Information and Communication Technologies and Feminization U Hypothesis: Empirical Analysis for Turkey ... 1
Gülsüm Akarsu, Ondokuz Mayıs University, Turkey

Chapter 2
Using Digital Technology to Enhance Medical Education and Improve the Quality of Patient Care in a Changing Global Healthcare System ... 20
Mitchell Alan Kaplan, Independent Researcher, USA

Chapter 3
Social and Economic Maxims of ICT in Education ... 38
Sonia Bhandari, Rajdhani College, New Delhi, India & University of Delhi, India
Aarti Bhandari, Lilawati Vidya Mandir, Delhi, India

Chapter 4
Impact of Information and Communication Technology on Economic Growth: Evidence From Arabian Peninsula Region Countries ... 59
Mehmet Hilmi Özkaya, Uşak University, Turkey
Abdullatif Almukaddam, Taiz University, Yemen
Maeen Alhuwesh, Taiz University, Yemen

Chapter 5
Exploring the ICT Diffusion and Economic Growth Nexus: Evidence From BRICS Nations ... 73
Oğuzhan Yelkesen, Bandirma Onyedi Eylul University, Turkey

Chapter 6
ICT Penetration and Entrepreneurial Activity: An Empirical Analysis for High Income Countries ... 92
Lina Karabetyan, Independent Researcher, Turkey

Chapter 7
A Detailed Analysis of the Digital Divide and Its Impact on the Development of Countries ... 104
Özlem Yorulmaz, Istanbul University, Turkey

Chapter 8
The Impact of Digitalization on Innovation: Evidence From Eurozone Countries 122
Funda Hatice Sezgin, Istanbul University-Cerrahpasa, Turkey

Chapter 9
The Journey of the Tourism Industry From Digital Marketing to Metaverse Network 134
Betül Garda, Selcuk University Social Science Vocational School, Turkey

Chapter 10
Examining Cryptocurrencies Within the Framework of Sustainability 151
Tolgahan Tuglu, Institute of Banking and Insurance, Turkey
Canan Dağıdır Çakan, Faculty of Financial Sciences, Turkey
Mehmet Hanifi Ateş, Institute of Banking and Insurance, Turkey
Aleyna Uca, Institute of Banking and Insurance, Turkey

Chapter 11
The Impact of Digital Transformation on Personnel Number in the Banking Sector 171
Gokce Tekin Turhan, Nisantasi University, Turkey

Chapter 12
The Impact of Knowledge Management on Organizational Learning in Universities 181
Gamze Sart, Istanbul University-Cerrahpasa, Turkey

Chapter 13
The Effect of Knowledge Sharing on Innovative Behavior in Organizations: A Case of Academic Members 196
Pınar Tokal, Netkent Mediterranean Research and Science University, Cyprus

Chapter 14
Assessment of University Students' Industry 4.0 Conceptual Awareness Levels 209
Gamze Sart, Istanbul University-Cerrahpasa, Turkey

Chapter 15
Business Transformation Projects: The Role of a Transcendent Software Engineering Concept (RoTSEC) 220
Antoine Toni Trad, IBISTM, France

Chapter 16
In Search of Cooperative Compliance: International Compliance Assurance Programme (ICAP) and Turkey 252
Hakkı Odabaş, Yıldırım Beyazıt Üniversitesi, Turkey
İmran Arıtı Erdem, Yıldırım Beyazıt Üniversitesi, Turkey

Compilation of References 277

About the Contributors 313

Index 316

Detailed Table of Contents

Preface xv

Chapter 1
Information and Communication Technologies and Feminization U Hypothesis: Empirical Analysis for Turkey 1
Gülsüm Akarsu, Ondokuz Mayıs University, Turkey

The use and dissemination of information and communication technologies contribute to the socio-economic development of countries. Due to information and communication technology improvements, remote working can help to engage different social groups in the labour force. This chapter aims to analyse the effect of information and communication technologies on women's participation in the labour market, considering other essential factors and testing the validity of the Feminization U hypothesis. For the analysis, the author employed panel data on 12 regions of Turkey over 2013-2020. Findings indicate that female internet usage increases women's participation in the labour market, and the Feminization U hypothesis is invalid. Policymakers should develop measures to improve information and communication technologies services to encourage female labour force participation.

Chapter 2
Using Digital Technology to Enhance Medical Education and Improve the Quality of Patient Care in a Changing Global Healthcare System 20
Mitchell Alan Kaplan, Independent Researcher, USA

The use of digital technology as an effective method of educating the emerging generation of physicians and improving the quality of services that patients receive has become an expanding trend in recent decades. The application of alternative methods of instruction in higher education, such as online degree courses, learning apps on smartphones, and computer-generated virtual reality models that simulate actual clinical situations, combined with frequent use of digital assessment tools, have made diagnosis and treatment of medical conditions online more precise and less time-consuming. These technological breakthroughs have redefined the roles of providers and patients across the spectrum of care. This chapter will present a comprehensive analysis of the beneficial impact that digital technology is having on medical education and the delivery of healthcare services to patients. It will examine how incorporating technological advances such as telemedicine is changing the dynamic of patient care and transforming healthcare systems globally.

Chapter 3
Social and Economic Maxims of ICT in Education 38
Sonia Bhandari, Rajdhani College, New Delhi, India & University of Delhi, India
Aarti Bhandari, Lilawati Vidya Mandir, Delhi, India

ICT in education, in terms of terminology and structural composition, may carry out three basic components, namely education, information, and communication technology. ICT in education has its sole concern of identifying the most suitable and appropriate use of developed technology (both hardware and software) for serving the educational needs of society at a particular time and place. Education using ICT tools also aids in expanding educational opportunities to socially vulnerable section of people such as low-income groups, women, aged, and differently abled. This effectiveness of ICT in education depends on the presence or absence of the conditions and elements, facilitating or obstructing the proper flow of information and technological methods. Proceeding from simple to complex always provides an appropriate learning sequence. Hence, a wise teacher cares for a beautiful integration of psychological principles to logical organisation of sequence of activities in his teachings for social and economic growth of the community altogether.

Chapter 4
Impact of Information and Communication Technology on Economic Growth: Evidence From Arabian Peninsula Region Countries 59
Mehmet Hilmi Özkaya, Uşak University, Turkey
Abdullatif Almukaddam, Taiz University, Yemen
Maeen Alhuwesh, Taiz University, Yemen

The present study aims at investigating the impact of information and communication technology (ICT) on economic growth of countries in the Arabian Peninsula region through an econometric study of seven countries over the period 1996-2020. To achieve the objectives of the study, the panel data was based on three independent variables, namely the number of mobile cellular subscriptions per 100 people, number of fixed (landline) telephone subscriptions per 100 people, internet usage (% of population), and the economic growth rate as a dependent variable. The results show that within the framework of static analysis the fixed effects model is the most appropriate and is evidence of the existence of fixed individual differences between the countries of the Arabian Peninsula as regards the impact of ICT on economic growth. It also shows that the mobile phone index has a positive and significant impact, and a negative and significant impact on economic growth is shown to be related to the internet index.

Chapter 5
Exploring the ICT Diffusion and Economic Growth Nexus: Evidence From BRICS Nations 73
Oğuzhan Yelkesen, Bandirma Onyedi Eylul University, Turkey

In the last few decades, there has been a surge of interest in the effects of information and communication technologies (ICTs) on economic growth. Although ICT diffusion affects economic growth in many ways, the magnitude of ICT penetration is not time-invariant and depends on many factors. Using pooled mean group (PMG) panel ARDL model, this study aims to explore the nexus between ICT penetration and economic growth for BRICS nations over 2003-2020. The findings show that, in the short run, fixed broadband and mobile cellular subscriptions have a positive effect on economic growth, while both R&D expenditure and ICT goods import have a negative insignificant effect. In the long run, however, the effect of both R&D expenditure and ICT goods import turns out to be significantly positive, whereas fixed broadband and mobile cellular subscriptions have a negative effect on economic growth. The results have significant implications for policymakers and practitioners who are willing to harness ICTs to improve the economic and social conditions that would lead to higher economic growth.

Chapter 6
ICT Penetration and Entrepreneurial Activity: An Empirical Analysis for High Income Countries ... 92
Lina Karabetyan, Independent Researcher, Turkey

Entrepreneurial activity is a critical factor for performance of firms and economies. Therefore, firms and governments try to create an entrepreneurial environment given positive economic and social gains of entrepreneurship. Furthermore, governments stimulate the entrepreneurial activities through financial and legal incentives. This study investigates the mutual interaction between ICT penetration and entrepreneurial activity in 13 high income countries for the 2002-2018 period by a panel causality test which considers the cross-sectional dependence and heterogeneity. The results of panel level causality analysis reveal a bidirectional causal interaction between ICT penetration and entrepreneurial activity. However, the results of country level causality analysis uncover that interaction between ICT penetration and entrepreneurial activity differs among higher income countries depending on country specific characteristics.

Chapter 7
A Detailed Analysis of the Digital Divide and Its Impact on the Development of Countries............ 104
Özlem Yorulmaz, Istanbul University, Turkey

Information and communication technologies (ICT) lead to significant social and economic changes in the lives of individuals. In parallel with this change, new criteria are needed when evaluating nations' development levels. In this study, the Digital Economy and Society Index (DESI), the Women in Digital Index (WID), and Rural Digital Index (RDI) proposed by the European Commission were considered as the proxies of ICT. The effects of these indexes on human development and gender inequality were studied. Findings showed that there is a digital divide among EU countries. The effect of DESI on HDI is higher in developed countries than in developing countries. In addition, it was observed that the increase in the WID value affected countries with high gender inequality more, and the RDI has an impact on the development of countries.

Chapter 8
The Impact of Digitalization on Innovation: Evidence From Eurozone Countries 122
Funda Hatice Sezgin, Istanbul University-Cerrahpasa, Turkey

Today, the relationship between science and technology is one of mutual interaction, with technology developing on the basis of science, and developing technology opening up new fields of science, which in turn play a driving role in new technologies and innovative activities. The decisive role of science and technology in the process of economic development increases day by day and it is seen that all countries are in an increasingly intense search for science and technology policy. In the rapidly globalizing world economy, countries' conscious efforts and policy-making searches are intensifying on scientific and technological advances, which are among the main factors determining the competitive advantage of nations. In this study, the impact of digitalization on innovation for Euro Area countries is determined using panel regression. As a result of the analysis, it was determined that digitalization indicators have a statistically significant positive relationship on innovation.

Chapter 9
The Journey of the Tourism Industry From Digital Marketing to Metaverse Network...................... 134
Betül Garda, Selcuk University Social Science Vocational School, Turkey

Metaverse virtual tourism market, which is frequently mentioned as a result of this change, will appear as an alternative tourism type that allows people living in different countries to travel to the same country or different countries in the virtual world. Consequently, it is possible that the Metaverse virtual tourism market will become widespread as an alternative niche market to traditional tourism in the near future. For this reason, it is anticipated that the conceptual framework presented by the department about digital tourism marketing and metaverse network will contribute to the sector and future academic studies.

Chapter 10
Examining Cryptocurrencies Within the Framework of Sustainability.. 151
Tolgahan Tuglu, Institute of Banking and Insurance, Turkey
Canan Dağıdır Çakan, Faculty of Financial Sciences, Turkey
Mehmet Hanifi Ateş, Institute of Banking and Insurance, Turkey
Aleyna Uca, Institute of Banking and Insurance, Turkey

Cryptocurrencies have been attracting a significant amount of attention in the world since they were first launched in 2009. Pretending to be a decentralized finance solution, it brought out a new era in technology called blockchain. Even though the benefits did not come into action in daily routines for many to be aware of, the market and its variety kept growing. On the other hand, there are also a lot of concerns and unpredictability about the future of this technology. Especially the high energy consumption while generating blocks for mining cryptocurrencies and completing transactions is commonly being criticised. In this study, blockchain technology and the basics of mining and validation procedures such as proof of work (PoW) and proof of stake (PoS) processes will be explained, and the environmental effects of bitcoin mining will be investigated. In the perspective of environmental sustainability of cryptocurrencies, the improvement in usage of renewable energy and its side benefits will be overviewed for a better prediction on the blockchain technology future.

Chapter 11
The Impact of Digital Transformation on Personnel Number in the Banking Sector 171
Gokce Tekin Turhan, Nisantasi University, Turkey

Digital transformation affects the banking sector as it does every sector and enables significant innovations and changes in the banking and finance market. In this sense, it aims to ensure that banking transactions can be carried out quickly, securely, and at less cost, and to achieve the highest level of satisfaction in customer service. New solutions should continue to be developed to enable customers to benefit from digital banking products in the fastest way possible. There are also negative effects of digitalization. When evaluated within the framework of the banking sector, the increasing number of digital services and the number of customers using these services may lead to a decrease in the number of bank personnel. In the study, quantile regression was applied for the period of 2010.Q1-2020.Q4. This study examines the changes that may occur in the number of personnel as a result of the digitalization process in the banking sector and determines the decreasing effect of digitalization on the number of personnel.

Chapter 12
The Impact of Knowledge Management on Organizational Learning in Universities 181
Gamze Sart, Istanbul University-Cerrahpasa, Turkey

In today's information and technology-driven markets, knowledge has an important role to play in gaining and maintaining competitive advantage and improving organizational performance. Since knowledge is the most important competitive advantage tool, effective management of knowledge will increase organizational performance. The knowledge management process is a process that includes activities related to the systematic acquisition, transformation, use, and protection of knowledge in order to better achieve organizational goals. This process develops depending on technology, culture, organizational structure, and human factors. The aim of this study is to determine the effects of knowledge management on organizational learning for 650 academicians working at universities in Istanbul. As a result of the structural equation model, a positive 68.1% increasing relationship was determined between the knowledge management and organizational learning.

Chapter 13
The Effect of Knowledge Sharing on Innovative Behavior in Organizations: A Case of Academic Members 196
Pınar Tokal, Netkent Mediterranean Research and Science University, Cyprus

Knowledge sharing is the transfer of knowledge owned by individuals to other individuals within the organization within the scope of knowledge management. Knowledge sharing is vital for all organizations that aim for success. The concept of knowledge includes concepts such as innovation, development, research, learning, effectiveness, progress, and replacing what is wrong with what is right. Today, one of the common characteristics of developed countries is the updating of information and the effective use of the information network. Knowledge and the transfer and sharing of this knowledge between individuals affect the development of societies and countries, as well as businesses and organizational structures within businesses. The aim of this study is to determine the effect of knowledge sharing on innovative behavior for 890 academics. As a result of the structural equation model analysis, it is seen that knowledge sharing is positively and significantly related to innovative behavior (β=0.593; p<0.01).

Chapter 14
Assessment of University Students' Industry 4.0 Conceptual Awareness Levels 209
Gamze Sart, Istanbul University-Cerrahpasa, Turkey

The results of the intensive digital transformation in the world are manifest in all industries. Developments within the scope of Industry 4.0 in educational organizations change the structure and functioning of educational institutions on the one hand, while affecting and changing student competencies on the other. The aim of this study is to evaluate the Industry 4.0 conceptual awareness levels of 890 students from the Faculty of Engineering, Faculty of Economics, and Faculty of Education Sciences at Istanbul University-Cerrahpaşa in terms of gender, grade level, and faculty. As a result of the analysis, it was seen that there was a statistically significant difference in terms of gender, faculty, and grade level. It was determined that male students, engineering faculty students, and fourth year students had higher conceptual awareness of Industry 4.0.

Chapter 15
Business Transformation Projects: The Role of a Transcendent Software Engineering Concept (RoTSEC) 220
Antoine Toni Trad, IBISTM, France

The hyper evolution of technologies is a major problem for transformation projects because such projects take a long time to terminate. That is why there is a need to find transcendent technological artefacts for all technology generations. The role of a transcendent software engineering concept (RoTSEC) is central for implementation projects in general and is also especially crucial for business transformation projects (or simply projects) because transformation phases incur major changes in the existing sets of the archaically defined requirements, development, and platform models. Software engineering (SE) is probably the riskiest part of a project because it consists of many related complex factors and dependencies, for example, the need for SE's artefacts to interact between various actors like project management, business users, business architects, implementation developers, and other project actors.

Chapter 16
In Search of Cooperative Compliance: International Compliance Assurance Programme (ICAP) and Turkey.......... 252
Hakkı Odabaş, Yıldırım Beyazıt Üniversitesi, Turkey
İmran Arıtı Erdem, Yıldırım Beyazıt Üniversitesi, Turkey

International Compliance Assurance Programme (ICAP) of OECD, a brand-new application, is the international reflection of the tax compliance expectations rising over cooperative compliance. ICAP is an application in which the tax risks of MNE groups are evaluated by bringing together the tax administrations of more than one country in which they operate and the relevant MNE group, and it has the capacity to create positive outputs at the point of ensuring voluntary tax compliance based on the establishment of tax certainty and consistency. In this study, the charactetistics, composition, and function of ICAP is revealed with the concept of cooperative compliance by comparing similar concepts with ICAP. Based on these elements, the chapter discusses Turkey's future participation in the programme.

Compilation of References 277

About the Contributors 313

Index 316

Preface

Enormous developments have been made in the field of information and communication technologies (ICT) during the past four decades, and ICT has spread rapidly in the world with the contribution of globalization and become a significant part of our daily life. The ICT development and penetration are continuing to affect all aspects of the societies and have been also playing an interdisciplinary role as it has a significant impact on education, environment, economics, management, energy, health, and medical care since its adoption for the improvement of socio-economic and cultural circumstances. This study explores the economic and social implications of the ICT development and penetration from a multidisciplinary perspective.

Chapter 1 investigates the effect of information and communication technologies on women's participation in the labor market, considering other essential factors and testing the validity of the feminization U hypothesis in 12 regions of Turkey over the 2013-2020 period and reveals that female internet usage increases women's participation in the labor market, and the Feminization U hypothesis is invalid. On the other hand, Chapter 2 researches the role of digital technology in enhancing medical education and improving the quality of patient care in the global healthcare system.

Chapter 3 examines the social and economic maxims of ICT in education and suggests that use of ICT in education helps to expand the educational opportunities to socially vulnerable section of people such as low-income groups, women, aged, and differently abled and effectiveness of ICT employment in education depends on the presence or absence of the conditions and elements, facilitating or obstructing the proper flow of information and technological methods. Chapter 4 investigates the impact of information and communication technologies on economic growth in the countries from the Arabian Peninsula region over the 1996-2020 period and uncovers a positive and significant effect of mobile phone index on economic growth, but a negative effect of internet index and fixed telephone variable on economic growth.

Chapter 5 investigates the impact of ICT on economic growth in BRICS economies for the 2003-2020 period through pooled mean group (PMG) panel ARDL model and reveal that in the short run, fixed broadband and mobile cellular subscriptions have a positive effect on economic growth, while both R&D expenditure and ICT goods import have a negative insignificant effect. In the long run, however, the effect of both R&D expenditure and ICT goods import turns out to be significantly positive, whereas fixed broadband and mobile cellular subscriptions have a negative effect on economic growth. On the other hand, Chapter 6 analyzes the mutual interaction between ICT penetration and entrepreneurial activity in 13 high income countries for the 2002-2018 period through causality test and finds a bidirectional causal interaction between ICT penetration and entrepreneurial activity at panel level causality analysis, but the interaction between ICT penetration and entrepreneurial activity differs among higher income countries depending country specific characteristics at country level causality analysis.

Chapter 7 examines the effect of various ICT indicators including Digital Economy and Society Index (DESI), the Women in Digital Index (WID), and Rural Digital Index (RDI) on human development and gender inequality and discovers a digital divide among EU countries. Furthermore, the study reveals that the effect of DESI on HDI is higher in developed countries than in developing countries. and the increase in the WID value affects the countries with high gender inequality more, and the RDI has an impact on the development of countries. Chapter 8 investigates the impact of digitalization on innovation in the Eurozone countries through regression analysis and uncovers a positive impact of digitalization indicators on innovation.

Chapter 9 examines the process from digital marketing to metaverse in tourism industry and suggests that metaverse virtual tourism market will become widespread as an alternative niche market to traditional tourism in the near future. On the other hand, Chapter 10 explores the sustainability of cryptocurrencies and the environmental effects of bitcoin mining in terms of carbon dioxide emissions and reveals a causality from bitcoin hash rate to the total hydroelectric consumption in the United States and insignificant causality from Bitcoin hash rate to the total consumption of electricity obtained from solar energy and a causality from the hash rate to the amount of carbon dioxide in the atmosphere and reaches the conclusion that bitcoin miners' use or non-use of renewable energy sources in the United States affects the total renewable energy production.

Chapter 11 examines the impact of digitalization process on personnel number in the banking sector over the 2010.Q1-2020.Q4 duration through quantile regression and discovers a negative impact of digitalization on the number of personnel.

Chapter 12 analyzes the impact of knowledge management on organizational learning in sample of 650 academicians working at universities in Istanbul through structural equation model and reveals a positive relationship between knowledge management and organizational learning. Chapter 13 examines the effect of knowledge sharing on innovative behavior in organizations in sample of 890 academics through structural equation model and discovers a positive relationship between knowledge sharing and innovative behavior.

Chapter 14 evaluates the university students' Industry 4.0 conceptual awareness levels in sample of 890 students from the Faculty of Engineering, Faculty of Economics and Faculty of Education Sciences at Istanbul University-Cerrahpaşa in terms of gender, grade level and faculty and reveals a statistically significant difference in terms of gender, faculty and grade level and male students, engineering faculty students and fourth year students have higher conceptual awareness of Industry 4.0. On the other hand, Chapter 15 explains the role of a transcendent software engineering concept in business transformation projects. Lastly, Chapter 16 explains e characteristics, composition and function of International Compliance Assurance Programme (ICAP) by comparing similar concepts with ICAP.

Lina Karabetyan
Independent Researcher, Turkey

Chapter 1
Information and Communication Technologies and Feminization U Hypothesis:
Empirical Analysis for Turkey

Gülsüm Akarsu
https://orcid.org/0000-0002-4877-1969
Ondokuz Mayıs University, Turkey

ABSTRACT

The use and dissemination of information and communication technologies contribute to the socio-economic development of countries. Due to information and communication technology improvements, remote working can help to engage different social groups in the labour force. This chapter aims to analyse the effect of information and communication technologies on women's participation in the labour market, considering other essential factors and testing the validity of the Feminization U hypothesis. For the analysis, the author employed panel data on 12 regions of Turkey over 2013-2020. Findings indicate that female internet usage increases women's participation in the labour market, and the Feminization U hypothesis is invalid. Policymakers should develop measures to improve information and communication technologies services to encourage female labour force participation.

INTRODUCTION

Information and communication technologies (ICTs) have spread rapidly all over the World during the last 30 years (Valberg, 2020). For example, as shown in Figure 1, the percentage of internet users in the world population has increased from 0.05% in 1990 to nearly 60% in 2020 globally (The World Bank, 2022a). The use and dissemination of ICTs contribute to the socio-economic development of countries through providing rapid information dissemination, easy access to information, cross-border knowledge, and technology transfer, creating job opportunities, increasing productivity, enabling online services related to education, health, finance, and government, among many others and remote working style-based

DOI: 10.4018/978-1-6684-6620-9.ch001

Figure 1. Percentage of individuals using the internet in the population and female labor force participation (FLFP) rate
Source: Own construction using data from World Bank World Development Indicators (The World Bank, 2022a, 2022b)

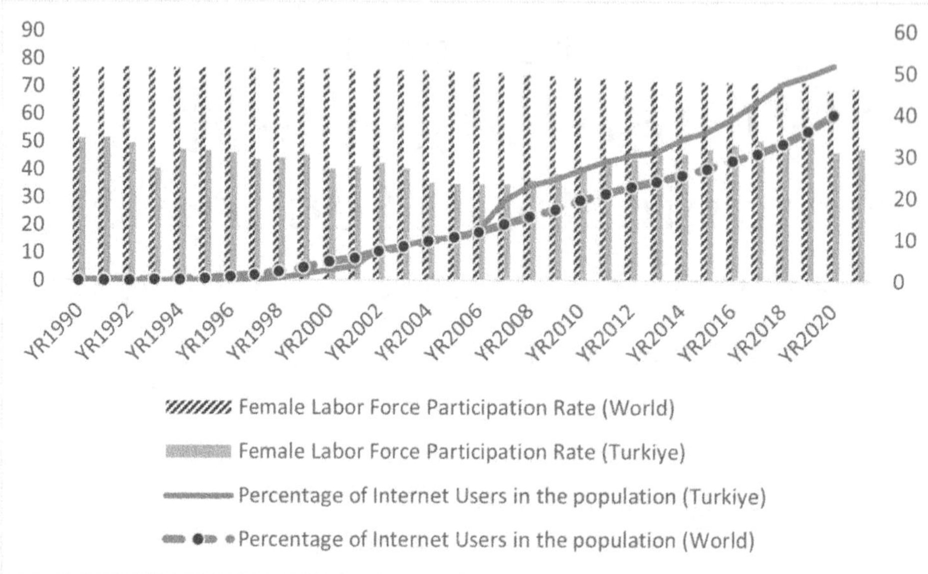

employment opportunities, digitalising economy, and globalising skills (Tüzemen, Barış-Tüzemen, & Çelik, 2021; Valberg, 2020; Watson, Corliss, & Le, 2018). Remote working has been widely discussed not only starting with the emergence of the Covid-19 pandemic but because of technological improvements which has made it possible. However, the discussion has intensified following the pandemic. Due to Information and Communication Technology (ICT) improvements, remote working can help engage different social groups in the labour force. One such group is women. In societies, social norms, the structure of the country, and gender-based labour division may impede women's active involvement in formal economic activities in addition to many other factors (Tüzemen et al., 2021). ICT use can increase female labour force participation (FLFP) by reducing the effect of these social, cultural, and religious obstacles, removing time and mobility constraints, and increasing productivity and output (Valberg, 2020; Watson et al., 2018). It can also reduce the impact of glass ceilings by providing innovative solutions for better positions related to women's employment and, therefore, ensure gender-balanced career development and fewer earning inequalities (Brussevich et al., 2018). There are many benefits of ICT use that induce engagement of women in the labour force, such as easy information access, job creation, lower transaction costs, and promotion of inclusiveness, transparency, flexibility, and innovation in the labour market (Ngoa & Song, 2021; Suhaida, Nurulhuda, & Yap, 2013; Valberg, 2020). However, in many G20 countries, there is evidence of a significant digital gender gap which limits benefits obtained from digitalisation and ICT use (Organisation for Economic Co-operation and Development [OECD], 2018).

Moreover, there are various factors affecting women's participation in the labour market shown by previous literature, such as economic development, financial development, noneconomic factors (for example, religion, governance, social norms, democracy, culture), household appliance ownership, education level and the fertility rate among many others (Ngoa & Song, 2021). Economic growth was shown to affect FLFP by previous studies. Some studies showed a U-shaped relation, while others found that the relation is inverted U-shaped (Ngoa & Song, 2021). The U-shaped relation is known as the feminisation U hypothesis, in which the structural changes, fertility changes, and gender-based educational differences are shown to be reasons (Uberti & Douarin, 2022). According to this hypothesis, as economies have become more service sector-oriented and the percentage of industry share declines, FLFP increases due to non-stigma job creation (Uberti & Douarin, 2022). This stigma is related to some societies' gender roles attached and cultural norms. Based on the labour intensiveness of economic activities, the gender wage gap may change, affecting FLFP (Uberti & Douarin, 2022). As economies grow, become capital-intensive, and economic structure changes, one may expect higher FLFP because of higher and more gender-equal wages offered to women if gender-neutral higher educational opportunities are provided. On the other hand, higher national income may defer women to work as there will be less need for women to participate in the labour force.

FLFP may contribute to economic growth and welfare improvement of countries by providing economic independence and freedom to women and improving women's well-being (Valberg, 2020). The United Nations Women (2022) have discussed that ensuring gender equality and empowerment of women is integral to all of the Sustainable Development Goals (SDGs) and contribute to achieving these goals of the 2030 Agenda for Sustainable Development. Moreover, various actions have been taken to reduce the gender gap in the labour market. For example, by considering "full economic and social participation of women" as a fundamental goal in their agenda in 2012, G20 economies have committed to reducing the gender gap in the labour force participation rate by 25% in 2025 in 2014 (OECD, 2018). However, there is a tendency for women to be less active in the labour market compared to men, although due to various factors such as a decline in fertility rates, better health conditions, infrastructural improvement and higher educational attainments of women, this trend has begun to change in favour of women as discussed by Valberg (2020). Figure 1 illustrates that between 1990 and 2021, the FLFP rate has slightly declined from 51% to 46% globally (The World Bank, 2022b). Reaching the desirable level of women's participation in the labour market is essential for all countries' sustainable development. For Turkiye, similar to global trends, as demonstrated in Figure 1, the FLFP rate has declined over the last 30 years. This rate decreased to 32% in 2021 from its 1990's level of 34% (The World Bank, 2022b). On the other hand, following the global trends but with higher momentum in Turkiye, the percentage of internet users has increased from 0,009% in 1993 to 78% in 2020 (Figure 1) (The World Bank, 2022a). Nevertheless, among the G20 countries, Turkiye is one of the countries having a higher gender gap in internet usage, with 16% in 2017 (OECD, 2018).

The main aim of this study is to investigate the effect of ICT usage on FLFP by considering additional factors and testing the validity of the Feminization U hypothesis. For this investigation, panel data on 12 regions of Turkey was employed over 2013-2020. The author shows the favourable effect of female internet usage on FLFP and the invalidity of the Feminization U hypothesis for Turkiye.

The chapter is organised as follows. After this brief introduction, the background section discusses both the theoretical studies and empirical literature on the feminisation U hypothesis and the impact of ICT usage on FLFP. In the methodological issues and empirical results section, the author presents the model, method, information on data and empirical results. Lastly, solutions, recommendations and future research directions are followed by the conclusion section.

Background

It is crucial to analyse the determinants of FLFP as it contributes to the socio-economic development of countries, including economic growth, inequality decrease, poverty reduction, and gender equality. By analysing these factors, one may determine the factors which can enhance FLFP and guide policymakers to take necessary actions. There are many theoretical contributions to explain the factors for FLFP, such as the Theory of Time Allocation, Theory of Marriage, Economics of Marriage, and Theory of Investment in Human Capital, as noticed by Suhaida et al. (2013). According to the first one, as leisure is a normal good, the wage increase will increase the labour supply as leisure time becomes relatively expensive. In the second one, marital status determines FLFP. Economics of Marriage presumes the determinants as marital status, spousal help, religiosity, culture, income, and education. Education of women, husbands, and all close relatives are taken as factors influencing FLFP in the Theory of Investment in Human Capital.

According to Ngoa and Song (2021), related to the analysis of FLFP, the literature dates back to the 1960s. They mentioned the study performed by Mincer (1963) as the first study dealing with this issue. In the literature, there are various studies analyzing the women's participation in the labor market and its determinants (for example, Adshade & Keay, 2010; Aldan, 2021; Chen, Qin, & Zhao, 2021; Genre, Salvador, & Lamo, 2010; Jayachandran, 2021; Lv & Yang, 2018; Morris, 2022; Nam, 1991; Nishikawa & Higuchi, 1981; Osundina, 2020; Saure & Zoabi, 2014; Schwarz, 2012; Vega & Elhorst, 2017; Ward & Pampel, 1985). They showed the importance of various economic, socio-demographic and cultural factors, such as economic growth, exchange rate, poverty, investment, real wages, working hours, taxation, trade liberalisation, trade openness, export orientation, foreign direct investment, private and public health care expenditures, government subsidies, transportation, access to electricity, electrification, ICTs, economic structural changes, informal sector, unemployment, urbanisation, demand for female labor, economic crises, public assistance, workers' remittances, employment of foreign maids, home appliance ownership, land restrictions, natural resources, geography, income inequality, spatial mismatch, migration, social norms, laws, democracy, institutions, fertility, health status, infant mortality, marital status, age, educational level, household's economic status, children number, family structure, ethnicity, racial differences, caregiving time, old and youth dependency ratios, social networks, religion, wars, terrorism, political participation of women, etc. However, as Ngoa and Song (2021) suggested, there is not a sufficient number of studies investigating the effect of ICT usage and dissemination. This may be related to the digital divide at all levels of geography and gender, as there are inequalities in accessing available technology. For example, "Globally, women are 23% less likely than men to access the Internet" (Tüzemen et al., 2021, p. 127). Therefore, studies may focus on first technology access to analyse its effect. According to Ngoa and Song (2021), there are two channels through which ICT use may enhance FLFP.

One channel is financial development. ICT dissemination may increase financial access and reduce information asymmetry and transaction costs by decreasing moral hazard and adverse selection impacts. Financial development is expected to influence labour markets through its effects on economic growth. Also, as the financial inclusion gap decreases between men and women, women's entrepreneurship and labour force participation has been shown to increase, as suggested by Ajide (2021) and Watson et al. (2018). As economies are financially developed, this may lead to structural changes. These structural changes may also increase FLFP. For example, digital platforms facilitate access of small and medium-sized enterprises to global markets and the functioning of small household businesses (Watson et al., 2018). Another channel is women's education. ICT use leads to easy access to educational resources, which may lead to higher educational attainment by women (Valberg, 2020). Therefore, FLFP may increase due to easy access to educational resources through ICT, providing them with better skills, qualifications, and abilities. Valberg (2020) mentioned three major channels for ICTs' effect on FLFP: information, new jobs and time-saving. Information channel contributes to job searching and matching, reduction of discrimination in the labour market resulting from asymmetric information, and elimination of structural constraints (for example, related to access to educational resources). Online job platforms reduce search costs for women by providing up-to-date information related to jobs (Valberg, 2020).

For a panel of 48 African countries over 2001-2017 using a dynamic panel data model estimated by system generalised method of moments, Ngoa and Song (2021) showed that the favourable effect of ICT usage (mobile phone and internet usage) on FLFP increases with women's education level and financial development. Their findings further indicate positive impacts of female education, female population, real GDP per capita, financial development, trade openness, and noneconomic factors (the rule of law, democracy, female political participation, and political stability). In contrast, the adverse effect of fertility rate was shown. Based on sector-level analyses, among industry, agriculture and service sectors, the industrial sector was found to benefit from ICT use at most to increase FLFP. Valberg (2020) also found positive effects of ICTs' use on women's participation in the labour force, which further increased with the country's economic development level for a panel of 156 countries over 1991-2014, but with a lower impact in developing countries and higher effect of internet use compared to mobile phone. Watson et al. (2018) showed that internet use contributed to 80% of the FLFP in Indo-Pacific countries from 2000 to 2016. They also found favourable impacts of free information flow, internet freedom, GDP per capita, urbanisation, agriculture's share, and government expenditure share. Baliamoune (2021) showed the favourable impact of ICTs' use interacted with trade openness on FLFP using panel data on 91 developing and emerging countries over 1990-2018. The author also reported nonlinear effects (U shaped) of trade openness and access to electricity. For 42 Sub-Saharan African countries between 2004 and 2014, S. A. Asongu and Odhiambo (2022) found that the favourable impact of ICT use can be obtained when the fixed broadband subscription rate is above 9.187 per 100 people. Using the panel data on the same group of countries over the same period, S. A. Asongu and Odhiambo (2020) calculated minimum threshold levels of ICTs usage to increase FLFP by considering income inequality. Favourable impacts of ICTs use and digitalisation were also reported and discussed by Alves and Steiner (2017), Asongu, Amankwah-Amoah, Nting, and Afrifa (2021), Atasoy, Banker, and Pavlou (2021), Dettling (2017), Efobi, Tanankem, and Asongu (2018), Goyal (2011), Krieger-Boden and Sorgner (2018), and Viollaz and Winkler (2022). On the other hand, Samargandi, Al Mamun, Sohag, and Alandejani (2019) showed that the undesirable impact of ICT diffusion could be eliminated by financial development.

For 12 African countries over 2005-2016, Ajide (2021) found a U-shaped relation between financial inclusion and FLFP. Thus, financial inclusion increases women's labour force participation after some level of financial inclusion. Also, there are statistically significant positive moderating impacts of information and communication technology use and trade openness. Democracy was shown to have a statistically significant adverse effect on FLFP. All other control variables were found to have a statistically insignificant impact, such as inflation, foreign direct investment, remittances, female unemployment, financial crises, and economic growth. Financial inclusion, trade openness, and ICT infrastructure improved FLFP. The author also shows that the results are robust by using different dependent variables (the female employment rate) and the GLS estimation technique except for inflation, growth, and democracy, which were found to statistically significantly and positively affect the female employment rate. In contrast, the female unemployment rate decreases the female employment rate statistically significantly.

However, Tüzemen et al. (2021) showed no relation between ICTs usage and FLFP in Turkey using the ARDL model between 1988 and 2018.

Uberti and Douarin (2022) showed the validity of the Feminization U hypothesis, especially for the countries whose ancestors were based on plough agriculture, analysing global panel data on 169 countries over 1990–2013. Tam (2011), Suh (2017) and Choudhry and Elhorst (2018) also present supportive results for a panel of 130 countries over 1950-1980, South Korea over 1980-2014 and a panel of 40 countries over 1960-2005, respectively. Some studies, such as Gaddis and Klasen (2014) and Klasen (2019), empirically find little or no support for the Feminization U hypothesis.

METHODOLOGICAL ISSUES AND EMPIRICAL RESULTS

This section discusses methodological issues and presents empirical results.

Methodology and Data

The author gives information related to the methodology and data in this section.

Methodology

This study used panel data techniques for the investigation of ICT usage's impact on FLFP in 12 regions[1] of Turkiye (NUTS-1 level) between 2013 and 2020. The author employed the dynamic panel data model given in equation (1) for each region i=1,…,12 and year t=2013,...,2020 considering persistency in FLFP and regional heterogeneity;

$$\begin{aligned} \boldsymbol{flfp}_{i,t} = {} & \alpha_i + \beta_1 \boldsymbol{flfp}_{i,t-1} + \beta_2 \boldsymbol{lrealgdppc}_{i,t} + \beta_3 \boldsymbol{lrealgdppc}_{i,t}^2 \\ & + \beta_4 \boldsymbol{femaleinternet}_{i,t} + \beta_5 \boldsymbol{inflation}_t + \beta_6 \boldsymbol{totalagedependency}_{i,t} \\ & + \beta_7 \boldsymbol{financialaccess}_{i,t} + \beta_8 \boldsymbol{netmigrationrate}_{i,t} + \beta_9 \boldsymbol{lgini}_{i,t} + \beta_{10} \boldsymbol{unemployment}_{i,t} + \boldsymbol{u}_{i,t} \end{aligned} \quad (1)$$

$u_{i,t}$ indicates the error term and $u_{i,t}$~$iidN(0, \sigma 2)$. The region-specific fixed effects (FE) are shown by αi. Equation (1) explains FLFP rate (flfp) as a function of the lagged value of FLFP rate (flfp-1), real income (lrealgdppc), square of lrealgdppc (lrealgdppc2), the proportion of female internet users (femaleinternet), the square of femaleinternet (femaleinternet2), inflation rate (inflation), the total age dependency ratio (totalagedependency), financial access (financialaccess), net migration rate (netmigrationrate), Gini coefficient (lgini), and unemployment rate (unemployment).

A dynamic model is employed to consider the persistency in FLFP. Country and time fixed effects are added to control for possible time-invariant heterogeneity among countries and common shocks (such as changing trends and economic crises) over time. Therefore, by using this model, one can eliminate serial correlation and cross-sectional correlation in error terms. Suppose the coefficient on lagged flfp ($\beta 1$) is statistically significant and close to 1. In that case, it shows high persistency in FLFP, implying persistency in norms restricting women's participation in the labour market (Uberti & Douarin, 2022). A priori, one may expect the following effects of these factors on the FLFP rate. According to the feminisation U hypothesis, as economies grow in the initial stages of economic development due to the higher industry share, women's participation in the labour market declines. However, FLFP increases due to economic structural changes towards a more service sector-oriented economy. If $\beta 2<0$ and $\beta 3>0$, then the validity of the feminisation U hypothesis can be shown. In some economies, this U-shaped relation may not occur because higher economic growth may not increase the need for an additional labour force after a certain level of economic growth. FLFP may not be affected.

ICT improvements may lead to a higher level of FLFP by decreasing the effect of various impediments related to social norms, the structure of the country, and gender-based labour division, as examples (Tüzemen et al., 2021). Also, by providing easy information access, job creation, and promotion of inclusiveness, transparency, flexibility, and innovation in the labour market, ICT may contribute to the increase in the FLFP rate (Ngoa & Song, 2021; Suhaida, Nurulhuda, & Yap, 2013). However, the advanced technological and technical systems may be biased towards male employment as they can access more straightforward technological advancements than women, thus increasing male labour force participation. Therefore, this study considers the proportion of female internet users as an ICT improvement to overcome the so-called gender digital divide (Valberg, 2020). One may expect $\beta 4>0$. Price adjustments can be considered by including the inflation rate (Ajide, 2021). Moreover, as the cost of living increases, this may also increase FLFP to compensate for real wage declines ($\beta 5>0$). The total age dependency ratio considers both old and youth dependency ratios. As the total age dependency ratio increases, one may expect lower flfp ($\beta 6<0$) b$_{e}$cause in most societies and most of the time, women are responsible for the child and elderly care.

There are two effects for financial development: income and substitution (Ajide, 2021). Because of financial development, countries' incomes increase, and due to this income effect, flfp may decrease ($\beta 7<0$). $_{I}$f women's wages increase due to financial development, flfp may increase ($\beta 7>0$) be$_{c}$ause of the substitution effect. Migration from rural areas may also increase FLFP ($\beta 8>0$) to $_{m}$eet the labour demand of industry (Tüzemen et al., 2021). Income inequality may increase flfp ($\beta 9>0$) to d$_{e}$crease this inequality. Nevertheless, it may also decrease it ($\beta 9<0$) becau$_{s}$e of competition in the job market. Male labour force participation may increase to reduce the income gap between the poor and rich in societies because men can earn higher wages and have more job opportunities than women. Unemployment may increase or decrease flfp ($\beta 10<0$ or $\beta 10_{>0}$). Due t$_{o}$ high unemployment, to compensate for income losses, the flfp may increase. However, as higher unemployment decreases the chance of finding a well-paid and decent job, the flfp may decline, as suggested by Ajide (2021). Its increasing effect is called an added-worker effect, while its decreasing impact is named the discouraged-worker effect (Tüzemen et al., 2021).

In the analysis, various panel data techniques were employed. First, the presence of cross-sectional dependency was checked. Second, panel unit root tests were performed to determine the series' integration order. If all of them are integrated of order 1, then panel cointegration tests are employed. The analysis can continue with an estimation of models if all series are stationary or cointegrated. For nonstationary series, series is differenced until they become stationary in the absence of a cointegrating relationship. Under the lack of regional heterogeneity, the OLS estimation method can be employed if there is no misspecification error. In the presence of regional heterogeneity and any misspecification error, such as cross-sectional dependency, heteroscedasticity or autocorrelation problem, the Feasible Generalised Least Square Dummy Variable (FGLSDV) estimation method is suitable. For dynamic panel data models, one can employ either GMM-based estimators developed by Arellano and Bond (1991) and Blundell and Bond (1998) or Bias corrected Least Square Dummy Variable (BCLSDV) dynamic panel data estimator

Table 1. Summary statistics

Variables	Mean	Standard Deviation	Coefficient of Variation (%)	Minimum	Maximum
flfp	32.154	5.454499	16.96	14.9	43.5
realgdppc	17564.3	7101.561	40.43	8609.943	36798.36
femaleinternet	52.047	15.59741	29.97	18.80322	85.6464
inflation	10.859	3.314302	30.52	7.494407	16.22052
totalagedependency	49.108	7.389913	15.05	40.1	69.65
financialaccess	8.3333	7.050444	84.61	1.599342	28.45915
netmigrationrate	-1.1051	8.89766	-805.14	-21.12	30.54
gini	0.36384	0.029347	8.07	0.302	0.451
unemployment	10.3677	3.688973	35.58	3.6	22.4

Source: Author's calculation

introduced by Bun and Kiviet (2003), Kiviet (1999) and Kiviet (1995). Bruno (2005) showed that for unbalanced panel data with a small cross-sectional dimension (N), BCLSDV outperforms Least Square Dummy Variable and GMM-based estimators. In this study, BCLSDV was employed as the number of regions is 12.

Information on Data

This study employed annual balanced panel data on 12 regions of Turkiye (NUTS-1 level) over 2013-2020. Data set includes FLFP rate (15 years and above, %), chained volume Gross Domestic Product (Value, Base year = 2009), total population, proportion of female internet users (by statistical regions level 1, %), Consumer Price Index (2003=100), the total age dependency ratio (%), distribution of bank branches by regions (%), net immigration rate (%), Gini coefficient (by equivalent household disposable individual income), and total unemployment rate (15 years and above, %). The author obtained data from the Turkiye Statistical Institute (TURKSTAT) database and The Banks Association of Turkiye.2 The only inflation variable is region-invariant but time-varying. Real GDP per capita and Gini coefficient were transformed using a natural logarithm. Table 1 presents summary statistics. On average, over the years and across the regions of Turkiye, FLFP rate, real GDP per capita, the proportion of female internet users, inflation rate, total age dependency ratio, regional distribution of bank branches, net migration rate, Gini coefficient and unemployment rate are 32.15%, 17564.3, 52.1%, 10.9%, 49.1%, 8.3%, -1.11%, 0.36, 10.37%, respectively. The coefficient of variation shows that the net migration rate has the highest time and regional variability.

According to the time averages of variables given in Table 2, significant regional differences exist in all of them except income inequality. TR1 (Istanbul) region has the highest realgdppc, femaleinternet, financialaccess, and Gini as this region is the most developed region of Turkiye. The FLFP rate is the highest in the TR9 (Eastern Black Sea) region, at nearly 40%. TRC (Southeastern Anatolia) region recorded the highest total dependency ratio and unemployment rate. TRA (Northeastern Anatolia Region) region sent the highest number of migrants, whereas TR2 (Western Marmara Region) received

Table 2. Time averages of variables except for inflation

Region	(1)	(2)	(3)	(4)	(5)	(6)	(7)	(8)
TRA	31.59	11075.27	37.73	57.14	1.73	-15.76	0.36	6.88
TAB	28.25	9765.56	37.44	56.67	2.23	-8.79	0.37	12.18
TRC	19.50	9987.19	37.50	66.70	4.79	-5.55	0.38	17.94
TR1	34.71	34236.55	69.03	40.33	27.29	-0.69	0.42	13.19
TR2	34.86	20472.01	56.56	43.02	4.49	8.91	0.36	7.41
TR3	37.06	19607.40	56.91	43.80	14.82	3.59	0.36	10.26
TR4	32.68	23610.19	60.36	43.68	9.17	7.46	0.33	9.80
TR5	31.00	24512.25	64.29	43.91	12.07	3.01	0.38	10.25
TR6	31.43	15714.08	54.96	49.00	11.27	0.26	0.39	12.09
TR7	28.41	14760.09	52.34	49.75	4.09	-3.14	0.35	10.98
TR8	36.45	13312.40	48.63	47.83	4.94	-2.24	0.34	7.10
TR9	39.91	13718.50	48.82	47.49	3.11	-0.32	0.33	6.35

(1) flfp, (2) realgdppc, (3) femaleinternet, (4) totalagedependency, (5) financialaccess, (6) netmigrationrate, (7) gini, (8) unemployment.
Source: Own calculation

the highest number of migrants. Migrant-receiving (migrant-sending) regions are TR2, TR3, TR4, TR5 and TR6 (TRA, TRB, TRC, TR1, TR7, TR8, and TR9).

To analyse the regional differences and change over time, Figure 2 and Figure 3 illustrate the regional FLFP rate and female internet usage proportion for 2013 and 2020, respectively. In Figures, darker colours indicate higher women's internet usage proportion in a region, whereas bigger circles show a higher FLFP rate. These figures also show the regional differences. However, this difference in FLFP and the proportion of women's internet usage changed over time. TR1 (Istanbul) region has the highest female internet among the regions. Also, the ratio increased from 52.8% in 2013 to 85.6% in 2020. The FLFP rate was the highest in the TR3 (Aegean) region in 2013, with 38.3%, but this rate declined to 34.1% in 2020. TR9 (Eastern Black Sea) region recorded the highest flfp in 2020, with a 38.4% increase from its rate in 2013 of 36.3%. TRC (Southeastern Anatolia) region has the lowest FLFP rate, with 14.9% and 20% in 2013 and 2020, respectively. Female internet usage proportion is the lowest in TRB (Central Eastern Anatolia) region, with 18.8% for 2013. However, TRC (Southeastern Anatolia) region's proportion became the lowest (52.2%) in 2020.

Table 3 illustrates the pairwise correlations among the variables. The correlations show a strong negative linear relationship between flfp and totalagedependency. However, there may be collinearity between lrealgdppc and other variables because pairwise correlations indicate strong linear relations between lrealgdppc and femaleinternet, totalagedependency, and financialaccess. The variance inflation factor (vif) was calculated as 5.59, indicating the absence of severe collinearity problem among the explanatory variables; however, after excluding all explanatory variables except lrealgdppc and femaleinternet, vif declines to 2.27. Therefore, the author employed the model in equation (2);

$$\boldsymbol{flfp}_{i,t} = \alpha_i + \beta_1 \boldsymbol{flfp}_{i,t-1} + \beta_2 \boldsymbol{lrealgdppc}_{i,t} + \beta_3 \boldsymbol{lrealgdppc}^2_{i,t} + \beta_4 \boldsymbol{femaleinternet}_{i,t} + \boldsymbol{u}_{i,t} \quad (2)$$

Figure 2. Female internet usage proportion and FLFP rate for regions of Turkiye for 2013
Source: Own construction using TURKSTAT data

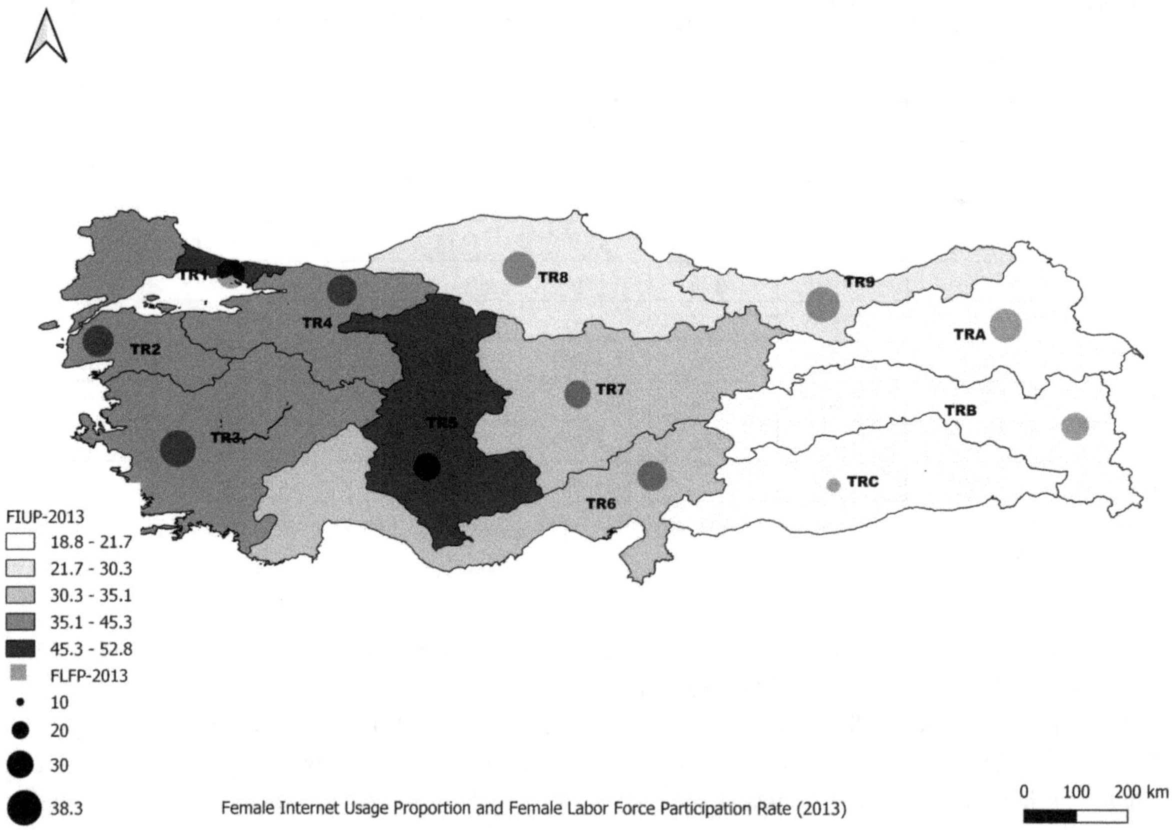

Results

In series, there is evidence of cross-sectional dependency. Because of the short time dimension (T=8), by assuming a common unit root process for each region, as suggested by Levin, Lin, and Chu (2002), Levin et al. (2002) (LLC) test was applied to the cross-sectionally demeaned series to eliminate cross-sectional dependency. In Table 4, the LLC unit root test indicates that all series are stationary. Therefore, the estimations can be performed without further tests for cointegration and data transformation.

1. Generalized least squares for heteroskedastic panels with cross-sectional correlation and common AR(1) coefficient for all panels was estimated as 0.2019. The utest algorithm showed that H_0 of Monotone or inverse U shape could be rejected against H_a of U shape with a t-value of 1.39 (P value = 0.0843).
2. Interval for good estimates for lagged flfp was obtained as 0.576-0.908 based on the OLS estimation of the pooled dynamic model and within the estimation of fixed effects dynamic panel data model. Results show that the coefficient estimate of lagged flfp is within this interval. Bias correction was initialised by Blundell and Bond estimator. First-stage regression results indicate that there is not any second-order serial correlation as the Arellano-Bond test for AR(2) in first differences has a z value of 1.14 with Pr > z = 0.256. Sargan's test result showed the validity of overidentifying

Figure 3. Female internet usage proportion and FLFP rate for regions of Turkiye for 2020
Source: Own construction using TURKSTAT data

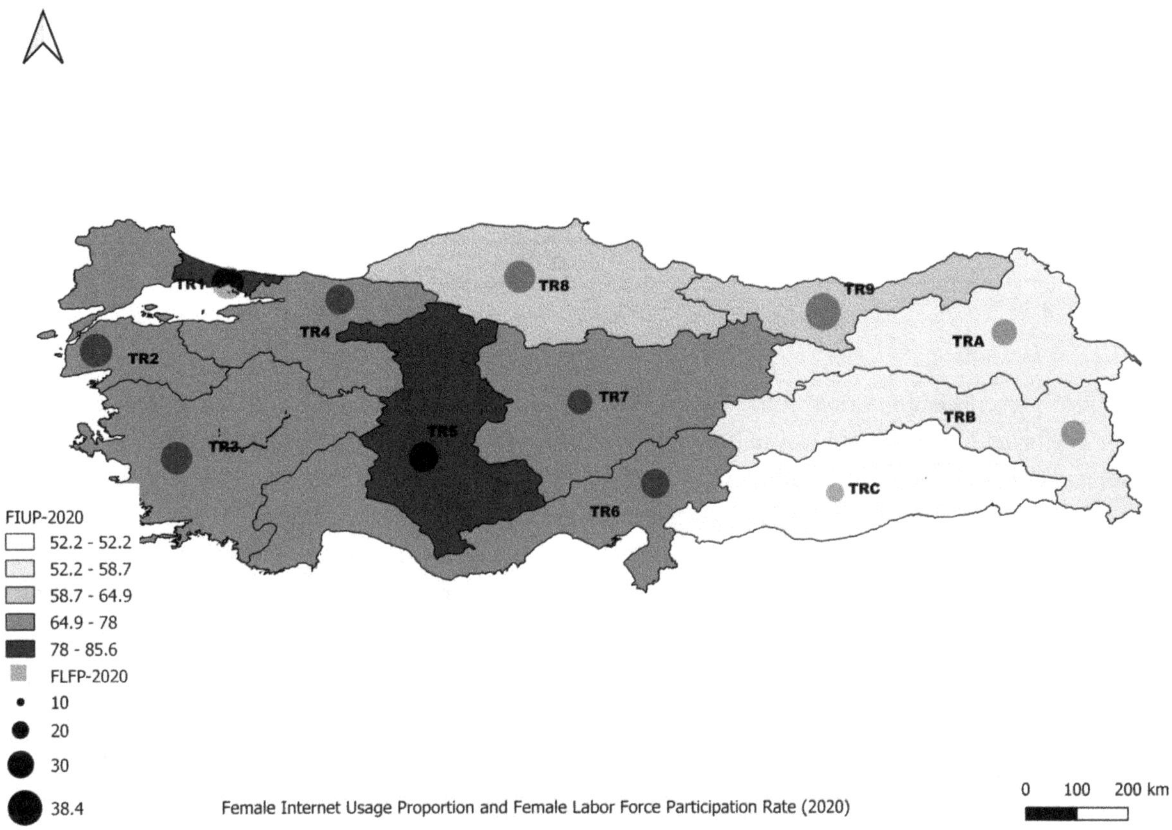

Table 3. Correlations

	(1)	(2)	(3)	(4)	(5)	(6)	(7)	(8)	(9)	(10)	(11)
(1)	1.00										
(2)	0.41	1.00									
(3)	0.41	1.00	1.00								
(4)	0.35	0.75	0.74	1.00							
(5)	0.32	0.73	0.73	0.99	1.00						
(6)	0.14	0.14	0.14	0.62	0.62	1.00					
(7)	-0.71	-0.85	-0.84	-0.59	-0.55	-0.03	1.00				
(8)	0.21	0.80	0.81	0.50	0.52	0.00	-0.57	1.00			
(9)	0.25	0.51	0.50	0.39	0.33	0.08	-0.55	0.26	1.00		
(10)	-0.18	0.24	0.25	0.12	0.17	-0.05	0.04	0.53	-0.13	1.00	
(11)	-0.60	0.01	0.02	0.24	0.26	0.30	0.37	0.27	-0.07	0.37	1.00

(1) flfp, (2) lrealgdppc, (3) lrealgdppc², (4) femaleinternet, (5) femaleinternet², (6) inflation, (7) totalagedependency, (8) financialaccess, (9) netmigrationrate, (10) lgini, (11) unemployment.

Source: Own calculation

restrictions with a $\chi^2(26)$ value of 21.07 ($Pr> \chi^2 = 0.738$). t values in parentheses were calculated based on bootstrapped standard errors.

Table 4. Cross-sectional dependency (CD) test and LLC panel unit root test

Variables	CD Test	LLC Test
flfp	10.78***	-3.3145***
lrealgdppc	21.19***	-6.2552***
femaleinternet	22.43***	-13.9562***

Source: Own calculation

Table 5 presents estimation results. To compare estimation results of static and dynamic models, the author performs four different estimations for the static pooled model, static fixed effects panel data model and dynamic model. According to the tests, there are statistically significant region-specific and time-fixed effects. However, as there is a misspecification problem in the static fixed effects panel data model, such as serial correlation, heteroscedasticity and cross-sectional dependency in the residuals, this model was estimated using FGLSDV given in column (3) of Table 5. Further, to consider dynamic

Table 5. Results of estimations

Variables	(1)	(2)	(3)[1]	(4)[2]
$flfp_{-1}$				0.80***
				(9.48)
lrealgdppc	219.68***	-162.62***	-158.81***	-16.31
	(17.11)	(-11.12)	(-6.51)	(-0.27)
$lrealgdppc^2$	-10.95***	8.40***	8.40***	1.17
	(-16.42)	(10.58)	(6.97)	(0.38)
femaleinternet	-0.002	0.07	0.06	0.13**
	(-0.05)	(1.35)	(1.15)	(2.34)
Constant	-1066.27***	814.58***	779.67***	
	(-17.09)	(11.04)	(6.24)	
R^2	0.2676	0.4672		
LM_p	107.6***	38.8***		
LR_H	68.4***	365.2***		
Individual Fixed Effects		62.1***		
Time-period Fixed Effects		6.1***		
Joint Fixed Effects Test		68.8***		
Pesaran's CD test	8.8***	-1.9*		

Source: Own calculation

Notes: *, **, *** indicate p<0.1, p<0.05 and p<0.01. The region-specific and time-period fixed effects' coefficient estimates are not presented to save space. In parentheses, t values are presented. (1) Pooled OLS, (2) Individual and time FE, (3) FGLSDV, (4) BCLSDV.

effect explicitly; therefore, serial correlation in the residuals, column (4) of Table 5 presents results for dynamic panel data estimated by BCLSDV.

Results show that female internet usage increases FLFP statistically significantly, and the Feminization U hypothesis is not valid as the coefficient estimates are statistically insignificant. Lagged FLFP rate is close to 1 and statistically significant, implying persistency in women's labour force participation. According to Valberg (2020), due to leapfrogging, Feminization U-hypothesis may not be relevant to developing countries as these countries can develop without passing the industrialisation stage; therefore, the development process of these countries may be directly from agriculture to information economy. This finding is also in line with Klasen (2019). In contrast to Tüzemen et al. (2021), the results of this study showed a statistically significant effect of ICT usage on FLFP for Turkiye. This difference may be due to the data, variables and method used. By considering regional differences in Turkiye, this study shows the importance of ICTs in FLFP. This finding is similar to the results of Alves and Steiner (2017), Atasoy, Banker, and Pavlou (2021), Dettling (2017), Efobi, Tanankem, and Asongu (2018), Ngoa and Song (2021), Valberg (2020), Viollaz and Winkler (2022), and Watson et al. (2018). Also, according to Brussevich et al. (2018), in countries with a low level of FLFP, women's engagement is mostly through jobs using higher levels of ICTs. Therefore, this result can be expected for Turkiye, which has a 32% FLFP rate lower than World'sWorld's rate of 46% in 2021.

SOLUTIONS AND RECOMMENDATIONS

The results showed the importance of ICT use and dissemination to engage women in the labour force. Therefore, policymakers should focus on the gender digital divide to encourage women's active participation in the labour market. As highlighted by the report prepared by Organisation for Economic Co-operation and Development (OECD) in 2018, concrete actions should be taken to ensure the digital inclusion of women and girls and to prevent discrimination against women caused by social norms and stereotypes. According to Watson et al. (2018), gender-specific policy actions can be classified into four groups: access to digital technologies, digital skill improvement and entrepreneurship, norms, and data collection for gender-specific ICT use and women's contribution to the economic output in the digital economy to develop and evaluate policies. ICT usage of women should be increased by informing and mentoring them about the use and importance of ICTs for their career development, training programs, financing their personal development and access to ICTs, and encouraging them to participate in the labour force using innovative solutions. Collaboration between government, businesses, and non-governmental organisations, incorporating gender equal digitalisation process into the strategies of all the sectors, setting female ICT usage targets, policy impact analysis, respecting the rule of law, protection of property rights, governmental regulations, technical and financial assistance for inclusion of women into the digital entrepreneurship may also help design policies, evaluate their results, ensure higher quality infrastructural improvement and investment, and digital inclusion (Watson et al., 2018). However, these policies should be supported by other social and economic policies, such as education, to have desirable and long-lasting impacts, as suggested by Valberg (2020). Moreover, because of the automatisation risk of many jobs performed by women due to technological advancements, Brussevich et al. (2018) suggested active labour market policies for this transition process, various support schemes for displaced workers, the adaptation of social security schemes, upskilling and reskilling of women, fiscal, institutional, regulatory and legal measures including tax deductions for training and infrastructural investment for the reconciliation of work and family life, increasing women's participation in science, technology, engineering, and mathematics sectors, peer mentoring, reducing glass ceiling (i.e., gender gaps in managerial positions) and tackling digital divide.

FUTURE RESEARCH DIRECTIONS

Similar analysis can be performed for different countries at the regional level considering regional differences and using spatial data analysis. Other innovative solutions can be considered for the increase in FLFP, and future studies may focus on these innovations. The channels that increase FLFP through ICTs may be analysed explicitly to understand the direct and indirect impacts of ICTs. Moreover, as different sectors have differentiated dynamics, a sectoral level analysis may provide much more detailed conclusions and policy recommendations.

CONCLUSION

New technologies have changed how people work (Brussevich et al., 2018). ICT use and dissemination can engage different social groups in the labour force, such as women. Therefore, in this chapter, the author analysed the impact of internet use on FLFP for Turkiye and tested the validity of the Feminization U hypothesis employing regional level data between 2013 and 2020. Results show that the relation between women's participation in the labour market and internet usage is inverted U-shaped, and the Feminization U hypothesis is valid. Therefore, digital inclusion is vital for FLFP. Policy measures should be developed to simultaneously increase ICT services to encourage women's participation in economic activities. Lastly, as discussed by Roche (2016), although there are various benefits of ICTs use, there are also some disadvantages due to employment losses as a result of artificial intelligence's replacement of humans in some jobs, cybercrimes, cyberterrorism, loss of privacy and individual autonomy, and cyberattacks to infrastructures related to transportation, finance, electricity, and communication. Therefore, necessary preventive actions, such as ensuring cybersecurity, should be taken globally.

REFERENCES

Adshade, M., & Keay, I. (2010). Technological and Organisational Change and the Employment of Women: Early Twentieth-Century Evidence from the Ohio Manufacturing Sector. *Feminist Economics, 16*(1), 129–157. doi:10.1080/13545700903382711

Ajide, F. M. (2021). Financial Inclusion and Labour Market Participation of Women in Selected Countries in Africa. *Economics and Culture, 18*(1), 15–31. doi:10.2478/jec-2021-0002

Aldan, A. (2021). Rising Female Labor Force Participation and Gender Wage Gap: Evidence From Turkey. *Social Indicators Research, 155*(3), 865–884. doi:10.100711205-021-02631-9

Alves, E. E. C., & Steiner, A. Q. (2017). Globalisation, Technology and Female Empowerment: Breaking Rights or Connecting Opportunities? *Social Indicators Research, 133*(3), 859–877. doi:10.100711205-016-1395-1

Arellano, M., & Bond, S. (1991). Some Tests of Specification for Panel Data - Monte-Carlo Evidence and an Application to Employment Equations. *The Review of Economic Studies, 58*(2), 277–297. doi:10.2307/2297968

Asongu, S. A., Amankwah-Amoah, J., Nting, R. T., & Afrifa, G. A. (2021). Information Technology and Gender Economic Inclusion in Sub-Saharan Africa. *Journal of Global Information Technology Management, 24*(2), 120–133. doi:10.1080/1097198X.2021.1914497

Asongu, S. A., & Odhiambo, N. M. (2020). Inequality and gender inclusion: Minimum ICT policy thresholds for promoting female employment in Sub-Saharan Africa. *Telecommunications Policy, 44*(4), 101900. doi:10.1016/j.telpol.2019.101900

Asongu, S. A., & Odhiambo, N. M. (2022). *Enhancing ICT for Female Economic Participation in Sub-Saharan Africa.* AGDI Working Paper. African Governance and Development Institute Research Department.

Atasoy, H., Banker, R. D., & Pavlou, P. A. (2021). Information Technology Skills and Labor Market Outcomes for Workers. *Information Systems Research, 32*(2), 437–461. doi:10.1287/isre.2020.0975

Baliamoune, M. (2021). *Trade, Infrastructure, and Female Participation in Labor Markets.* Research Paper. Policy Center for the New South.

Becker, G. S. (1965). A Theory of the Allocation of Time. *Economic Journal (London), 75*(299), 493–517. doi:10.2307/2228949

Becker, G. S. (1973). A Theory of Marriage: Part I. *Journal of Political Economy, 81*(4), 813–846. doi:10.1086/260084

Becker, G. S. (1974). A Theory of Marriage: Part II. *Journal of Political Economy, 82*(2, Part 2), 511–526. doi:10.1086/260287

Blundell, R., & Bond, S. (1998). Initial conditions and moment restrictions in dynamic panel data models. *Journal of Econometrics, 87*(1), 115–143. doi:10.1016/S0304-4076(98)00009-8

Bruno, G. S. F. (2005). Estimation and inference in dynamic unbalanced panel-data models with a small number of individuals. *The Stata Journal, 5*(4), 473–500. doi:10.1177/1536867X0500500401

Brussevich, M., Dabla-Norris, E., Kamunge, C., Karnane, P., Khalid, S., & Kochhar, K. (2018). *Gender, Technology, and the Future of Work.* Retrieved August 21, 2022, from https://www.imf.org/-/media/Files/Publications/SDN/2018/SDN1807.ashx

Bun, M. J. G., & Kiviet, J. F. (2003). On the diminishing returns of higher order terms in asymptotic expansions of bias. *Economics Letters, 79*(2), 145–152. doi:10.1016/S0165-1765(02)00299-9

Chen, M. X., Qin, L. Y., & Zhao, C. Z. (2021). What Are Reasons of the Significant Change in the Chinese Female Labour Force Participation Rate? Analysis on Influencing Factors Based on the Macroscopic Trend. *Transformations in Business & Economics*, *20*(3c), 405–428.

Choudhry, M. T., & Elhorst, P. (2018). Female labour force participation and economic development. *International Journal of Manpower*, *39*(7), 896–912. doi:10.1108/IJM-03-2017-0045

DeSerpa, A. C. (1971). A Theory of the Economics of Time. *Economic Journal (London)*, *81*(324), 828–846. doi:10.2307/2230320

Dettling, L. J. (2017). Broadband in the Labor Market: The Impact of Residential High-Speed Internet on Married Women's Labor Force Participation. *Industrial & Labor Relations Review*, *70*(2), 451–482. doi:10.1177/0019793916644721

Efobi, U. R., Tanankem, B. V., & Asongu, S. A. (2018). Female Economic Participation with Information and Communication Technology Advancement: Evidence from Sub-Saharan Africa. *The South African Journal of Economics*, *86*(2), 231–246. doi:10.1111aje.12194

Evans, A. W. (1972). On the Theory of the Valuation and Allocation of Time. *Scottish Journal of Political Economy*, *19*(1), 1–17. doi:10.1111/j.1467-9485.1972.tb00504.x

Gaddis, I., & Klasen, S. (2014). Economic development, structural change, and women's labor force participation: A reexamination of the feminisation U hypothesis. *Journal of Population Economics*, *27*(3), 639–681. doi:10.100700148-013-0488-2

Genre, V., Salvador, R. G., & Lamo, A. (2010). European women: Why do(n't) they work? *Applied Economics*, *42*(12), 1499–1514. doi:10.1080/00036840701721547

Goyal, A. (2011). Developing women: Why technology can help. *Information Technology for Development*, *17*(2), 112–132. doi:10.1080/02681102.2010.537252

Jayachandran, S. (2021). Social Norms as a Barrier to Women's Employment in Developing Countries. *IMF Economic Review*, *69*(3), 576–595. doi:10.105741308-021-00140-w

Kiviet, J. F. (1995). On Bias, Inconsistency and Efficiency of Various Estimators in Dynamic Panel Data Models. *Journal of Econometrics*, *68*(1), 53–78. doi:10.1016/0304-4076(94)01643-E

Kiviet, J. F. (1999). Expectation of expansions for estimators in a dynamic panel data model; some results for weakly exogenous regressors. In C. Hsiao, K. Lahiri, L.-F. Lee, & M. H. Pesaran (Eds.), *Analysis of Panels and Limited Dependent Variable Models* (pp. 199–225). Cambridge University Press. doi:10.1017/CBO9780511493140.010

Klasen, S. (2019). What Explains Uneven Female Labor Force Participation Levels and Trends in Developing Countries? *The World Bank Research Observer*, *34*(2), 161–197. doi:10.1093/wbro/lkz005

Krieger-Boden, C., & Sorgner, A. (2018). Labor market opportunities for women in the digital age. *Economics-the Open Access Open-Assessment E-Journal, 12.*

Levin, A., Lin, C. F., & Chu, C. (2002). Unit root tests in panel data: Asymptotic and finite-sample properties. *Journal of Econometrics*, *108*(1), 1–24. doi:10.1016/S0304-4076(01)00098-7

Lv, Z. K., & Yang, R. D. (2018). Does women's participation in politics increase female labor participation? Evidence from panel data analysis. *Economics Letters, 170*, 35–38. doi:10.1016/j.econlet.2018.05.013

Mincer, J. (1962). Labour force participation of married women: A study of labour supply. Aspects of labour economics. In *A conference of the universities: National bureau committee for economic research* (pp. 63–105). Princeton University Press.

Morris, T. (2022). Re-examining female labor supply responses to the 1994 Australian pension reform. *Review of Economics of the Household, 20*(2), 419–445. doi:10.100711150-021-09561-y

Nam, S. (1991). Determinants of Female Labor-Force Participation - a Study of Seoul, South-Korea, 1970-1980. *Sociological Forum, 6*(4), 641–659. doi:10.1007/BF01114405

Ngoa, G. B. N., & Song, J. S. (2021). Female participation in African labor markets: The role of information and communication technologies. *Telecommunications Policy, 45*(9).

Nishikawa, S., & Higuchi, Y. (1981). Determinants of Female Labor-Force Participation. *Japanese Economic Studies, 9*(2), 62–87. doi:10.2753/JES1097-203X090262 PMID:12313523

Organisation for Economic Co-operation and Development. (2018). *Bridging the digital gender divide: include, upskill, innovate*. Retrieved August 21, 2022, from https://www.oecd.org/going-digital/bridging-the-digital-gender-divide.pdf

Osundina, O. A. (2020). Sustainable development: Does improvement in education and health of women improve female labour force participation rate? *Sustainable Development, 28*(1), 13–24. doi:10.1002d.1961

Roche, E. M. (2016). Information and Communication Technology Still a Force for Good? *Journal of Global Information Technology Management, 19*(2), 75–79. doi:10.1080/1097198X.2016.1172952

Samargandi, N., Al Mamun, M., Sohag, K., & Alandejani, M. (2019). Women at work in Saudi Arabia: Impact of ICT diffusion and financial development. *Technology in Society, 59*, 59. doi:10.1016/j.techsoc.2019.101187

Saure, P., & Zoabi, H. (2014). International trade, the gender wage gap and female labor force participation. *Journal of Development Economics, 111*, 17–33. doi:10.1016/j.jdeveco.2014.07.003

Schwarz, P. (2012). Tax disincentives and female employment in Organisation for Economic Co-operation and Development (OECD) countries. *Journal of European Social Policy, 22*(1), 17–29. doi:10.1177/0958928711425267

Suh, M. G. (2017). Determinants of Female Labor Force Participation in South Korea: Tracing out the U-shaped Curve by Economic Growth. *Social Indicators Research, 131*(1), 255–269. doi:10.100711205-016-1245-1

Suhaida, M. A., Nurulhuda, M. S., & Yap, S.-F. (2013). Access to ICT as Moderating Factor to Women's Participation in the Labor Force: A Conceptual Framework. *International Journal of Trade. Economics and Finance, 4*(4), 197–201. doi:10.7763/IJTEF.2013.V4.285

Tam, H. (2011). U-shaped female labor participation with economic development: Some panel data evidence. *Economics Letters*, *110*(2), 140–142. doi:10.1016/j.econlet.2010.11.003

The United Nations Women. (2022). *Women and the Sustainable Development Goals (SDGs)*. Retrieved August 7, 2022, from https://www. unwomen.org/en/news/in-focus/women-and-the-sdgs

The World Bank, World Development Indicators. (2022a). *Individuals using the Internet (% of population)*. Retrieved August 7, 2022, from https://databank.worldbank.org/source/world-developmentindicators

The World Bank, World Development Indicators. (2022b). *Labor force participation rate, female (% of female population ages 15+) (modeled ILO estimate)*. Retrieved August 7, 2022, from https://databank.worldbank.org/source/world-developmentindicators

Tüzemen, S., Barış-Tüzemen, Ö., & Çelik, A. K. (2021). The relationship between information and communication technologies and female labour force participation in Turkey. *Economics and Business Review*, *7*(4), 121–145. doi:10.18559/ebr.2021.4.7

Uberti, L. J., & Douarin, E. (2022). The Feminisation U, cultural norms, and the plough. *Journal of Population Economics*.

Valberg, S. (2020). ICT, Gender, and the Labor Market: A Cross-Country Analysis. In D. Maiti, F. Castellacci, & A. Melchior (Eds.), *Digitalisation and Development Issues for India and Beyond* (pp. 375–405). Springer Nature. doi:10.1007/978-981-13-9996-1_15

Vega, S. H., & Elhorst, J. P. (2017). Regional labour force participation across the European Union: A time-space recursive modelling approach with endogenous regressors. *Spatial Economic Analysis*, *12*(2-3), 138–160. doi:10.1080/17421772.2016.1224374

Viollaz, M., & Winkler, H. (2022). Does the Internet Reduce Gender Gaps? The Case of Jordan. *The Journal of Development Studies*, *58*(3), 436–453. doi:10.1080/00220388.2021.1965127

Ward, K. B., & Pampel, F. C. (1985). Structural Determinants of Female Labor-Force Participation in Developed Nations, 1955-75. *Social Science Quarterly*, *66*(3), 654–667.

Watson, T., Corliss, M., & Le, M. (2018). Digitalisation and Women's Workforce Participation in the Indo-Pacific. *Australian Journal of Labour Economics*, *21*(1), 45–74.

KEY TERMS AND DEFINITIONS

Digital Inclusion: All activities guarantee equal access to and use of information and communication technologies to participate in digital economic and social services and activities.

Economics of Marriage: Analysing the marriage-related issues using economic analysis methods starting with Becker's (1973, 1974) seminal theory of marriage.

Female Labour Force Participation: The percentage of women in the total labour force shows the activeness degree of women in the available labour supply for producing goods and services in the economy.

Feminisation U Hypothesis: Hypothesis showing the U-shaped relation between women's participation in the labour market and economic growth in such a way that as economies grow, female labour force participation declines in the industrialised economies; however, as economies become service sector-oriented, then women's participation increases in the labour market.

Gender Digital Divide: A term used to reflect gender differences in the usage of digital technologies. In most cases, digital technology usage by women is lower than men.

Glass Ceiling: A term coined by Marilyn Loden in 1978 for the discrimination against women in their career advancements. In a broader context, it refers to difficulties and invisible barriers, such as social barriers, including norms and biases, preventing minorities' promotion to higher management positions in their jobs.

Information and Communication Technologies: The combination of computing and communication hardware and software technologies which are used to acquire, process, disseminate, retrieve, store, transmit, and manipulate information digitally and form electronic input and includes a heterogeneous set of techniques, systems and devices or electronic platforms, intelligent machines and computer networks, for example, the radio, television, telephones, mobile phones, computers, Internet, wireless networks, software, and all digital media applications. Various areas employ these technologies, such as education, health, public services, and libraries.

Leapfrogging: A concept used in industrial organisation and theory of development to explain the development stages of developing countries. Catching up with developed countries' economic growth by skipping some growth stages which developed ones previously took.

Panel Data Analysis: Statistical analysis using panel data which is two-dimensional data obtained by collecting observations on the set of cross-sectional units over time.

Theory of Investment in Human Capital: Analysing the effect of various investments related to human capital, such as education and health, on the well-being of individuals, for example, future income and consumption. There are multiple views on the usefulness of human capital, such as the views of Becker, Gardener, Schultz/Nelson-Phelps, Bowles-Gintis and Spence.

Theory of Time Allocation: In the model introduced by Becker (1965), under budget and time restrictions, the utility maximisation of consumers is realised with choices on commodities produced by using market goods and time. The model was further improved by DeSerpa (1971) and Evans (1972), considering additional restrictions and redefinition of commodities.

ENDNOTES

1 TRA: Northeastern Anatolia Region, TRB: Central Eastern Anatolia Region, TRC: Southeastern Anatolia Region, TR1: Istanbul Region, TR2: Western Marmara Region, TR3: Aegean Region, TR4: Eastern Marmara Region, TR5: Western Anatolia Region, TR6: Mediterranean Region, TR7: Central Anatolia Region, TR8: Western Black Sea Region, TR9: Eastern Black Sea Region.

2 https://www.tbb.org.tr/tr/bankacilik/banka-ve-sektor-bilgileri/istatistiki-raporlar/59.

Chapter 2
Using Digital Technology to Enhance Medical Education and Improve the Quality of Patient Care in a Changing Global Healthcare System

Mitchell Alan Kaplan
Independent Researcher, USA

ABSTRACT

The use of digital technology as an effective method of educating the emerging generation of physicians and improving the quality of services that patients receive has become an expanding trend in recent decades. The application of alternative methods of instruction in higher education, such as online degree courses, learning apps on smartphones, and computer-generated virtual reality models that simulate actual clinical situations, combined with frequent use of digital assessment tools, have made diagnosis and treatment of medical conditions online more precise and less time-consuming. These technological breakthroughs have redefined the roles of providers and patients across the spectrum of care. This chapter will present a comprehensive analysis of the beneficial impact that digital technology is having on medical education and the delivery of healthcare services to patients. It will examine how incorporating technological advances such as telemedicine is changing the dynamic of patient care and transforming healthcare systems globally.

TECHNOLOGY IN MEDICAL EDUCATION AND CLINICAL TRAINING

In recent decades, the rapid expansion of various types of digital technology at hospitals and ambulatory care centers around the United States has facilitated significant changes in how physicians and other primary care professionals are trained and how patient services are delivered across the spectrum of medical specialties. Incorporating the advantages of technology as an essential part of the educational

DOI: 10.4018/978-1-6684-6620-9.ch002

process is opening doors to a wide range of instructional possibilities for medical school faculty who want to share their accumulated knowledge and clinical expertise with a growing global audience of medical students, interns, and residents who will become the next generation of practicing physicians. During periods of national crisis, such as that which occurred after the September 11, 2001, terrorist attack on the United States, medical schools and other institutions of higher education have sought to protect the safety of their students and faculty by transitioning their in-house instructional and training programs to online access in response to these catastrophic circumstances. The COVID-19 global pandemic fostered a similar reaction from medical institutions across the country, seeking to find remote venues to educate and train students while protecting them from this deadly disease. A joint study conducted by Professor of Psychiatry and Medical Education Jacob Appel and his third and fourth-year medical students at Mount Sinai School of Medicine, published in *Academic Medicine* in 2020, provides evidence supporting the efficacy of this approach to medical education and training. Data analysis indicates that implementing restrictive COVID-19 policies is one of the primary factors fueling the exponential growth of remote education and training programs for medical students using the distance learning approach known as telemedicine. The analysis revealed that using telemedicine platforms in medical education effectively facilitates the acquisition of basic medical knowledge and helps improve students' decision-making skills. For example, the research confirmed that using technology-assisted techniques such as three-dimensional computer-generated simulation models greatly enhances medical students' ability to absorb and understand essential human anatomy and physiology knowledge. The study disclosed that incorporating technological innovations such as video conferencing into the academic curriculum of medical institutions is a highly effective method of fostering the sharing of case discussions, short instructional presentations, and real-time case evaluations with students studying medicine at educational institutions located in underserved countries where access to knowledge about the benefits of remote care would otherwise be unavailable. The investigation also showed that technology-assisted simulation is a highly effective tool instructors can use to improve students' clinical assessment and diagnostic capabilities through engagement in virtual situations that mirror actual crisis events they will encounter when they become physicians

Based on these findings, Dr. Appel and his team concluded that incorporating telemedicine-based instructional methods such as remote lectures, online examinations, and computer-assisted preclinical skills training has significant positive benefits for fourth-year medical students who want to complete their studies at an accelerated pace. The investigators contend that integrating remote methods of instruction into the traditional medical school curriculum accelerates the process of academic learning, allowing more medical students to graduate early and enter hospital-based internship and residency programs where they will learn the applied knowledge and gain the practical experience necessary to become board-certified service providers (Appel et al., 2020).

Similar outcomes in other studies affirm the positive effects of technology-assisted instruction on medical education. Research by Professor Phyllis A. Guze of the University of California School of Medicine documented in a 2015 article in Transactions of the American Clinical and Climatological Association online posits that the expanding trend toward the use of digital learning techniques in medical education is a direct response to changing social and economic conditions taking place within the infrastructure of healthcare institutions across the United States and abroad. Dr. Guze argues that such internal changes at all levels of the U.S. healthcare system create a broad set of new educational challenges for medical school faculty that must be addressed effectively through technological intervention. Dr. Guze points to several critical factors as responsible for mitigating this emerging trend; key among them are changes in society's expectations of physician competencies, medical ethics, and patient safety concerns, combined with the explosion of knowledge about advances in medicine that have triggered the need for physicians to engage in lifelong learning. Professor Guze conjectures that each of these essential factors is a component of what is currently driving the movement to transition the delivery of patient services away from the traditional hospital setting into the realm of ambulatory care. She argues that this transition allows doctors to deliver treatment services to patients in a more cost-efficient and less time-consuming atmosphere. Dr. Guze points out that exposure of the emerging generation of tech-savvy medical students to a technology-enriched instructional environment in medical school dramatically enhances their educational experience and professional potential as future service providers. She believes that using different technology-based instructional techniques has a transformative impact on students' ability to achieve crucial academic milestones. These milestones include the following:

- The acquisition of the most up-to-date medical knowledge which strengthens their basic clinical competency in their area of specialization.
- The improvement of students' capacity to make critical medical decisions while raising their level of professional ability to coordinate their skills.
- Enables medical students to develop psychomotor skills and improve perceptual variation abilities.
- Allows students to practice their crisis intervention skills and enhances their team coordination training, making them better prepared to deal collectively with challenging clinical situations.

On the basis of these results, Dr. Guze concludes that the most significant challenge facing faculty engaged in the medical education system is adapting technology to meet the changing educational and training expectations of the current cohort of medical students. She contends that incorporating various forms of technology as a routine part of the medical school's instructional curriculum constitutes one of the most progressive methods of turning the learning experience into a more collaborative, empowering, and personally rewarding one for students (Guze, 2015).

Further evidence of the instructional advantages that technology can bring to faculty attempting to meet the changing educational and professional training demands of today's medical students is provided in a joint study conducted by Professor Norman Berman of Geisel School of Medicine at Dartmouth University and his colleagues in 2016. The study examines the benefits of integrating virtual reality, a computer-generated simulation technology, into medical students' educational and training milieu. Findings reveal that the integration of patient-centered virtual technology into the medical school curriculum broadens the scope of the learning experience by enabling students to master complex scientific information that forms the foundation of their clinical education and promotes the development of clinical reasoning skills essential to the diagnosis and treatment of patients with a variety of medical conditions. Dr. Berman and his colleagues believe that using patient-centered virtual technology as conceptual support has the power to transform medical education across the boundaries of specialization. The researchers posit that the multi-institutional sharing of virtual patient data and other resources creates high-quality educational content that medical school instructors can draw upon in their lectures and reality-based clinical demonstrations. They also contend that V.P. data helps establish national standards for integrating various types of technology into the instructional curriculum of medical institutions (Berman et al., 2016).

Similar evidence that supports Professor Berman's findings is revealed in the results of a 2019 study by Dr. Jack Pottle, co-founder and chief medical officer at Oxford Medical Simulation in Great Britain. Dr. Pottle's research focused on evaluating the effects of integrating technology into the medical school curriculum based on the instructional perspective that computer-enhanced simulation techniques represent the gold standard of medical education and clinical training. Results show that incorporating computer-generated simulation technology is one of the most efficient methods of delivering experience-based instruction to medical students. He argues that the feedback learners receive from immersive engagement in VR-created clinical situations increases the amount of knowledge students retain from their clinical training sessions. For example, Pottle contends that studies show that V.R. immersion in surgical training programs decreases injury, increases the speed of surgical procedures, and improves overall surgical outcomes. Many surgical training programs presently use VRS technology as a standard instructional tool for their surgical residents, with excellent professional results.

Pottle's research points out that although technologies such as screen-based learning, 360 video demonstrations, and interactive virtual reality presentations powered by artificial intelligence offer the current cohort of medical students many positive advantages that go far beyond what they would gain from more traditional teaching methods, it is not without its drawbacks. On the positive side, the expansive integration of VRS technology into medical education enables students to:

- Access simulation-based learning modules outside the classroom setting on a flexible schedule
- Engage in repeatable visual education scenarios that highlight medical errors that they can learn from, subsequently improving their professional performance
- Complete their institutional requirements using fewer instructional resources, thus significantly decreasing the financial cost of their formal education and experiential training

On the negative side, results show that VRS technology is not the perfect solution to improving learning outcomes for medical students in all clinical situations. Data analysis indicates that VRS immersion techniques do not apply to certain experiential situations, such as training students in the intricacies of specific complex medical procedures like abdominal palpation, which usually requires hands-on exposure to a three-dimensional physical representation of the abdomen rather than a computer-generated simulation of the core of this anatomical structure on a screen. The analysis also reveals that interactions with virtual images manufactured through artificial intelligence are no substitute for face-to-face encounters between physicians and patients, especially in sensitive clinical situations such as breaking bad news about an individual's medical condition to a family. The investigation concludes that despite the host of socioeconomic advantages that widespread use of VRS can offer medical schools in terms of cost-effectiveness and improved deliverability of services, many medical educators acknowledge that this type of technology cannot replace the hands-on experience and expertise of a skilled instructor able to train students how to handle patients in clinical circumstances that involve life and death decisions (Pottle, 2019).

TRANSFORMING MEDICAL SERVICES THROUGH DIGITAL TECHNOLOGY

The delivery of services in the traditional U.S. healthcare system is undergoing one of the most phenomenal overhauls in modern times, thanks to the expanding utilization of information and communication technologies. Socioeconomic disparities that were once considered barriers to equal access to treatments that affect health outcomes of children and the elderly are systematically eliminated by integrating digital technology as a routine part of medical practice. Research summarized in a book chapter written by

Dr. Najeeb-Al-Shorbaji, President of E-Health Development Association in Jordan in 2021, documents several compelling reasons why patient access to essential healthcare services is jeopardized. The most transparent of these reasons are:

- Lack of knowledge about the existence of services offered by medical institutions
- Patient acquisition of physical or mental health conditions that result in disability
- Geographic distance from institutional facilities that provide essential healthcare services
- COVID-19 pandemic-imposed social distancing restrictions and lockdowns that limit access to care

The investigation describes several methods by which technology impacts provider ability to deliver medical services. Paramount among them are:

- The transitioning of health information about patients from hard copy into the realm of digital record keeping over the internet
- The expansion of the use of electronic treatment devices controlled by artificial intelligence
- The growing use of sensory technologies that communicates vital data about the medical condition of patients to physicians and other primary care professionals
- The use of wearable communication devices that enable physicians to collect critical medical information about a patient's condition at the bedside or in outpatient clinic settings
- The use of the internet as a tool to educate physicians about what type of treatment options are available to assist patients with specific medical conditions

The researcher believes that applying remote health solutions through digital technology empowers physicians with practical alternatives for reaching out to patients to provide treatment services and professional guidance that can resolve their health concerns effectively. He further contends that the routine adoption of digital technology as a compatible component of a changing service delivery system will incentivize global nations to implement resolutions that support digital access to health services (Al-Shorbaji, 2021).

Similar outcomes were uncovered in a 2018 joint study by investigators at the World Health Organization, John Hopkins University School of Medicine, and the Center for Health Policy and Inequalities Research at Duke University Medical School. The research focused on evaluating the efficacy of providing equal access to high-quality medical services using digital health strategies to overcome human resource constraints. Investigators utilized popular search engines such as Medline, Pubmed, EMBASE, and CINAHL to evaluate the findings of 92 peer-reviewed research articles cited in the medical literature. From those, researchers selected 24 abstracts for inclusion in the final evaluation. Published results from the selected studies examined the practical benefits of technological interventions in the healthcare industry. Topics covered include the assessment of specific interventions tied to digital approaches to health service delivery, technical methods of enhancing provider education and training, integration of mobile diagnostics and disease management techniques in the traditional clinical environment, mobile strategies for upgrading clinical decision-making, and patient referral coordination systems. It also evaluates the strength of specific technological interventions for improving compliance with treatment protocols by enhancing the essential communication dynamic between service providers and patients. Analysis of findings reported in selected studies reviewed in the assessment revealed that frontline healthcare workers, particularly physicians who use mobile devices such as cell phones to manage patient's medical conditions through online education and follow up via health-related text messages and email reminders on patients' mobile phones about components of their treatment plan such as medication routines and appointment scheduling elicited the most dramatic change in consumer behavior associated with positive treatment outcomes

A peer review of studies about the advantages of introducing mobile technology into the doctor-patient relationship cited by the authors confirms that this approach offers patients practical provider alternatives for delivering care and educational strategies that can assist them in managing their illness. Published results from the series of articles documented in the context of this assessment show that investigators used research methodologies such as randomized controlled trials, prospective cohort studies, and double-blind protocols to analyze the effectiveness of communication technologies like short messaging services (SMS) on physician capacity to collect accurate patient information remotely and train students in essential medical procedures. Data analysis determined that integrating mobile devices such as cell phones and laptop computers into clinical settings strengthens interns' and residents' ability to acquire experiential knowledge. and improves their capacity to apply fundamental skills across disciplinary boundaries. The investigations also affirm that implementing communication technologies in healthcare benefits service providers and patients. Utilization of these specialized technologies as a routine part of medical care:

- Allows service providers to increase their adherence to prescribed guidelines of treatment
- Enables physicians to convey accurate information to patients about their medical condition, which creates a better understanding of disease management and improves treatment outcomes
- It provides an incentive for primary care providers to upgrade the effectiveness of their job performance and allows supervisors to improve staff oversight

Evidence supporting these findings is highlighted in a meta-analysis of cross-cultural studies conducted by research teams in Libera, Kenya, Korea, Iran, and Tanzania between 2010 and 2013. Investigators used randomized controlled clinical trials to evaluate how the use of SMS technology for remote data collection promotes each of the outcomes previously mentioned. A brief synopsis of the effectiveness of incorporating digital technology into global healthcare systems is documented in the summation of international studies below:

- A team of researchers in Liberia evaluated the effectiveness of a three-day prenatal education program designed to train healthcare professionals to instruct midwives in the community to use short messaging service technology (SMS) to collect essential clinical data on pregnant women in remote areas of the country. The investigators recruited 99 traditionally trained midwives practicing in community settings to participate in the evaluation of this technological intervention. The assessment revealed that midwives who completed the SMS training program reported a significant increase in their professional knowledge and skill acquisition post-intervention (Lori et al., 2012).
- A research team in Kenya utilized randomized cluster trials to evaluate the beneficial effects of SMS reminders on two groups of healthcare professionals. Findings indicated that providers in the experimental group that received the remote messages increased their adherence to malaria treatment guidelines by 24 percent above providers in the control group (Zurovac et al., 2011).

- A team of investigators in Korea evaluated the impact of SMS communication and other web-based technology on patients' capacity to manage their symptoms of diabetes. Findings suggest that patients in the experimental group utilizing web-based technologies to monitor their diabetic symptoms achieved a statistically higher degree of blood sugar control than patients in the control group (Kim et al., 2010).
- Researchers in Iran implemented a study using a controlled clinical trial to evaluate the effectiveness of SMS communication technology on patient compliance with diabetes education and management programs. Findings suggest that service providers in the experimental group who used the technology on mobile devices like smartphones to send messages to educate patients about physical exercise routines, diet, medication, and self-monitoring of blood sugar reported higher levels of compliance with the treatment and management of their disease compared to patients in the control group whose providers did not use this technological intervention for educational purposes (Goodarzi et al., 2012).
- A controlled clinical trial conducted by researchers in Tanzania suggests that provider use of digital decision-making devices significantly enhances their adherence to integrated management of childhood illness protocols. Findings indicate that routine use of mobile health supports improves communication between service providers and patients and upgrades supervisors' ability to evaluate the staff's job performance more objectively and efficiently (Mitchell et al., 2013).
- Similar outcomes were disclosed by a team of researchers performing a meta-analysis of technological studies published between 1996 and 2017. The assessment evaluated the efficacy of using digital technology to enhance patient compliance with HIV/AIDS treatment protocols. The review suggests that M-Health and E-Health technologies on cell phones and laptop computers increased the adherence rate of AIDS patients to antiretroviral therapy and follow-up clinic attendance. It also means that routine use of these technologies dramatically improves the level of self-care of patients contracting the AIDS virus and significantly reduces the risk of re-infection (Daher et al., 2017).

The research findings summarized above provide evidence of the benefits of establishing digital technology as a national standard for the future of medical education and experiential training. The analysis of global studies confirms that introducing technological innovation into clinical settings is one of the most efficient methods of streamlining the delivery of patient care across the broad scope of medical needs. The following sections focus on specific areas of medical specialization where technology has significantly contributed to the treatment outcomes of patients with life-threatening illnesses and psychiatric disorders.

USING TECHNOLOGY IN THE TREATMENT OF CANCER PATIENTS

Using technology to enhance the survival rates of patients with advanced cancer is fast becoming the standard intervention model implemented by physicians specializing in treating debilitating, life-threatening illnesses worldwide. Results of epidemiological research by the World Health Organization cited in an article released online last July indicate that between 2000 and 2019, the spread of cancer became the leading cause of mortality among adults of both sexes before age 70 in 112 of 183 countries and a primary barrier to increasing human life expectancy on a global scale. International public health data analyzed in 2020 disclosed that 58.3 percent of global mortality from cancer occurred in Asia. The Indian Council of Medical Research projects that by 2025 the nation is expected to see a 12 percent increase in the number of cancer cases adding 1.56 million people to the global burden of the disease (Priyadarshini & Bhargava, 2021).

Prior research by investigators in China and India has produced data on global cancer rates that support the above findings. Data from two epidemiological studies sponsored by the World Economic Forum released in 2019 revealed that approximately 23.6 million newly diagnosed cases of advanced cancer and 10 million cancer-related deaths were reported in the United States and abroad. Results indicated the global incidence of cancer rose an average of 26.3 percent, and the average mortality rates for patients in the final stages of the disease increased by 21 percent, a comparative rate much higher than what was reported in 2010 (Liu & Bhargava, 2022).

Based on these statistics, the investigators argue that a substantial proportion of the recent uptick in late-stage cancer cases and mortality is directly attributable to government-backed constraints on the healthcare system linked to the COVID-19 pandemic. The studies found that delaying access to outpatient cancer diagnostic, treatment, and preventive services because of severe pandemic-related public health restrictions imposed by local health departments has had a devastating impact on the help-seeking behavior of thousands of cancer patients, causing massive numbers to forgo in-house diagnostic testing and hospital administered treatment interventions until the latter stage of the disease process when these measures are least effective.

Researchers contend that dire circumstances mitigated by the COVID-19 health crisis have had a significant impact on the medical community's continuing battle against cancer, which has led many oncologists and other allied service providers on the frontlines of cancer diagnosis and treatment to incorporate digital technologies into their medical practice as a means of delivering comprehensive outpatient care to those suffering from the disease. Drs. Liu and Bhargava state that one of the essential lessons medical experts have learned from the pandemic is that digital technology is a crucial component of the team approach to the care of patients with cancer (Ibid, 2022).

Technology has made significant strides that have bridged the communication gap between service providers and patients through access to critical information that enhances their knowledge about the benefits of cancer therapies. Scientific breakthroughs driven by technological innovations, like genetic sequencing and computer-assisted screening techniques driven by artificial intelligence, have made it possible for doctors to detect, diagnose, and treat specific early-stage and advanced cancers in adults and children more precisely, thereby raising the probability of positive therapeutic outcomes for patients afflicted by the disease. Physicians recognize that computer-assisted diagnostic screening tools guided by artificial intelligence are a significant game-changer in the war against cancer. Providers conjecture that integrating Al technology into clinical care for cancer can reduce health disparities, especially in institutional settings with limited healthcare resources. They contend that incorporating Al technology into multifaceted treatment protocols for cancer can:

- Aid in the genomic characterization of cancer-related tumors;
- Accelerate the development of new medications used to treat the disease;
- Upgrade the speed of medical decisions that improve therapeutic outcomes for patients.

In addition, the investigators posit that Al-based algorithms have the potential to pave the way for scientists to identify genetic mutations and aberrant protein interactions that are often present at the early stage of many types of cancer. They acknowledge that because this type of medical technology is so versatile, its utilization is highly adaptable to the medical needs of biomedical researchers who want to bring its many advantages into healthcare facilities worldwide safely and effectively (Priyadarshini & Bhargava, 2021).

Diagnostic screening tools powered by artificial intelligence are just one of the many forms of cutting-edge technology introduced by the collaborative partnership between the Fourth Industrial Revolution for Sustainable Transformation of Cancer Care Project and Microsoft Corporation's Global Health Initiative. The collaboration between these two research giants has enhanced the treatment options for patients with cancer. Two types of breakthrough technology that are making significant inroads in the battle against cancer are machine learning (ML) and natural language processing (NLP). These technologies assist cancer specialists worldwide in developing treatments tailored to the specific medical needs of the individual undergoing cancer therapy. For example, one application, the *Inner Eye*, pairs machine language with computer-generated visualization to give radiologists detailed information about the progress of a patient's tumor. This technology is under investigation by researchers at Addenbrooke Medical Center in Cambridge. They are using it to develop artificial intelligence models from in-house medical data capable of distinguishing between cancerous tumors and healthy tissue on patient scans. A related technological advance developed through a collaborative partnership between the British Columbian Cancer Research Centre in Vancouver and Microsoft Canada is called *Single Cell Genomics,* which allows doctors to view a computer simulation of the genomes of individual cancer cells in real time. Researchers believe this kind of computer-augmented detail will enable cancer specialists to combine specific treatment options that will allow them to accurately predict how individual cells inside a patient's tumor will react to chemotherapy (Ibid, 2022).

Another computer-generated technological advance with similar potential developed by Microsoft is the Bio Model Analyzer (BMA). This cloud-based visual simulation tool allows biologists to model how cells interact and communicate with one another and monitor the connections they make. Cancer researchers view BMA technology as a potent therapeutic instrument in the diagnosis and treatment of cancer because it allows physicians to detect the disease early and aids them in developing models that can predict which medications will be most effective for patients to be treated with at different stages of the illness and determine at what point they may become resistant to their therapeutic effects. In partnership with AstraZeneca, Microsoft is using BMA technology to study drug interactions and resistance in patients with certain types of leukemia. Together, they have launched Project Hanover, a technology-based scientific initiative that combines the disciplines of biomedicine and computer science to connect fragmented pieces of relevant medical information about the genetic composition of tumors so that doctors can apply the most effective therapies to cancer patients. In conclusion, the technologies in this review account for just some of the many ways companies like Microsoft are changing the face of cancer care for thousands of patients in the global community. Cancer researchers believe that computer systems supported by artificial intelligence will be the technology that will eliminate the socioeconomic burden of cancer for millions who suffer from this life-threatening illness around the world (Ibid, 2022).

USING TECHNOLOGY TO TREAT PATIENTS WITH PSYCHIATRIC DISORDERS

Technology is revolutionizing the provision of mental health services to individuals with psychiatric disorders in the United States and other countries. Research by the National Institute of Mental Health

(NIMH), released online, shows that recent advances in digital technology and information sharing are opening up new frontiers in the field of mental health support and data collection that are changing the dynamic of interaction between patients and service providers in the context of the therapeutic relationship. Like other life-altering medical conditions discussed earlier in this chapter, the expanding utilization of sophisticated sensor-driven apps on smartphones and digital tablets is creating new ways for patients to access support services when experiencing psychological distress.

These apps have the technical capability to monitor patients' typical behavior patterns and send a remote distress signal to service providers when a therapeutic intervention such as peer counseling is required to head off the onset of an emotional crisis. NIMH researchers posit that mental health apps offer practitioners a valuable opportunity to assess the complex social and psychological relationships influencing the individual's psychiatric illness. The online digital assessment tools provide therapists with the necessary clinical information that will enable them to develop targeted treatment options tailored to the management of specific behavioral aberrations linked to the patient's illness. The app technology also allows researchers to collect remote data, determining the degree of incidence of mental health disorders in the general population (NIMH, 2019).

Mental health experts posit as society becomes more comfortable discussing psychiatric disorders like depression, anxiety, and other behavioral abnormalities, app technology will take on a more significant role in treating these conditions that impact the lives of so many Americans. According to an article published on Digital Health Today, a web-based global platform for leadership in digital health, app technologies provide clinicians with four ways to influence how mental health disorders are treated. One of the fastest growing of these is virtual reality (V.R.), a mobile headset tool that psychotherapists use to desensitize patients with post-traumatic stress disorder (PTSD) and other stress-related psychiatric illnesses through a virtual recreation of the personal triggers that underly the behavioral symptoms causing the patient's distress. V.R. technology has proven to be a highly effective method of teaching patients coping skills to manage their illnesses effectively. In addition, several popular mental health apps and websites, like Calm, Headspace, Talkspace, Pacifica, and Moodnotes, provide a safe technology-based environment online where individuals can go to learn practical techniques for managing their mood swings and access support services from a licensed clinician. Two of these top-rated remote apps, Calm and Pacifica, use guided meditation, breathing programs, and other techniques to improve mindfulness which helps patients build resilience to feelings of anxiety. Other mental health apps like Moodnotes help patients with mood disorders identify troublesome patterns in their behavior and raise self-awareness, so they can implement appropriate measures to correct them. Patients use these apps in conjunction with online psychotherapy with a clinician trained to offer insight and social support as part of the treatment process (Kendall, 2018).

Researchers evaluating the efficacy of delivering mental health services through digital technology believe that this method of providing care has several advantages that can benefit both patients and professionals. A few of the potential benefits of technology-based care are:

- Convenience: Technology-based mental health appointments do not require the patient to take time off from work or coordinate mass transportation and childcare logistics. Patients can schedule virtual meetings with their therapist on a more flexible schedule that fits in with their daily routine and generally requires less advance notice.

- Anonymity: Technology-based care enables patients with psychiatric disorders to seek and receive treatment services from a mental health professional privately without the involvement of other people.
- Lower cost: Some mental health apps can be downloaded online for free. In contrast, others are highly inexpensive, making remote treatment services accessible to a more significant proportion of the general population than traditional therapeutic alternatives.
- Twenty-four-hour service: Technology-based care allows psychotherapists to provide treatment intervention and monitoring services to patients around the clock.
- Consistency: Technology enables psychotherapists to provide the same intervention services to all patients that use it.
- Broader reach: The technology makes mental health services more readily available to a wider proportion of consumers across the United States, particularly those in rural areas where access to care in emergencies is limited or nonexistent.
- Fewer barriers: Technology-based mental health services eliminate social barriers by reducing the stigma that causes patients with psychiatric disorders to be hesitant about seeking care in the traditional system.
- Support: Mental health professionals recognize that technology-based treatment services can augment traditional care by reinforcing coping skills learned in psychotherapy and extends social support and monitoring for patients outside the therapist's office.
- Advances in technology: The growing demand for treatment services by patients with psychiatric disorders is the primary motivator encouraging many mental health professionals nationwide to transition their professional practice to remote access (Ibid, 2022).

Although web-based counseling services are proving to be of significant benefit to patients with psychiatric disorders, some mental health experts have raised critical concerns about the potential drawbacks of providing supportive care using this method. Many argue that technology-based mental health services can create social and economic barriers that inhibit individuals from seeking assistance in a crisis. Immediate reservations that patients and professionals have about the delivery of intervention services via remote technology are summarized below:

- Access: Mental health professionals contend that a significant obstacle to delivering treatment services through web-based technology is that some patients, particularly those in economically disadvantaged communities, may have reduced access to internet connections or other digital devices that would enable them to receive care remotely.
- Quality issues: Mental health professionals argue that another potential limitation affecting the quality of treatment services patients receive through remote access is inconsistent technological sophistication and service rates available in many regional locations.
- Cost: While some mental health experts argue that web-based intervention services offer patients with psychiatric disorders a less expensive alternative to traditional in-house care, in contrast, others contend that continuous upgrades and maintenance of digital equipment, web-based platforms, and patient/provider networks can have a detrimental impact on the cost of online psychotherapy, making services more expensive and thereby unattainable for many patients, especially those without health insurance.

- Privacy: Critics of the extended utilization of online mental health services by patients with psychiatric disorders have voiced concern that video cameras and other virtual surveillance raise significant confidentiality issues that impact the professional relationship between patients and service providers. They believe that video recording equipment in many home environments infringes on professional privacy, making some patients hesitant to share sensitive personal information with a provider in situations where they might not feel secure.
- Insurance coverage: Mental health professionals acknowledge that the recent expansion of technology-driven healthcare services during the COVID-19 pandemic has significantly changed the policy provisions of the health insurance industry so that essential services can remain accessible to the general public during this crisis. However, as public restrictions imposed by the COVID-19 outbreak begin to ease, many providers are concerned that flexible payment provisions for online clinical services may not be permanent, creating an economic barrier that further compromises patient access to care (Ibid, 2019).

While the emerging era of mental health technology offers patients and professionals some promising alternatives that can significantly broaden the scope of traditional therapeutic intervention, some critics remain skeptical about the effectiveness of the care that patients receive utilizing this approach. Many have voiced concern about the regulatory challenges that online therapy presents to state licensing boards, whose quality assurance standards vary from state to state. Finding ways to address these issues adequately has become a significant focus of the mental health community in partnership with the software development industry. Researchers and software development engineers are combining their professional skills and talents to address critical issues associated with delivering mental health services via the internet. Some of the specific areas of app development and evaluation receiving a significant amount of collaborative professional attention are:

- The use of self-management apps that enable patients to set medication reminders and develop strategic tools that can assist them in managing the psychological symptoms of stress, anxiety, and sleep disorders related to their condition.
- The use of certain specialized apps can assist patients with cognitive remediation to improve their thinking skills significantly.
- The use of skills-training apps containing online educational tools that can assist patients in learning new coping and cognitive processing skills to manage their illness more effectively.
- The use of illness management and social support apps enables patients to connect with a peer counselor directly or send necessary clinical information to a service provider online to get the help they need to manage their behavioral symptoms during a crisis.
- Using data collection apps powered by smartphone sensors enables service providers to gather passive information that allows them to monitor patterns of physical movement and social interaction that patients have with other people via text messaging and phone calls to determine their state of mind throughout the day. The apps alert caregivers to changes in a patient's behavior patterns that signal mood swings such as depression, mania, or psychotic episodes requiring professional intervention.

Researchers examining the impact of mental health apps on treating and managing patients with psychiatric disorders have determined that this technology holds significant promise in terms of its capacity to transform how intervention services are delivered. The results of a groundbreaking study by Dr. Patricia Arean, Professor of Psychiatry at the University of Washington, and her colleagues provide evidence that substantiates this claim. The research team used various strategies, like placing ads on city buses, in newspapers, and on social media platforms like Twitter, along with advertisements on popular internet websites such as GoogleAdwords and Craigslist, to recruit patients willing to participate in the study, which utilized a controlled trial approach to treatment intervention and clinical assessment.

Those interested in joining the trial received specific instructions from investigators that guided them to the study's website, where they answered asked a set of questions about their mobile device use. They were also asked to view a video presentation about informed consent and answer more detailed questions that indicated they had a general understanding of the goals and details of the research.

To determine eligibility, investigators administered a short quiz to patients to assess which ones met the inclusion criteria to be part of the sample. A total of 2,900 patients with symptoms of clinical depression were selectively chosen for inclusion five months into the recruitment process. Those selected for enrollment were randomly assigned to one of three technology-based depression interventions and required to download a treatment app to their mobile device after receiving instructions. Patients enrolled in the 12-week Trial were asked to use their treatment app daily and complete self-assessments of their progress, which was subsequently monitored on customized dashboards by members of the research team. Each app in the pilot test contained a different therapeutic intervention designed to manage the clinical symptoms of depression. Researchers also utilized the treatment apps to collect supplementary data on various aspects of patients' social behavior, such as the number of text messages sent on their smartphones and the type of interactions they exhibit when they visit family and friends. Patients in the clinical trial received financial compensation from the investigators for their participation. Rates of completion for patients enrolled in the three-month clinical assessment for the treatment apps were as follows:

- Sixty-six percent of patients completed one month of the Trial;
- Fifty percent of patients completed two months;
- Forty-one percent of patients completed the entire three months of the assessment.

When investigators compared attrition rates for patients in the app-based clinical trial with other studies, they found that those in the technology-driven assessment had a lower dropout rate than patients involved in research that used a more traditional therapeutic approach that involved at least some face-to-face human contact. Findings determined that patients who used the three app-based intervention systems to manage their symptoms of depression effectively had a lower incidence of mood disturbances and less psychiatric disability over time compared to those treated with conventional forms of psychotherapy. The research team concluded the outcome of the controlled clinical trial utilizing app technology provides a window into the future of delivering comprehensive mental health services to a broader population of patients that need access to psychiatric care. They contend that the results of technology-based studies such as this will open the door to new horizons, upgrading the quality of psychiatric services patients with depression and related disorders receive from the mental health system (Anguera et al., 2016).

SUMMARY AND CONCLUSIONS

The research findings highlighted in this chapter document the many profound benefits that digital technology can bring to global healthcare systems. Evaluations of the substantive contributions of technological innovation to improving medical education and services are very promising. Experts acknowledge that the multifaceted global expansion of digital health systems makes a vast difference in how patients with various medical conditions seek and access initial treatment services and receive provider follow-up care. To date, institutional and public policy decisions to incorporate technology into the multiple levels of education and training for medical students and the delivery of necessary services to patients have been supported by the following systemic objectives;

- Expand the availability of preclinical instruction and post-clinical experiential learning for the global population of college students that want to become practicing physicians
- Improve accessibility to critical care services by patients as needed through the adoption of digitization that allows providers to diagnose, treat, and prevent the onset of illness more effectively than traditional methods
- Lower the economic burden of the extreme cost of treatment services that often creates a prohibitive barrier for patients seeking adequate medical care
- Give providers improved access to the most up-to-date information about the patient's medical history so that appropriate therapies can be developed to address current health concerns in a timely and efficient manner
- Develop universal standards for government oversight and provider monitoring of the quality of therapeutic services offered to patients through digital systems of care
- Implement methodological strategies for collecting reliable patient data that support the rationale for transforming traditional systems of delivering care into a digital model

Although a significant proportion of the digital conversion of healthcare systems is still in its formative stage, policymakers and healthcare professionals worldwide believe that the long-range goals for system-wide change in service delivery outlined above will become considerably more attainable over the next several decades. As the integration of M-Health and E-Health technologies become more commonplace in the delivery of medical care, the expectations and responsibilities of patients and providers will undergo a radical shift that will permanently alter the traditional dynamic of the doctor/patient relationship.

While some health experts and policy professionals argue that the increasing acceptance of digital technology as a significant part of the changing healthcare system will facilitate the need to implement more rigid national standards to restrict corporate takeover of healthcare, in contrast, others conjecture that the expansion of this technology will serve to provide all subdivisions of the industry with new opportunities to gain the competitive edge needed to improve the quality of patient-centered services related to achieving long-term business goals. With this prospect in mind, it is not surprising that many healthcare analysts believe that whatever the outcome of this debate, one thing is sure the future of delivering the full range of quality care services to patients with varying degrees of medical need rests squarely on the shoulders of providers who understand the advantages of using technology to make the long-term goals of the system a reality (Ricciardi et al., 2019).

REFERENCES

Al-Shorbaji, N. (2021). *Improving Healthcare Access through Digital Health: The Use of Information and Communication Technologies. Healthcare Access.* doi:10.5772/intechopen.94675

Anguera, J. A., Jordan, J. T., Castaneda, D., Gazzaley, A., & Arean, P. A. (2016). Conducting a Fully Mobile Randomized Clinical Trial for Depression: Access, Engagement, and Expense. *British Medical Journal Innovations, 2*, 14-21. Retrieved from https://innovations.bmj.com/content/bmjinnov/2/1/14.full.pdf

Appel, J., Race, J., Yang, E., & Jumreornvong, O. (2020). Telemedicine and Medical Education in the Age of COVID-19. *Academic Medicine, 95*(12), 1838–1843. doi:10.1097/ACM.0000000000003711 PMID:32889946

Berman, N. B., Durning, S. J., Fischer, M. R., Huwendiek, S., & Triola, M. M. (2016). The Role of Virtual Patients in the Future of Medical Education. *Academic Medicine, 91*(9), 1217–1222. doi:10.1097/ACM.0000000000001146 PMID:26959224

Cancer. (n.d.). In *Merriam-Webster Dictionary*. Retrieved July 24, 2022, from https://www.merriam-webster.com/dictionary/cancer

Clinical Trial. (n.d.). In *Merriam-Webster Dictionary*. Retrieved July 24, 2022, from https://www.merriam-webster.com/dictionary/clinical%20trial

Daher, J., Vijh, R., Linthwaite, B., Dave, S., Kim, J., Dheda, K., Peter, T., & Pai, N. P. (2017). Do digital innovations for HIV and sexually transmitted infections work? Results from a systematic review (1996-2017). *BMJ Open, 7*(11), e017604. doi:10.1136/bmjopen-2017-017604 PMID:29101138

Digital Technology. (n.d.). In *Dictionary.com*. Retrieved July 24, 2022, from https://www.dictionary.com/browse/digital-technology

Distance Learning. (n.d.). In *Merriam-Webster Dictionary*. Retrieved July 24, 2022, from https://www.merriam-webster.com/dictionary/distance%20learning

Goodarzi, M., Ebrahimzadeh, I., Rabi, A., Saedipoor, B., & Jafarabadi, M. A. (2012). Impact of Distance Education via Mobile Phone Text Messaging on Knowledge, Attitude, Practice and Self-efficacy of Patients with Type 2 Diabetes Mellitus in Iran. *Journal of Diabetes and Metabolic Disorders, 11*(10), 10. doi:10.1186/2251-6581-11-10 PMID:23497632

Guze, P. A. (2015). Using Technology to Meet the Challenges of Medical Education. *Transactions of the American Clinical and Climatological Association, 126*, 260–270. https://www.ncbi.nlm.nih.gov/pmc/articles/PMC4530721/ PMID:26330687

Kendall, D. (2018, May). Four Ways Technology is Helping Mental Health. *Digital Health Today*. Retrieved from https://digitalhealthtoday.com/blog/technology-and-mental-health/

Kim, C. S., Park, S. Y., Kang, J. G., Lee, S. J., Ihm, S. H., Moon, G. C., & Yoo, H. J. (2010). Insulin Dose Titration System in Diabetes Patients Using a Short Messaging Service Automatically Produced by a Knowledge Matrix. *Diabetes Technology & Therapeutics*, *12*(8), 663–669. doi:10.1089/dia.2010.0031 PMID:20615108

Liu, C., & Bhargava, R. (2022). *6 Experts Reveal the Technologies Set to Revolutionize Cancer Care.* World Economic Forum, Project on Health and Healthcare, Beijing China and The Fourth Industrial Revolution for Health Project, India. Retrieved from https://www.weforum.org/agenda/2022/02/cancer-care-future-technology-experts/

Lori, J. R., Munro, M. L., Boyd, C. J., & Andreatta, P. (2012). Cell Phones to Collect Pregnancy Data from Remote Areas in Liberia. *Journal of Nursing Scholarship*, *44*(3), 294–301. doi:10.1111/j.1547-5069.2012.01451.x PMID:22672157

Medical assessment. (n.d.). *Workplacetesting.com.* Retrieved July 24, 2022, from https://www.workplacetesting.com/definition/17/medical-assessment-occupational-health-and-safety

Medical Education. (n.d.). In *The Free Dictionary by Farlex.* Retrieved July 25, 2022, from https://medical-dictionary.thefreedictionary.com/medical+education

Mitchell, M., Hedt-Gauthier, B. L., Msellemu, D., Nkaka, M., & Lesh, N. (2013). Using Electronic Technology to Improve Clinical Care – Results from a before-after cluster trial to evaluate assessment and classification of sick children according to Integrated Management of Childhood Illness (IMCI) protocol in Tanzania. *BMC Medical Informatics and Decision Making*, *13*(1), 95. doi:10.1186/1472-6947-13-95 PMID:23981292

National Institute of Mental Health. (2019, September). *Technology and the Future of Mental Health Treatment.* Retrieved July 18, 2022, from https://www.nimh.nih.gov/health/topics/technology-and-the-future-of-mental-health-treatment

Orton, M., Agarwal, S., Muhoza, P., Vasudevan, L., & Vu, A. (2018). Strengthening delivery of health services using digital devices. *Global Health, Science and Practice*, *6*(suppl 1), S61–S71. doi:10.9745/GHSP-D-18-00229 PMID:30305340

Pottle, J. (2019, October). Virtual Reality and the Transformation of Medical Education. *Future Healthcare Journal, 6*(3), 181-185. Retrieved from https://www.ncbi.nlm.nih.gov/pmc/articles/PMC6798020/?report=classic

Pryadarshini, K., & Bhargava, R. (2021). *These Al Projects are Improving Cancer Screening and Outcomes.* World Economic Forum, Fourth Industrial Revolution for Health, India and Worldwide Health Microsoft Asia. Retrieved from https://www.weforum.org/agenda/2021/07/ai-projects-improving-cancer-screening-outcomes

Psychiatric Disorder. (n.d.). In *The Free Dictionary by Farlex.* Retrieved July 24, 2022, from https://medical-dictionary.thefreedictionary.com/Psychiatric+disorders

Quality of Care. (n.d.). *World Health Organization*. Retrieved July 24, 2022, from https://www.who.int/health-topics/quality-of-care#tab=tab

Ricciardi, W., Barros, P. P., Bourek, A., Brouwer, W., Kelsey, T., & Lehtonen, L. (2019). Expert Panel on Effective Ways of Investing in Health (EXPH), How to govern the digital transformation of health services. *European Journal of Public Health, 29*(3), 7–12. doi:10.1093/eurpub/ckz165 PMID:31738442

Telemedicine. (n.d.). In *Merriam-Webster Dictionary*. Retrieved July 24, 2022, from https://www.merriam-webster.com/dictionary/telemedicine

Virtual reality. (n.d.). In *Merriam-Webster Dictionary*. Retrieved July 24, 2022, from https://www.merriam-webster.com/dictionary/virtual%20reality

Zurovac, D., Sudoi, R. K., Akhwale, W. S., Ndirtu, M., Hamer, D. H., Rowe, A. K., & Snow, R. W. (2011). The Effect of Mobile Phone Text-message reminders on Kenyan Health Workers' Adherence to Malaria Treatment Guidelines: A Cluster-randomized trial. *Lancet, 378*(9793), 795–803. doi:10.1016/S0140-6736(11)60783-6 PMID:21820166

ADDITIONAL READING

Boruff, J. T., & Storie, D. (2014). Mobile devices in medicine: A survey of how medical students, residents, and faculty use smartphones and other mobile devices to find information. *Journal of the Medical Library Association: JMLA, 102*(1), 22–30. doi:10.3163/1536-5050.102.1.006 PMID:24415916

Eggermont, S., Bloemendaal, P. M., & Van Baalen, J. M. (2013). E-learning anytime, anyplace anywhere on mobile devices. *Perspectives on Medical Education, 2*(2), 95–98. Advance online publication. doi:10.100740037-013-0045-4 PMID:23670682

Fan, S., Radford, J., & Fabian, D. (2016). A mixed-method of research to investigate the adoption of mobile devices and Web2.0 technologies among medical students and educators. *BMC Medical Informatics and Decision Making, 16*(1), 43. doi:10.118612911-016-0283-6 PMID:27094813

KEY TERMS AND DEFINITIONS

Cancer: A term used by medical oncologists to define a pathological state in which the human organism develops progressive tumors that can spread throughout the body and lead to death (Merriam-Webster.com, 2022).

Clinical Trial: A descriptive term used by medical researchers to determine the effectiveness of a therapeutic agent under controlled conditions with the consent of human subjects (Merriam-Webster.com, 2022).

Digital Technology: A generic term coined by scientists and engineers to define the process of computer-generated information sharing that facilitates online communication between individuals via text messages, emails, and internet-based social media platforms (Dictionary.com, 2022).

Distance Learning: Higher education professionals define this teaching method as an internet-based alternative to traditional classroom learning that does not require face-to-face interaction between students and teachers and has greater flexibility (Merriam-Webster.com, 2022).

Medical Assessment: A term used by physicians and other healthcare professionals to define the evaluation of a patient's state of health utilizing specific diagnostic tests (Workplacetesting.com, 2022).

Medical Education: A term used by medical school faculty to define the instructional process involved in educating and training future service providers in the art of medicine (Thefreedictionary.com, 2022).

Psychiatric Disorder: A term mental health professionals use to define the pathological manifestations of a psychological syndrome characterized by disruptive behavioral symptoms that can affect an individual's mental state (Thefreedictionary.com, 2022).

Quality of Care: A term used by the World Health Organization to determine the degree to which health services administered to individuals or entire populations increase the likelihood of the desired outcome (WHO.com, 2022)

Telemedicine: A term used to describe the computer-enhanced process of delivering healthcare services to patients remotely utilizing two-way virtual or visual communication on mobile devices like smartphones, laptop computers, and tablets (Merriam-Webster.com, 2022).

Virtual Reality: A term scientists use to define an artificial environment produced by computer generation that individuals experience through sensory stimulation created by the device (Merriam-Webster.com, 2022).

Chapter 3
Social and Economic Maxims of ICT in Education

Sonia Bhandari
Rajdhani College, New Delhi, India & University of Delhi, India

Aarti Bhandari
Lilawati Vidya Mandir, Delhi, India

ABSTRACT

ICT in education, in terms of terminology and structural composition, may carry out three basic components, namely education, information, and communication technology. ICT in education has its sole concern of identifying the most suitable and appropriate use of developed technology (both hardware and software) for serving the educational needs of society at a particular time and place. Education using ICT tools also aids in expanding educational opportunities to socially vulnerable section of people such as low-income groups, women, aged, and differently abled. This effectiveness of ICT in education depends on the presence or absence of the conditions and elements, facilitating or obstructing the proper flow of information and technological methods. Proceeding from simple to complex always provides an appropriate learning sequence. Hence, a wise teacher cares for a beautiful integration of psychological principles to logical organisation of sequence of activities in his teachings for social and economic growth of the community altogether.

INTRODUCTION

ICT in education is a method of using information and communication technology to build, enhance, empower, motivate and engage learners inclusively by facilitating the attainment of basic skills and lifelong learning for all. Educators, parents and students have long complained about the one-size-fits-all model of education through conventional schooling, but COVID-19 has changed the way the education was imparted globally. Education is a complex ecosystem and is galloping towards the future with a speed of change in top gear as things we imagined once to be fiction are all turning into reality. Integrating ICT appropriately in education helps to expand access to education, strengthen the relevance and quality of

DOI: 10.4018/978-1-6684-6620-9.ch003

education to different masses from various economic and social backgrounds (Kanematsu and Barry, 2015). The cost-effectiveness of edtech sector allows learner to overcome the paywall between them and premium educators, and virtual nature of this learning erases geographical constraints. The allocation and management of ICT resources provides more creative solutions to different types of learning inquiries. ICT enables teacher-learner community to communicate, share, and work collaboratively anywhere, any time. Better navigation of resources, animated graphics and interactive platforms allow for a more engaging educational experience. However, edtech industry making use of ICT in education still has its own challenges and drawbacks. Information and communication technology growth is dwarfed by inadequate infrastructure, poor accessibility and conditions of digital divide. Skills like problem solving, logical reasoning, collaboration and communication are essential in organisations today. When new age technologies are pervading all our interactions, can we afford to be ignorant about them? As we hurtle towards an era of digital economy, positioning oneself with combination of domain knowledge and digital skill will set them apart. The educational institutions are tying up with edtech players to introduce ICT based programs at school and college level making an effort to integrate it with core curriculum. For fulfilling the expectations of classroom, distance and correspondence education, ICT provides learners with access to all types of programmed learning modules from beginner to advanced levels with ease through computers and personal digital machines. For the management of affairs of educational practices in an effective, fruitful and fecund way, one must approach with certain professional responsibilities and applications (Jo Shan Fu, 2013). In this cause, educational institutions and related stakeholders are strongly advised to integrate communication & technology among all learning levels and across all of the learning spaces. When this is accomplished, learners are able to attain higher levels of cognition in context of a particular activity. The time is ripe for all the concerned stakeholders in the education sector to join hands and tap the umpteen opportunities knocking at our door for successful implementation of social and economic maxims of ICT in education.

BACKGROUND

Throughout the journey of life, learning is a continuing activity where people change their assumptions and beliefs by seeking rational knowledge, which departs from traditional approaches and practices within a society. Be willing to seek out new sources of knowledge and implementing them in an educational structure plays the role of remodelling of community, both socially and economically. The youth of today is brimming with potential and advanced skills aspiring to have opportunities for a better life. It is worthy to consider Information and Communication Technology (ICT) as a powerful tool for educational change and reform due to its easy accessibility and powerful presentation across the globe (Antonio, Gomes, and Marcelino, 2015). Information and communication technology (ICT) includes computer hardware, the internet and application software, and numerous other electronic automated delivery systems used in today's education field. With the advent of science and technology era, developing skills for using ICT in education will be an indispensable prerequisite for the teachers as well as the learners to radiate a sense of self-esteem and communicate better with different sections of the world. Today‘s society is a world of information explosion that is taking place at a expeditious pace. Even an educated and knowledgeable person discern as if he or she is capable enough to cope up with such an information explosion. Information Technology is nothing but the rapidly increasing amount of published information and its methodical management to help reduce disproportionate learning and behavioural effects within the society.

Information technology helps building communication networks in a society and is defined as the acquisition of facts and figures, its processing and manipulation, storage statistics and dissemination of detailed knowledge. This distribution of processed knowledge can be in the form of a vocal note or speech, pictorial illustrations, textual or animated particulars through any guided and secure telecommunication medium. On the other hand, communication technology implies the transmission or reception of data or information, knowledge comprehension, skills, awareness and understanding needed to exchange information either verbally or non-verbally. It is the activity of designing, processing and coding of information, sending it via a medium or a communication system and decoding this information at the receiver's end, converting it back to its written, oral or gesture form. This is very much dependant on constructing and maintaining correct choices of communication systems for connecting geographically dispersed regions and serving the purpose of real time access to information around the world.

Information and communications technology (ICT) in education is the science of social-cognitive processing of information along with its communication solutions and provisions that potentially support teaching, learning and various other developmental activities in education (Chris Abbott, 2003). Utilizing ICT in education tends to expand access to educational pedagogy with an objective of:

1. Preparing pupils for state-of-the-art learning experiences
2. Reinforcing their credence and values while sharing the experience
3. Relating the experience with the behavioural conditioning and thus stimulating further learning.

Even though efforts to transform a nation into a digitally-enabled knowledge-based economy have been taking shape for a few years now, but one should be aware of the promises and pitfalls in such re-vamping policies. With the advent of recent research developing newer and more efficient technologies in education with ICT, great ameliorations can be brought in learner's behaviour and response that can have a magnificent impact in society both socially and economically. Assimilating ICT in education has an impact on following key sectors resulting in the advancement of a community:

1. Progress in science and technology for raising the standard of living and improving the quality of life.
2. Modernisation, improvement and remodelling of community services, such as health, banking, transport, and other district services.
3. Transformation and innovation, by fostering and encouraging creativity with better management of human resource.

WHAT MAKES ICT IMPORTANT IN EDUCATION

There are different kinds of learning such as cognitive, reflective, logical, emotional, social, physical, sequential, and spiritual learning. Good communication skills are essentially needed to understand information more accurately and quickly. Education helps people become better citizens and gives the community an ability to change their coming future. Hence, while learning with the incorporation of ICT, there must be a focus on relationships of things between the educational system and the required learning environment. ICT has changed the way we communicate with each other and the manner in which we manage our social lives. Technological changes of any kind in the field of education will require better governing principles and tools to be safe, secure, and effective. However, the factors responsible for the progress of information and communication technology in education also cannot be overlooked. The numerous factors causing the development of educational technology are:

1. Rising costs of infrastructure for development related to transport, energy and social networking opportunities.
2. Population explosion of different culture and demography
3. Problems related to health and economy
4. Shortage of trained educator staff, unemployment and lack of government initiatives

ICT in education promotes self-regulated learning. Self-regulation abilities include self-monitoring, self-reinforcement, and goal setting. With a self-regulated learning, one is able to control and understand one's learning environment. It is a self-paced graduated learning with the purpose of developing cognitive, behavioural, emotional features of the learner. For the attainment of efficient and effective use of ICT, communication between the teacher and student is essential (Bridget Somekh, 2007). But here, communication is more than "teacher talks and student listens".

The importance of ICT learning gives teacher an opportunity to adopt a paraphrase policy, styling the learners to be more accommodating, tolerant and forbearing. With inclusion of ICT and paraphrasing, listeners must accurately summarize what a speaker has said before being allowed to respond. Paraphrasing is more than a classroom exercise and can be the first step in communicating with the students other than classrooms. With ICT, a teacher should use empathetic listening to allow the students to find a solution and stay motivated. Hearing the intent and emotions behind what student says and reflecting them back with paraphrasing help the teacher keep the communication lines open. At the same time, ICT enables the innovative use of vast educational resources and the re-establishment and improvement of learning methods leading to the advancement of existing social and cultural conception. A few factors that highlight the importance of the role of ICT in education are given below –

Globalization and Education Reforms

Globalization and technological change conjunction have accelerated global literacy rate over the past few years creating a new economic shift worldwide. This technological contribution is very much evident in respect of the proliferation of our public sectors such as transport, petro-chemicals, banks, hospitals, factories, agriculture and offices. But, our classrooms and educational institutions have still remained inconsiderate to the technological inputs and their influences on the learning abilities of the pupils. At the same time, the effective integration of ICTs into an educational system being a complex, and multifaceted process requires not just technology, but enough initial capital to be able to reach out to the masses. The prolonged pandemic has caused a paradigm shift in the learner's behaviour and their requirements. The necessity of getting the technology available and accessible for all, indeed we need to look at the issues of development of curriculum and pedagogy, Institutional readiness, teacher competencies and long term financing, among others.

Education Technology and Community Revolution

The evolution and progression of information and communication technology has heralded a kind of revolution in our social, occupational and educational world. These advancements in educational technology are an outcome of use of broad range of technological solutions, emphasis on individual and self-regulated learning together with a sequel approach to education system. Individually paced calibration, elevation and incorporation of ICT in education for learners from different socio-economic backgrounds helps them gain deep understanding of an era of choices that they may use to revolutionize our society for the benefit of all living beings. The learners today, look for value propositions in the form of durability, high-computing capability, energy-efficiency, and an immersive experience that allows them to multitask.

ICT Inclusion and Quality of Human Learning

The very purpose of education using characteristic technological reforms is to facilitate and improve the quality of human learning by providing an inclusive environment (Woolfolk, 2010). Inclusion in education refers to all socio-economic learner groups being able to access and gain equal opportunities to education and learning. The aim of inclusion is to embrace a community irrespective of race, disability, gender, and economic status. It is concerned with achieving the goals—of maintaining internal discipline, the ability to quickly and successfully accept change and adapt effectively in response to the changing environment. Inclusion with ICT in education believes in the virtues of innovation, determination, passion, time management, quality, perseverance, value centricity, ingenuity, integrity, responsibility and honesty. From facing new challenges in this domain, to solving the varied problems of education successfully, various media of mass communication have been developed over the past decades. With ICT, the set of elements and efforts needed to guide for suitable child learning processes, and modern testing and evaluation techniques are essential for a comprehensive human learning. As such both quantitative augmentation and qualitative growth of education can be facilitated and stimulated with the help of educational technology to upgrade our society.

HOW ICT IS USED IN EDUCATION

ICT helps to interact, stimulate, motivate and engage with the learner. There are various advantages linked with the use of ICT in education as described below:

1. **Easy Access:** It is easy to get a quick access to information by connecting to the internet and surfing through relevant data.
2. **Smooth Availability of Updated Facts and Figures:** With the comfort of sitting at any indoor or outdoor place, the desired information can be conveniently accessed frequently. This helps the students to learn the updated information or any arduous content repeatedly. Teachers too can benefit and keep themselves in alignment of the latest teaching learning strategies and related technologies.
3. **Connecting Geographically and Culturally Dispersed Population:** ICT as a collaborative technology offers vivid functionalities for co-ordination of group work. With the advancement of ICT, libraries of solutions and best practice learning tools are available as education is no longer

restricted within four walls of the educational institutions. Students from different parts of the world can learn together by using online and offline platform cum resources.

4. **Catering to the Individual Differences:** ICT can contribute in communication and knowledge construction as per individual needs of the learners capabilities and interest. To consider the needs of every student in the class has always been a challenge for the teacher. But a collaborative learning through ICT encourages and enables students to work and learn together, explore a problem and find its solution to complete the task.
5. **Wider Range of ICT Tools:** With the advent of ICT, different means of verbal and non-verbal communication are being introduced in the teaching-learning process. Offline learning, online learning, and hybrid learning are some of the resources that can be used in educational institutions. Collaborative learning, individualized and group learning strategies can enhance the quality of a society's emotional, cognitive and economic wellbeing.
6. **Wider Learning Opportunities for Pupils:** Application of latest ICT in education has provided various types of educational experiences to the teachers as well as the learners. Many options to the learners for a self-guided learning opportunity with courses of their choice are readily available. At the same time, scores of online courses are within easy reach of them to select as per their aptitude and interest.

A few innovations and strategies that can be incorporated with the use of ICT to improve teaching –learning process, enhance networking with resourceful persons, boost advancements of industrialized communities and foster the science of application of knowledge to practical use are stated below.

Approach to Instructional Design

For focus in our approach to instructional design, one must follow three rules of behaviourism, pragmatism and constructivism. Following the system approach along with the knowledge of what, how and where to integrate ICT tools and techniques, we can knit everything together in a teaching-learning process. One must understand the circumstances surrounding the learning situation in deciding which approach to learning is most appropriate (Latwal, Kommers, Mahajan, and Sharma, 2020). It is necessary to modify and accommodate constructivist values to the problems that require highly prescriptive solutions. Compared with a human instructor, control of technology is less apprehensive and adaptive to suit the learner's environment. Once a lesson plan of integration is implemented and executed, it is less likely to change unless recorded for student's reactions or responses. This is where instructional design plays an important role in bridging effectively pedagogy and technology together. Subject contents have to be well organized, structured and tailored for a goal achievement. At the same time, strategies for teaching have to be well-thought-out, self-directed and participative. Instructional design can help the educators make a rational and logical use of technology, thereby assuring a successful integration. Quality of course is ensured through instructional design supporting all the phases of content development. This designing and practice is important in order to accomplish and achieve learning of cognitive tasks as well as motor skills.

Nature of Instructional Design Technology

It studies the effect of science, psychology, and technology in a continuous, dynamic, progressive, and effective manner. New concepts like programmed learning, micro and macro teaching, simulated teaching, outmanoeuvre the modes of traditional teaching. Instructional design technology cannot replace a teacher but can be used successfully in a school system with a goal-oriented and self-directed teaching learning mechanism.

Characteristics of Instructional Design Technology

Instruction designing allows an educator develop a systematic application of scientific knowledge to the practical tasks. Integrating ICT with instructional design can help a teacher to structure the course according to the learner's needs and pace. This implementation is helpful in achieving cognitive objectives as well as is able to meet the shortage of effective teachers and controlling the individual differences in an educational institution.

Implementation of Instructional Design Technology

Instructional technology has several different aspects which includes the division of a practical task into separate sections and subsections. The distinct process of designing instruction involves:

1. Understanding the application of learning theories and methodologies needed to design instruction
2. Making the selection of materials, tools and techniques to design and implement the learning concept.
3. Manner in which the instruction design would be communicated and delivered to the learners for a particular goal-attainment.
4. Evaluation and procurement of design approach in terms of its effective and efficient use for an enhanced learning experience.

Assist Students in Accessing Digital Information

ICT is used as an efficient and effective tool by teachers and students to discover topics of interest, solve problems in a self-regulated manner, and provide solutions to these problems. ICT makes knowledge acquisition more accessible and attainable with strategic planning and futuristic approach (Meadows and Leask, 2012). Concepts of learning difficulties in areas of different demographic and socio-economic status are dealt with while engaging students in the application of ICT. Assisting and supporting student-centred and self-directed learning creatively is required in accessing digital information efficiently and effectively. The purpose is to build new knowledge and creative abilities through accessing, selecting, organizing, and interpreting information and data. Based on learning through ICT, students are more capable of using information and data from various sources, and act critically in analysing the quality of the learning materials.

Produce a Creative Learning Environment

Preservation of cultural and environmental ecosystems alongside empowerment of local communities, especially women and other marginalised groups is at the heart of ICT teaching-learning process. ICT involves purpose-designed applications that are meant to bridge the gap between the learner's current level of knowledge and skills with respect to the comprehensions required to perform a certain task. With blend of ICT in teaching-learning process, we can provide innovative ways to meet a variety of learning needs. Students not only acquire knowledge better together, but also share diverse learning experiences from one another by expressing themselves and reflecting on their learning behaviours. This can be in the form of some alternative reading applications such as e-books offering a reading-aloud interface, relevant vocabulary-building activities encouraging discussions and collaborations, games related to reading skills and vocabulary acquisition, and more to embrace creative classroom management. Therefore, students may have the opportunity to analyze problems, create visual goals, encourage autonomy and explore ideas to develop concepts. They may further evaluate ICT learning solutions where more opportunities are offered to develop critical and higher-order thinking skills. ICT helps students focus on higher-level concepts hence improving the quality of teaching and learning process. Therefore, the whole learning process creatively and innovatively enriches student's learning skills and broadens their consciousness and awareness beyond what they already know.

Distance Versus Traditional Classroom Learning

Distance learning help learners having social anxiety, health issues or no commuting means available. Distance learning appears to have significant positive effects of effective curriculum learning (W. Mark, 2006). But, such courses are likely to break down a learning experience when they are delivered as if they were traditional courses. ICT can be used to promote collaborative learning in a distance-learning environment with the following purposes:

1. To broadcast material of different subjects as a source of information for strengthening the concept of education for all
2. To facilitate communication in between economically weaker sections of society and students with special needs
3. To use visuals, animations and illustrations to develop spatial awareness and psycho-motor skills among the learners
4. To use the online applications as a discussion forum for supporting collaborative learning and school of thought.
5. To disseminate videos, online tutorials, and conducting webinars to involve wide range of students from distant geographic areas having different customs and cultures.
6. Combining conventional classroom learning with e-learning for developing a hybrid learning experience as per the exigencies and demands of the learner.
7. Processing and recording administrative and assessment data for corrective and controlled evaluation.
8. For the professional growth of teachers by allowing them to exchange and share ideas among multitude of trained executives in their field.

ICT in Adult Education

Much research has been conducted for the effective and useful integration of ICT in hybrid classrooms set ups. However, its integration into an adult learning environment has not been fully examined yet. Conventionally, adult education is mainly concerned with developing basic educational skills such as literacy and numeracy in many countries. Therefore, adult education teachers usually do not see the obligation for ICT integration. Indeed, many view ICT as an interruption in learning the basic skills of reading and writing. But, looking at the current scenario, where a majority of learners, irrespective of age, have access to ICT it is no longer an issue as to whether ICT should be used in the education system. The learners need to be prepared for a competitive, cost-effective and antagonistic market for their career choices. One must focus on making the learners of today being comfortable with the use of technology and not afraid of it. This is also one of main objectives for ICT use in today's classroom. As educational goals may be influenced by local and national employment trends of a nation, altering the learner's expectations through learning experience. A country's economy is dependent on the employment opportunities it offers, but with rapidly evolving technological needs and trained manpower, any country's economy may fall behind other countries if they do not integrate ICT effectively in their education system approach. Also, for many adults, returning to education after a long gap can make them apprehensive about a structured learning environment. Using interactivity, ICT and creative instruction design, it could be viewed as a mechanism to conquer this fear. This may allow and support more learners to participate in their learning activity with enthusiasm than they would do in a formal education forum and advance for a social and communal change within the society.

IMPORTANT CHARACTERISTICS NEEDED TO DEVELOP LEARNING WITH ICT

Information and communication technology can significantly impact a student's learning when teachers are digitally literate and understand how to integrate it for advancement in a society's habits, customs and standards of living. The mental conditions of learning for socio-economic development of a community are indentified as a guided step by step process. These events of instruction, which correlate to and address the conditions of learning, are listed below as follows:

1. Draw learners attention and interest
2. Highlight the importance and objectives of information learning to rethink, recreate and renew
3. Attempt recall of prior knowledge influencing new knowledge build up
4. Present, plan and diversify material of visual or auditory nature to communicate ideas for learning
5. Provide mentoring for cognitive, personal and emotional potential development
6. Draw out performance based initiatives and goals
7. Assess performance and goal-oriented achievements
8. Provide productive and meaningful feedback for better learning outcomes

Accommodating and Catering to Differences Between Learners

From time to time, instruction designers try to discern different learning practices and present the learner with materials which they anticipate would be better for the overall learning experience. The differences between learners are subjected to gender, age, education qualifications, cultural differences, or stimulated learning styles. But, providing for the essential needs of all potential learners means conditioning the design of the program to reflect those requirements. In some instances, educators provide different lesson experiences or representations of the content that they think will address student's desires for different forms of perceptual learning. On other occasions, the content is so developed as to adequately represent differences in gender, age, culture, and socio-economic demographics. But, these additions and alternatives within a specific learning environment add to the costs of the use of ICT tools and techniques. Hence, educators simply leave it to the learners to accommodate and make choices among various alternatives as per their own cognitive skills and areas of interest (Vrasidas and Glass, 2005).

Development of Technology-Based Learning Resources and Cost Flexibility

The development of technology-based learning environments is a costly and excessive business where exorbitant budgets will not allow designing flexible ICT tools for all aspects of learner's requirement. Analysis of content curation tools and software in depth is carried out as a part of educational technology to make the process of collecting relevant information easy. Well curated content is valuable to the teacher as well as the learner and thus, a wide range of presentation, control and feedback devices may be employed for learner's thoughts and opinions on that content. Moreover, it should be emphasised that along with teaching machines, stimulators and computers, techniques such as voicing learning experiences through curriculum development methods and task analysis are essential components too. In fact, as long as programmed learning plan and research is coordinated with these techniques, it is intertwined to be an essential art of educational technology.

There are two meanings attached to the definition of the term educational technology. While one meaning refers to the detailed application of principles of psychology in learning, the other deals with application of science and technology in practical teaching problems. For successful implementation of the education technology, school culture encompasses the vision, plans, norms and values that are bound by the school curriculum. Effective ICT integration in school culture depends on the perceptions and vision of the school leaders along with teacher's ICT skills. Social customs influence the patterned way of behaving that is characteristic of life in a society. In addition, school culture has a mediating role that influences learner's actions, beliefs, and attitudes (Lewin, Griffith, Assche, Anido, and McNicol, 2015). To achieve this goal, more professional and cost effective training sessions are required with a focus on increasing teacher's skills enabling them to overcome apprehensions associated with use of technology. Further, new teaching approaches and technical support should be offered to facilitate learning with computers. Overall, implementing effective teaching with technology amalgamation requires changes and modernization in teachers as well as learner's knowledge, beliefs, and school culture. Education unification with innovations in science and technology should offer conditions needed to optimize learning and promote the transfer of knowledge and skills for upliftment of a society.

Authenticity and Open-Ended Learning Environments

Authenticity is a matter of concern in open-ended learning environments which should be addressed in terms of the design and development of learning material and content. Learning environments refer to the educational approach, physical context or the cultural setting in which teaching and learning is meant to occur. Learning resources and technology need to reflect the potential uses of knowledge that pupils are expected to master. Open ended learning environments should provide a platform where learners feel safe, supported and inspired in their pursuit of knowledge in order to prevent the acquired knowledge from becoming inert. Rich contexts and tasks based on objective reality, rational reasoning and human intellect should be designed as authentic as possible and presented to the world outside school. In addition, teachers should stimulate pupils to engage in practices of authority, intuition and scientific empiricism for active knowledge construction. This calls for open-ended learning environments where children learn to react better in situations, make choices and feel more inspired to learn. Co-operation and interaction in the classroom environment are equally important in order to foster academic and social learning experiences instead of learning environments which focus on a mere transmission of facts (J.A. Belianca, 2010). Finally, for effective classroom practice, differentiated teaching is considered to be one of the key criterions. The objective of differentiation is to lift the performance of all students in terms of problem solving skills, and social relations since classes are of mixed ability.

BARRIERS IN USE OF ICT

A teacher is responsible to bring out the best of what is hidden in a student. Fortune or misfortune, nectar or poison are both hidden but the right kind of education will show the path to godliness and moral uprightness to a living being. As a teacher, one may find many student's behaviours unacceptable, unpleasant, or troubling when using ICT tools for class teaching. Widespread other barriers to the successful integration of ICT in teaching and learning can be identified as lack of computers and quality software, technical glitches and problems, teacher's attitudes and confidence in computer handling, poor funding and administrative reasons. Following discussion is the detailed description of barriers one encounters when using ICT and a few possible solutions to facilitate change as suggested-

Barriers and Solutions: Student Perspective

Student mobility, lack of focus and concentration, special needs, fear of failure and anxiety over standardized test results are the main challenges associated with ICT use. Children who are extremely anxious, withdrawn, shy, depressed, and hypersensitive, who cry easily and have little confidence have few social skills and hence very few friends. Children who have conduct disorders are aggressive, destructive, disobedient, uncooperative, distractible, disruptive and persistent. They have been corrected and punished for the same misbehaviour countless times. The most successful strategies for helping these children are the behaviour management approaches with very clear rules and consequences, consistently enforced. These emotional factors and personal resistances can be solved by providing more authentic group- and problem-based learning activities, and adequate learning support. Children having subpar technical skills face difficulties of access to ICT in classroom additionally troubled by an insufficient number of academic advisors and lack of timely feedback from instructors. Not reinforcing a child's effort to master basic cognitive skills may develop a sense of guilt or shame within the child. Autonomy vs shame and doubt marks the beginning of self-control and self confidence. Initiative adds to autonomy or the quality of undertaking, planning and attacking a task for the sake of being active. But if children are not allowed to do things on their own, they may come to believe that what they want to do is always 'wrong'. This further reduces their interaction with peers and instructors. Therefore, it is recommended to follow given strategies to facilitate the learning process with ICT incorporation:

1. More orientation and training sessions for students to develop ICT skills
2. Increased emphasis on the prominence of instructor access and effective administration
3. Expansion of webcasting and exclusively distributing online conferencing tools

In general, strengthening the skills, instincts and abilities of students in terms of educational processes and resources are required in order to lower student barriers and improve the effectiveness of ICT to survive, adapt and thrive in a fast-changing world.

Barriers and Solutions: Teacher Perspective

The goal of education must be life itself, not merely providing instruction in how to earn a livelihood. The means of a person's livelihood is only the existence of his body, but education will only be alive when it teaches life itself. To teach life is to teach knowledge of the self. Each learner carries the seed of that supreme development within him that needs to be cultivated by the teacher. When education moves towards real knowledge and real peace, a new being and a new humanity will be born. Teachers themselves identify the need for administrative support, adequate time, funding, and training as essential to facilitate changes in education with inclusion of information and communication technology (E.C. Sheninger, 2019). A few limitations can be listed as follows:

1. Lack of teacher collaboration and pedagogical support
2. Lack of clear goals for ICT use in education and low teacher expectations
3. Insufficient time to master new software, formal training and self-directed exploration, manage teaching materials or integrate ICT during classroom learning
4. Low software competence, limited knowledge and lack of experience or negative experiences with ICT use in the past among cooperating teachers
5. Habitual ways of teaching technical or operational skills rather than focusing on what and how the students should learn from existing pedagogical content knowledge
6. Pressure to improve scores but a lack of recognition and encouragement of the timely and effective use of ICT
7. Lack of specific and definite ideas on how integrate technology in instruction design and classroom management
8. Lack of motivation, uncertain technical and financial support restricting the possible benefits of using ICT in the classroom especially where pupil-to-computer ratios are poor
9. Fear of embarrassment and loss of status in front of pupils and colleagues

Barriers and Solutions: Educational Institute Perspective

Educational institution is a place where people of different ages gain education. Educational institutions such as schools, colleges and universities are responsible for the systematic transmission of knowledge, skill, social and cultural values within a formally organized structure. There are numerous challenges faced by institution administration for effective implementation of ICT in education. Lack of capital and expensive higher education, cost of acquiring, using and maintaining ICT resources, poor maintenance and inadequate facilities, obsolete and outmoded use of software and hardware, limitation of communication and transport facilities, lack of institutional support through leadership, planning and the involvement of authorities in implementing the change are a few listed factors (Beach and Hall, 2011). As a result, we face the problem of mass illiteracy when there is an uncontrollable increase in the population. However, these educational institutions play a very crucial role in the socialization of technologically simple societies and thus transmit a way of life, skills and values from one generation to the next. Hence, these educational institutions are encouraged to:

1. Provide technical support, professional development programs, workshops related to technology to update teacher's skills and knowledge
2. Support partnerships programs and conferences that allow teachers to reflect upon effective strategies for technology integration into instruction
3. Supplement the education curriculum with technology-enhanced materials and provide enough freedom for teachers in seeking curriculum materials
4. Provide effective, timely, and continuous training to improve management, integration and implementation of ICT in a technology-rich classroom
5. Encourage and construct a positive attitude about the significance of open ended learning with ICT

Barriers and Solutions: Administrative and ICT Infrastructure Perspective

An Institutional administration focuses more on the quantity of course content and student evaluation scores rather than on ICT usage. Lack of appropriate administrative support mandates improving examination results in terms standardized test. Educational administration distinguishes a candidate's ability to perform a particular task for the purpose of certification in terms of grades. This shifts the focus away from using ICT in engaging students for higher-order thinking activities. In most developing countries, ICT Infrastructure is also a limited resource with a lack of appropriate course content and instructional material programs. However, it is the availability of ICT tools, the establishment of disciplinary educational and socioeconomic principles and procedures for teaching assistance that are crucial for establishing a well-managed ICT-integrated class. Each school should indeed develop unique strategies to provide meaningful learning activities using ICT tools to both students and teachers during school hours. By emphasizing these elements, a learning process is more likely to be facilitated by engaging students in critical and analytical skills making our society more advanced and a better place to live.

ROLE OF TEACHER BEHAVIOUR AND INTERACTION ANALYSIS

The association between the quality of early teacher-child relationships and later school performance can be both strong and persistent. The role of a teacher is to identify, extract, and pull together available and pertinent heterogeneous methods and use appropriate computational principles, platforms and technologies to discover new relations and deliver insights into teaching-learning process. Students with significant behavioural problems in the early years are less likely to have problems later in school if their teachers are sensitive to their needs and provide frequent, consistent feedback. What energizes and directs a child's behaviour can be a part of intrinsic or extrinsic motivation (Anita Woolfolk, 9th Edition, 2010). The explanation could be drives, needs, incentives, fears, goals, social pressure, self-confidence interests, curiosity, beliefs, values, expectations and more. The teacher should be able to conceive, design, implement and manage the different information systems, services and processes by using principle of teaching with ICT tools effectively and efficiently. Intrinsic motivation is the natural tendency to seek out and conquer challenges as we pursue personal interests and exercise capabilities. When we are intrinsically motivated, we do not need incentives or punishments, because the activity itself is rewarding. When we do something in order to earn a grade, avoid punishment, please the teacher or some other reason that has very little to do with the task itself, we experience extrinsic motivation. Teaching can create intrinsic motivation by connecting to student's interest and supporting growing competence. Teachers must encourage and nurture intrinsic motivation, while making sure that extrinsic motivation supports learning. Indeed, teachers have been both criticized as ineffective and lauded as the best hope for young people. The main responsibilities and modalities of the teachers will be arranging for the computer aided learning technology through their course format, creating, elucidating and explaining the new assignment for efficient implementation of ICT in education.

Understanding Student's Learning With ICT

One of the most significant issues is to understand the process of student's congenial familiarization with technology if we want to understand how student's learning will happen with educational technology. As educators, we also have to recognize that each student has a different adoption rate, as far as computer knowledge is concerned. Observation can be a very efficient learning process and includes five elements: paying attention, retaining information or impressions, changing inhibitions, producing behaviours, and being motivated to repeat the behaviours and attitudes. Teachers have to deal with their observation skills, appreciate different adoption rates and take into consideration the differentiation in terms of the utilization of technology in the classroom. A teacher should understand that incentives can affect child's performance and constant discussion with their families regarding their adoption rates can be an acceptable and sustainable way of learning for their children.

Students Cognitive Processing Style and Abilities

The child's culture shapes cognitive development by determining what and how the child will learn about the world. Student's cognitive styles are used to determine the way individuals think, perceive and remember information. According to the above scheme, each individual, according to his/her personal cognitive abilities and learning style encompass language, imagination, perception and planning as the most appropriate learning strategy in order to achieve the desired learning outcome. Therefore, the role of the teachers is to provide educational technology activities of an appropriate level of difficulty based on their cognitive abilities such as memory learning, language use, reasoning, problem solving, decision making and intelligence. Teachers have to provide a plethora of learning strategies that are expressed in a variety of ways in order to adjust the curriculum to a variety of learning styles based on age, needs of kinesthetic learners, and for open ended platforms. In this process, the students should learn how to select the most appropriate learning style for their desired learning outcome allowing children to express and develop ideas of their own. Teachers are there to facilitate this process and to identify where the learning blockage occurs, in order to facilitate the learning process. They must deal with a wide range of student abilities and challenges: different languages, different home lives and different needs. The teacher must use new technologies and techniques appropriately to accomplish important goals. All students need to interact with teachers and peers in order to test their thinking, to be challenged, to receive feedback, and to watch how others work out problems. The teacher should be able to employ a qualitative analysis approach with a thematic framework to collect data of learning with ICT tools and to understand socio-psychological and ethical challenges in ICT enabled teaching-learning process. Disequilibrium is often set in motion quite naturally when the teacher or another student suggests a new way of thinking about something. But the disequilibrium must be kept "just right" to encourage growth. Student's beliefs about personal competence or self-efficacy should be future-oriented. Self-efficacy focuses on one's ability to successfully accomplish a particular task with no need for comparison. Greater efficacy leads to greater effort and persistence in the face of setbacks.

Structured vs. Spontaneous Learning

People develop abilities in a logical order. The degree of flexibility an individual wants during the learning process is how much structured and well- organized or spontaneous the individual wants the learning material to be for a better learning outcome. Through learning, an individual interact with their environment. Such changes make up a large part of a person's social development. Human development can be divided into a number of different aspects:

1. **Physical Development:** Deals with changes in the body
2. **Personal Development:** Used for changes in an individual's personality
3. **Social Development:** Changes in the way an individual related to others
4. **Cognitive Development:** Changes in thinking

All individuals inherit two basic psychological tendencies for understanding and interacting with the world. First of these tendencies is towards organization – combining, arranging, recombining, and rearranging of behaviours and thoughts into coherent systems. The second tendency is towards adaptation to the environment. Moreover, when people are learning a new behaviour, they will learn it faster if they are reinforced for every correct response. This reinforcement schedule can be continuous or intermittent depending on how persistently the child is responding when reinforcement is withheld. Teaching students to troubleshoot the problems on their own builds their analytical and decision making skills and hence result in autonomy. With a set of structured instructions to attain autonomy, the teacher is introducing the concept of spontaneous teaching to stay resilient and establishing strong foundations of virtuous learning habits.

Teacher Attitudes, Perceptions, and Confidence in Use of ICT

There is evidence that more qualified teachers make a difference in student learning. Well qualified teachers who establish positive relationships with their students appear to be a powerful force in those student's lives. Moreover, teacher's attitudes toward technology significantly predict teacher and student technology use. Teacher's constructivist teaching beliefs, teaching self-efficacy, and demonstration with computers reflect prospective ICT use for efficient learning process. A teacher needs to be a reflective thinker and practitioner, with a purpose of understanding his/her students, learning environment, motivating strategies, teaching practices and ready access of information. Furthermore, more confident pre-service teachers have stronger intentions to integrate technology into their future teaching practices in real classrooms. Although the teachers confronted their concern over the issues such of technology availability, accessibility, professional support, and classroom management but at the same time are more capable of and interested in using computers. However, till date many teachers are not able to translate the positive intentions with increased pedagogical knowledge into actual teaching, largely due to unfamiliar school environments. For this reason, it is concluded that teacher education programs need to increase awareness regarding benefits of integrating technology into student-centred learning approaches. ICT use should not be disregarded and its intentions in providing technology integration for pedagogical knowledge to foster student-centred learning must be disseminated. Self-determination is the need to experience choice and control in what we do and how we do it. Classroom environments that support student self-determination and autonomy are associated with greater student interest, sense of competence, creativity, conceptual learning, and preference for challenge. Controlling environments tend to improve performance only on the rote recall tasks.

Teaching Gifted Students

Teaching methods for gifted students should encourage abstract thinking (formal operational thought), creativity, reading of high level and original texts and independence. Gifted students tend to learn more when they work in groups with other high ability peers. In working with gifted and talented students, a teacher must be imaginative, flexible, tolerant and unthreatened by the capabilities of these students.

Gender Role Identity

Gender-role identity is the image each individual (boy or girl) has of himself or herself in terms of traits, beliefs or behaviours that a particular culture judges to be appropriate for them. Different treatment of the sexes and gender-role stereotyping continue in early childhood. As children grow, they have developed a gender schema that describes what clothes, games, toys, behaviours and careers are "right" for boys and girls. These ideas can be quite rigid and likely to create gender bias differences in the classroom. Unfortunately, Schools foster these gender biases in a number of ways. Most of the textbooks produced for the early grades portrayed both male and female in stereotyped roles. Whereas boys are depicted as more aggressive and argumentative, girls are portrayed as more expressive and affectionate. A teacher must make a thorough content analysis and present the learning material such that it helps in eliminating gender biases from the classroom. The teacher's duty is to remove the imbalances in terms of attention given to the boys and the girls in class. All teachers need to be sure that both girls and boys realize the value of advanced knowledge in opening doors to advanced education. The goal of incorporating ICT in education should be to create culturally compatible classrooms with elimination of racism, sexism, and ethnic prejudice while providing equal educational opportunities for all students.

SOCIAL AND ECONOMIC MAXIMS OF SUCCESSFUL TEACHING WITH ICT

Recent perspectives have called attention to two other aspects of learning that are critical – social and cultural factors. The process of adapting to innovation seems to be circular, and this cycle ideally reflects the changing modern technological world in all areas. Language is critical for the execution of ground-breaking innovation and cognitive development. It provides a means for expressing ideas and asking questions, and thus links the past and the future of a community.

It is apparent that ICT integration is inevitable and entails an evolving process, not a final product. To achieve successful integration of technology, one requires an effort from three sides: teachers, students, and school administrators besides precondition of possible infrastructure coherence. The fate of mankind is in the hands of education to help attain the perfection of self-knowledge and thus reshape and recreate a peaceful living world for everyone. The illumination of the whole world is useless if the self is in darkness. To begin this ascent to consciousness, it takes effort, courage, and determination to achieve successful teaching with ICT with an aim to provide authentic education for all learners.

Financing and Sustainability

Costs and sustainability are ongoing challenges for all of the educational institutions when attempting to bring in unhackneyed, contemporary and innovative resources such as ICT. These educational institutions endeavour and experiment to manage sustainability of their ICT activities in the following manner:

1. To acquire ICT resources along with pre requisite of technically trained staff
2. To control the finances related to ICT activities and obtain more funds or ICT resources

The sources of funding the educational institutions will rely on government programs that provide an infrastructure to support these activities. Indeed, another critical source is the community itself. Successful schools/colleges have developed righteous associations and concord with the surrounding community, and the communities in turn value the ICT initiatives of these institutions and help in terms of establishing their own small sources of revenue. The responsibility for social and moral change cannot rest solely on the shoulders of the teachers. Bringing about cultural and socio-economic changes in a society is a long- term, incremental process and is dependent on a broad range of factors, from leadership to funding to effective professional development. For the success of ICT in learning, schools/colleges should be encouraged to look at all the available options of financing that help create and sustain the conditions for a change in learning custom within a society. Effective reform requires succour investment and encouragement along with multifarious dimensions of the educational system, including physical and technical infrastructure, manpower resources, curriculum advancements, evaluation, and assessment standards. In the end, the success of teachers is vitally dependent on the conditions in which they work and opportunities for employment and career growth.

To Help Build Community Connect

Multicultural education is one response to the increasing diversity of the school population as well as to the growing demand for equality for all groups. The prevailing socio-economic status is a significant dimension of cultural difference in terms of wealth, power and prestige, often overpowering other differences such as gender and ethnicity. ICT should bridge the gap among all these cultural conflict sectors for living amicably, although not all cultural differences lead to clashes. ICT use should promote culturally relevant pedagogy that includes academic success, developing and maintaining cultural competence, and developing a critical consciousness to challenge the status quo. Connectivism learning theory is a relatively new concept in this regard where student are supposed to learn best if they are taught to navigate and create social networks of their own making the use of technology. For learning through connectivism, students should be able to combine thoughts, theories and general information in a meaningful manner.

Technology is an important part of this learning process as it gives a learner the opportunity to make choices about their learning and guiding principles mapping out knowledgeable content from a diversity of opinions. Principles of connectivism include the following factors:

1. Learning is a process of connecting through representatives from different social, cultural, and economic groups.
2. Critical thinking and ability to see connections between different fields, ideas, and concepts is a core skill
3. Accurate, up-to-date knowledge with analytical and decision-making skills
4. Necessity of nurturing and maintaining connections with a conception that what we know today might change tomorrow as a result of constantly evolving information circle

In a connectivist viewpoint, the new learning responsibilities unlike traditional teaching methods and other theories shift from the teacher to the learner. The educator's role and contribution is to guide students to become effective agents for their own learning and personal development. Connectivism relies heavily on the efficient use of information and communication technology by introducing more opportunities for digital learning—like online courses, webinars, social networks, and blogs. Connectivism in the e-learning classroom is implemented with an objective to enhance learning networks and engage more active learners to create their own learning experiences.

Learning Through Massive Open Online Courses (MOOCs)

MOOCs are free online courses designed to learn a new skill, bring advances in one's career or to deliver affordable and flexible educational experiences at large (J. Haber, 2014). These are available for everyone around the world and have dramatically changed the way the world learns. These are usually provided by the most trusted and eminent institutions for a variety of reasons such as supplemental learning, corporate training, career developments and mass awareness programs. These online courses with ICT incorporation offer certain advantages over traditional teaching methods which are given below as follows:

1. These are open access and help provide open ended learning platforms for all interested, regardless of location, resulting in a more diverse learner tally.
2. MOOCs bring to light top level educational professionals that would otherwise be unavailable to many of the world's population.
3. MOOCs assist and support techniques of data collection and analysis to closely monitor the success and failure of classroom and distance learning methods.
4. There are numerous benefits of learning with open educational resources such as its reuse, redistribution, and remix. Open educational platforms allow people to share, adapt, modify, translate, or collaborate for a change in teaching-learning processes.

Ethical Practices in ICT-Enabled Learning Process

The core of teaching-learning process consists of five basic values: truthfulness, responsibility, dignity, freedom, and communication. Ethics in education are important as they allow the system to run smoothly and protectively in the interest of both teachers and learners. A teacher's ethical stance will govern how they are assessing or instructing the students. Few professional standards and practices in ICT enabled learning methods are listed below:

1. **Copyright:** It is a type of intellectual property that protects original works of authorship. Copyright prevents others from using the original available work in certain ways, such as mimeographing or duplicating the work and placing it online.
2. **Plagiarism:** It is presenting someone else's work or ideas as your own, with or without their consent. All published and unpublished material, whether in manuscript, printed or electronic form, is covered under this definition. Plagiarism may be intentional or unintentional by incorporating someone else's thoughts and notions into your work without full acknowledgement to them.
3. **Creative Commons:** It is an internationally functional non-profitable organization that provides free licenses to content creators when making their work available to the public domain. These licenses give permission to avail oneself of the original work under certain conditions.

FUTURE RESEARCH DIRECTIONS

We have been practising the traditional classroom method of teaching for ages but ICT enabled education offers an all-encompassing learning environment that spans all languages, educational systems and fields of study. Information sharing between the teachers and students is made possible by the more efficient use of electronic gadgets that are supported by artificial intelligence and semantic architecture. During the pandemic COVID-19, conventional classroom teaching took a 360-degree turn and everything along with education went online. The edtech industry accelerated during this phase enabling the learners to become more flexible and independent in learning (A. Kingsley, 2021). Every student is different and so is the pace of their understanding. One of the most attractive advantages of ICT enabled education is the ability to access class material by anyone from anywhere and submit work online. With the help of modern-day technology, learners have gained enough courage and confidence to speak up their minds. ICT in education is effective in communicating with people from distant places and learn from their perspectives. It emphasis on ideas coupled with adaptable learning schedules. The future of teaching-learning process is all about ideas with correct decision making and not about physical judgements. The classroom of the future has evolved into a viable option for students looking to further their education at their own learning pace. At the same time, this is a wonderful opportunity for people of all ages to continue their education further, without having to leave their homes or jobs.

CONCLUSION

In a work environment that has been disrupted by new technologies like artificial intelligence, machine learning, data science etc. it is imperative that the workforce of today and the future is well equipped to adapt to the changing dynamics of the evolving market. In an era where technology is evolving at an overwhelming pace, staying relevant, connected and upgraded is the key to success for an individual both socially and economically. The advent of automation and its subsequent outcome is set to bring great economic benefits but will also jeopardize many traditional roles. The impact of ICT in education has come to terms with the fact that constant upgradation of skills is an absolute necessity to survive, grow and excel in teaching-learning process. The objective of ICT incorporation in education is to investigate a blended learning system with a careful study and review. The implementation of the hybrid learning model in education has benefited teachers, students and institutions in a significant way. An ecosystem approach that benefits all stake holders and synchronises in-school and after-school learning will not only scale-up access to quality education for all masses but also ensures creative enhancement of the learning objectives. The world has been witnessing an unprecedented phase of uncertainties and students of today are more likely to face complex and unexpected global, social, economic and environmental issues in the future. Hence, with fast changing automation of information and communication technologies, learners of today must remain prepared with entrepreneurial skills to adapt to new roles, responsibilities and challenges for the future workforce. At the same time, teachers must demonstrate the qualities of integrity, impartiality and identity to encourage creativity, innovation and collaboration among students. They must model strong characteristic traits among students such as hard work, perseverance, honesty, patience, respect and good ethical and moral behaviour. The teaching-learning process can only be a success when both parts endeavour to understand and respect the values and traditions of the diverse cultural and social representations within or outside a classroom. It will help learners to make their own identity and mitigate the existing and prevalent gaps in society for the overall progress of mankind both socially and economically.

REFERENCES

Abbott, C. (2003). *ICT: Changing Education.* Taylor & Francis. doi:10.4324/9780203400197

Antonio, J. M., Gomes, M. C. A., & Marcelino, M. J. (2015). *ICT in Education: Multiple and Inclusive Perspectives.* Springer Publication.

Beach, S. N., & Hall, L. R. (2011). *The Connected Educator Learning and Leading in a Digital Age.* Solution Tree Press.

Belianca, J. A. (2010). *21st Century Skills: Rethinking How Students Learn.* Solution Tree Press.

Fu, J. S. (2013). Article. *International Journal of Education and Development Using Information and Communication Technology*, *9*(1), 112–125.

Haber, J. (2014). MOOCs. MIT Press.

Kanematsu, H., & Barry, D. M. (2015). *STEM and ICT Education in Intelligent Environments.* Springer International Publishing.

Kingsley, A. (2021). *My Secret #EdTech Diary Looking at Educational Technology through a wider lens.* John Catt Educational Limited.

Latwal, G. S., Kommers, P., Mahajan, P., & Sharma, S. K. (2020). *Role of ICT in Higher Education: Trends, Problems and Prospects.* Apple Academic Press. doi:10.1201/9781003130864

Lewin, C., Griffith, D., Assche, F. V., Anido, L., & McNicol, S. (2015). *Re-engineering the Uptake of ICT in Schools.* Springer International Publishing.

Mark, W. (2006). *Laptops and Literacy.* Teachers' College Press.

Meadows, J., & Leask, M. (2012). *Teaching and Learning Using ICT in the Primary School*. Taylor & Francis. doi:10.4324/9780203137048

Sheninger, E.C. (2019). *Digital Leadership: Changing Paradigms for Changing Times*. Corwin Publication.

Somekh, B. (2007). *Pedagogy and Learning with ICT: Researching the Art of Innovation*. Taylor & Francis. doi:10.4324/9780203947005

Vrasidas, C., & Glass, G. V. (2005). *Preparing Teachers to Teach with Technology*. Information Age Publishing.

Woolfolk, A. (2010). *Educational Psychology* (9th ed.). Pearson Education Inc.

KEY TERMS AND DEFINITIONS

Educational Technology (Edtech): Application of principles of psychology in learning along with application of science and technology in practical teaching problems.

Extrinsic Motivation: When a learner does something in order to earn a grade and avoid punishment.

Inclusion: To include all socio-economic learner groups for having equal opportunities to education and learning.

Instructional Design: Strategies for teaching that integrate pedagogy and technology together to develop behaviourism, pragmatism, and constructivism in learners.

Intrinsic Motivation: Natural tendency to seek out and conquer challenges as we pursue personal interests.

Open-Ended Learning: Platform where learners feel safe, supported, and inspired to learn at their own pace.

Reinforcement Schedule: Use of continuous or intermittent reinforced behaviour for faster learning.

Self-Regulation: Abilities of self-monitoring, self-reinforcement, and goal setting for an effective and meaningful learning.

Chapter 4
Impact of Information and Communication Technology on Economic Growth:
Evidence From Arabian Peninsula Region Countries

Mehmet Hilmi Özkaya
Uşak University, Turkey

Abdullatif Almukaddam
Taiz University, Yemen

Maeen Alhuwesh
https://orcid.org/0000-0003-1527-2626
Taiz University, Yemen

ABSTRACT

The present study aims at investigating the impact of information and communication technology (ICT) on economic growth of countries in the Arabian Peninsula region through an econometric study of seven countries over the period 1996-2020. To achieve the objectives of the study, the panel data was based on three independent variables, namely the number of mobile cellular subscriptions per 100 people, number of fixed (landline) telephone subscriptions per 100 people, internet usage (% of population), and the economic growth rate as a dependent variable. The results show that within the framework of static analysis the fixed effects model is the most appropriate and is evidence of the existence of fixed individual differences between the countries of the Arabian Peninsula as regards the impact of ICT on economic growth. It also shows that the mobile phone index has a positive and significant impact, and a negative and significant impact on economic growth is shown to be related to the internet index.

DOI: 10.4018/978-1-6684-6620-9.ch004

INTRODUCTION

Information and communication technology (ICT) facilitate the exchange of information and transforms it into knowledge. This, no doubt, affects the modern societies and leads to a paradigm shift in the levels of economic and social development such as education, health and standards of living (Conole & Dyke, 2004). ICT including the collection, organization, storage, processing, transmission and dissemination of data (audio, video, and text or symbols) is done by using computer tools and telecommunications (Niebel, 2018; WB, 2002). In the last two decades, the world witnessed great and rapid transformations as a result of the expansion in the use of ICT, which directly affected various aspects of human life (Benlahbib, 2018). ICT has provided countries with great potentials to improve the economic situation and reduce disparities between developing and developed countries (Al-Amin & Benlahabib, 2018). The World Bank (WB) takes an optimistic view that ICTs hold great promise for reducing poverty, increasing productivity, promoting economic growth, and increasing accountability and governance (WB, 2012).

Studies done in this field have revealed that ICT is a key factor in the economic and social development of countries owing to its positive effects on economic growth, productivity and employment. A study conducted by the World Economic Forum (WEF) concluded that increasing the digitization of a country by 10% will lead to an increase of 0.75% in per capita GDP, and a decrease in the unemployment rate by 1.02% as digitization provided a boost of $193 billion to global economic output and created 6 million jobs globally in 2011 (Sabbagh, Friedrich, EL-Darwiche, Singh, & Koster, 2013). However, the countries of the world vary in terms of their use of ICT and the extent to which it is of benefit. This is related to its level of progress and aspiration transformed into an information society having a human objective and centering on development (Usman, Ozturk, Hassan, Zafar, & Ullah, 2021; Khireddine, Noureddine, & Bouzelifa, 2018).

To monitor the progress achieved by countries in transforming into an information society, a composite index of three sub-indicators containing 11 indicators[1] is used with the aim of monitoring and comparing developments in ICT between countries and over time (ITU, 2022). The first sub-indicator focuses on access and includes five indicators of infrastructure and access. The second sub-indicator focuses on usage and includes three indicators of the intensity of ICT use. The third sub-indicator focuses on technology skills and includes three indicators related to the skills needed to benefit from ICT, which is an indirect indicator and gives less weight when calculating the IDI compared to the other two indicators (ITU, 2022 a). In the Information Society Measuring Report 2018, it was indicated that more than half of the world's population is connected to the Internet, and there is a general upward trend in access to ICT, with the exception of fixed telephones. This, of courses, represents an important step towards a more comprehensive global information society. From another angle, the report indicates that the lack of ICT skills is one of the important obstacles to the population's access to the Internet, and this problem is more prominent in the case of developing and least developed countries (ITU, 2018).

Based on estimates made by the International Monetary Fund (IMF) for the year 2022, the countries of the Arabian Peninsula are classified into high-income countries where the average per capita income ranges between 23.4-84.5 thousand dollars, and low-income countries like Yemen, which is grouped under the lowest-income countries in the world and has an average per capita income of $891. The countries of the Arabian Peninsula also achieved high economic growth rates that ranged between 3.3-8.2%, with the exception of Yemen, which achieved a 1% (IMF, 2022; IMF, 2022a). They are classified among the countries with a high index in the IDI 2017 rank, where the index ranged between 6.43 and 7.60. Yemen, however, was not included in the report (ITU, 2022b).

In light of these findings, a question arises about the nature of the relationship between the rate of economic growth and the extent of access and use of ICT in the countries of the Arabian Peninsula. The question can be 'What is the nature of the impact of ICT on the economic growth of the countries of the Arabian Peninsula? In order to know this effect, the use of ICT and the effect of the indicators of access on economic growth were studied due to the direct impact of these two axes and the nature of the different economic and social life in the countries of the Arabian Peninsula.

Accordingly, this study aims at identifying the impact of ICT on economic growth of the countries in the Arabian Peninsula region through the access and use of indicators, besides the use of the econometric approach and based on the analysis of Panel Data models for 7 Arab countries, namely (Bahrain, AUE, Kuwait, Oman, Qatar, Saudi Arabia, Yemen) over the period 1996-2020. The hypothesis of the study is that there is a positive impact of ICT on the economic growth of these countries. To achieve the objective and test its hypothesis, the study is divided into five chapters: the first chapter is the introduction. The second chapter is devoted to literature review. The third chapter presents the data and methodology of the study. The fourth chapter analyzes and discusses the results. The fifth chapter is the sum-up of the study given in the form of conclusion.

LİTERATURE REVİEW

The impact of ICT expansion on economic growth has been widely discussed by researchers, as many empirical studies were conducted to find out the impact of ICT on economic growth. These studies arrived at inconclusive results. While some studies have revealed positive effects (Nguyen, Dinh, Ngoc, & Thuy, 2022; Kurniawati, 2022; Maneejuk & Yamaka, 2020; Bahrini & Qaffas, 2019; Majeed & Ayub, 2018; Stanley, Doucouliagos, & Steel, 2018; Saidi, Hassen, & Hammami, 2015; Vu, 2011), others have presented ambiguous or negative effects depending on the type of technology used and the degree of a country's economic development (Nabi, Tunio, Azhar, Syed, & Ullah, 2022; Niebel, 2018; Ishida, 2015; Pradhan, Arvin, & Normancd, 2015; Farhadi, Ismail, & Fooladi, 2012; Yousefi, 2011).

With regard to the studies on the economic situation in the countries of the Arabian Peninsula region, these countries were not studied independently, but within certain groups such as the Middle East and North Africa (MENA), the Organization of Islamic Cooperation (OIC), the League of Arab States (LAS), or the Organization of the Petroleum Exporting Countries (OPEC). Studies have reached different results. Though some of them revealed positive effects (Hussein, 2020; Sepehrdoust & Ghorbanseresht, 2019; Aghaei & Rezagholizadeh, 2017; Hodrab, Maitah, & Luboš, 2016), other studies revealed negative or inconclusive effects (Kacham & Kebir, 2020; Al-Amin & Benlahabib, 2018; Benlahbib, 2018). A study by (Hodrab, Maitah, & Luboš, 2016) found that there is a positive impact of ICT on the economic growth of Arab countries. Using the cluster analysis, a study by (Aghaei & Rezagholizadeh, 2017) found that there is a significant impact of investment in ICT on economic growth in the countries of the OIC. The study of (Al-Amin & Benlahabib, 2018) revealed that the Arab Gulf countries occupy advanced ranks among the Arab countries within the ICT index, while Yemen was in the last rank. The study, using the regression analysis, found that the Internet index has a positive impact on the economic growth of the Arab countries, with an explanatory power equivalent to 53%. However, there was no impact found on the fixed or mobile phone. The study of (Benlahbib, 2018) found that the Internet index has a negative and significant impact on the economic growth of developing countries, while the fixed-line phone index has a negative and insignificant impact. As for the study by (Bahrini & Qaffas, 2019), it was found that

Table 1. Dataset definition

Abbreviation	Variable	Data Source
EGR	Real economic growth rate	WB (2022); ICO (2022)
MCS	The number of mobile cellular subscriptions per 100 people	WB (2022)
FTS	The number of fixed telephone subscriptions per 100 people	WB (2022)
INT	Individuals using the Internet (% of population)	WB (2022)

the Internet and mobile phone indicators have positive effects on the economic growth of the MENA countries, but there was no effect found in terms of the fixed phone index. In a study by (Sepehrdoust & Ghorbanseresht, 2019), it was found that an increase in the ICT index by 1% leads to an increase of 0.05% in the economic growth of the OIC countries.

A study by (Kacham & Kebir, 2020) found that the Internet index affects negatively and insignificantly the economic growth of the Organization of Arab Petroleum Exporting Countries (OAPEC). The fixed and mobile indicators were found to have a positive and significant effect. Similarly, the study of (Yaba, Bouaziz, & Lnassari, 2021) found that the Internet index has a negative and insignificant impact on the economic growth of the Arab countries, while the mobile telephone index has a positive and significant impact. The study of (Bouallaga & Kebir, 2021) found that both the Internet and mobile phone indicators have positive and significant effects on the economic growth of Algeria and of some Arab countries. The fixed phone index,however, shows a negative and insignificant effect.

DATA AND METHOD

This study aims at measuring the impact of ICT on the economic growth in the seven countries of the Arabian Peninsula region through annual data for the period 1996-2020. To achieve this objective, an econometric model based on panel data was estimated. This model has been employed in many similar empirical studies such as (Kurniawati, 2022; Bouallaga & Kebir, 2021; Hussein, 2020; Bahrini & Qaffas, 2019; Aghaei & Rezagholizadeh, 2017). The data used in the study were obtained from the World Development Indicators issued by the WB as well as from OIC as shown in Table 1. The EViews10 package w+0 as used for econometric analyses.

The study sample consisted of 7 countries in the Arabian Peninsula region. They are Bahrain, United Arab Emirates,, Kuwait, Oman, Qatar, Saudi Arabia, and Yemen.. The mobile and fixed access indicators were used as a measure of the access sub-index, and the indicator of the Internet was used as a measure of the use sub-index .

The descriptive statistics of the data set are indicated in Table 2. The mean of economic growth rate was 3.76% of the total final economic growth in the sample, but it was the least changeable among the sample countries.The mean of the access to a mobile phone was 89.13%, and this rate was the most changeable. On the other hand, the mean of fixed-telephone access rate was 16.54%, which is less changeable compared to the economic growth rate. The mean of the internet usage rate, however, was 41.35%, which is also highly changeable from one country to another.

Table 2. Descriptive statistics for variables

Characteristic	EGR	MCS	FTS	INT
Mean	3.76	89.13	16.54	41.35
Median	3.72	86.27	16.39	28.97
Maximum	30.01	212.64	32.88	100
Minimum	-27.99	0.06	1.32	0
Std. Dev.	5.96	64.24	7.81	35.89
Observations	175	175	175	175

To find out the impact of the ICT access sub-index through the mobile and fixed phone indicators and the use sub-index through the Internet index on economic growth, this study estimated an econometric model as follows:

$$GDPG_{it} = \alpha_0 + \alpha_1 \mathrm{MCS}_{it} + \alpha_2 FTS_{it} + \alpha_3 INT_{it} + \vartheta_i + \varphi_t + \varepsilon_{it} \quad (1)$$

Whereas $GDPG_{it}$ is the Real economic growth rate of country i during the period t, MCS_{it} is the number of mobile phone subscribers per 100 people in country i during the period t, FTS_{it} is the number of fixed telephone subscriptions per 100 people in country i during the period t, INT_{it} are the individuals using the Internet (% of population) in country i during the period t, ϑi indicates the unobserved cross-sectional effects that vary from country to country but remain constant over the time period, φt refers to unobserved temporal differences between sample countries, and εit means the random error of the rate of economic growth in country i over the period t. It is assumed to be distributed symmetrically and independently $\varepsilon_{it} \sim \mathrm{IDD}\left(0, \tilde{A}_{\mu}^{2}\right)$,

i = Bahrain, Emirates, Kuwait, Oman, Qatar, Saudi Arabia, and Yemen,
t = 1996, 1996, 1997, ……, 2020

Panel data has a number of advantages. First it controls the heterogeneity of variance that may appear in the case of cross-sectional data or time series. Second, it gives better efficiency -- an increase in the degree of freedom and less linear multiplicity between variables. Third, its informational content is more than when using cross-sectional data or time series. Fourth, it takes into account what is known as the unbserved component in the sample whether cross-sectional or time. Fifth, it can take into account the time effects common to the sample countries which change over time.The panel data model is concerned with how to deal with sectional unobserved effects through the fixed effect model (as fixed effects), or by random effect model (as random effects) (Özkaya & Alhuwesh, 2017).

RESULTS AND DISCUSSION

In the empirical part of the study, first, the most appropriate models were explored in the case of static analysis, and then the cross-sectional dependency and heterogeneity in the sample countries were explored

Table 3. The results of static model

Dependent Variable: EGR			**Method: Panel Least Squares**			
Sample: 1996 2020	Periods included: 25		Cross-sections included: 7			
Total panel (balanced) observations: 175						
Method of estimation	pooled regression model PM		fixed effect model FEM		Random Effect Model REM	
Variable	**Coefficient**	**t-Statistic Prob.**	**Coefficient**	**t-Statistic Prob.**	**Coefficient**	**t-Statistic Prob.**
C (constant)	0.764103	0.4873	5.521392	0.061	1.406719	0.3852
MCS	0.03173	0.0295	0.046858	0.0011	0.041717	0.003
FTS	0.219329	0.0001	-0.043176	0.7727	0.174212	0.0236
INT	-0.083584	0.0015	-0.126257	0.0000	-0.102608	0.0001
Adjusted R-squared	0.1245		0.2473		0.1219	
F-statistic Prob.	0.000		0.0000		0.0000	
D-W stat	1.3016		1.5675		1.4542	
H test	--------		Chi-Sq. Prob.= 0.0185			
Chow Test	F. Prob.= 0.0000				--------	

as regards ICT indicators (mobile phone, fixed phone, and Internet) and the rate of economic growth in order to determine the most appropriate tests for the panel unit root and dynamic analysis of the model.

The static panel model was estimated. Three different models were estimated: Fixed effects model, random effect model, and the pooled regression model. The results presented in Table 3 were obtained.

As shown in Table 3 and by comparing PM with FEM through Chow's test, it was found that the F statistic is significant; thus, the null hypothesis that there is homogeneity between the parameters of the variables (common constant) is rejected accepting in this respect the alternative hypothesis. This means that there is a special constant for each country. Thus, the cross-sectional and time effects had to be included in the model (FEM), which is seen the most appropriate model. By comparing FEM with REM, through Hausman's test, it was found that the statistic $\chi 2$ is significant, and therefore the null hypothesis that the random effects parameter is consistent and efficient is rejected and the alternative hypothesis that the random effects parameter is inconsistent is accepted. This justifies our preference of FEM model.

Going by the results of FEM, it was found that the estimations of the parameters of the fixed telephone and the internet variables came with a negative sign, violating the hypothesis of the study. The positive impact of the mobile phone variable is considered to be low, as an increase of 1% in the number of mobile phone users leads to an increase of 0.05% in the economic growth rate of the countries of the Arabian Peninsula region. The negative impact of the Internet variable is also considered to be low, as an increase of 1% in the number of internet users leads to an increase of 0.13% reduction in the economic growth rate of the countries of the aforesaid region. The negative impact of the fixed-line variable, however, was not statistically significant. The value of the adjusted R-squared indicates that 25% of the changes in the economic growth rate are explained by the independent variables of the model, and 75% is connected with other variables. Concerning DW statistical value, it is found to be equal to 1.7, thus the model does not contain the autocorrelation problem. As for the F-statistic, it is found to be significant at the level

Table 4. Cross-sectional dependence tests results

Test	Stat	P-Value
LM	61.4611	0.0000
LM_{adj}	6.2433	0.0000
CD	3.9930	0.0001

of 1%, and therefore the model is considered to be good and appropriate for estimating the relationship between the economic growth rate and ICT variables in the countries of the Arabian Peninsula region.

Despite the advantages of the fixed effect model, it is criticized for its assumption that the slope parameters of the model are equal while expressing the cross-sectional or temporal effects through the common constant, which leads to the problem of biased estimator of the homogeneous slope parameters. To avoid this problem, a dynamic analysis of the panel model was conducted. It imposes a homogeneity constraint for the long-run parameters and allows the variance in the short-run parameters (Özkaya & Alhuwesh, 2017). For the dynamic analysis of the model, cross-sectional dependency and homogeneity tests must be conducted followed by the unit root test to verify the degree of integration of the study variables in order to choose the most appropriate model for estimation (Sart, Bayar, Corpădean, & Gavriletea, 2022).

In order to correctly define the unit root test method, cross-sectional dependence in residuals should be considered. If there is no cross-sectional dependence in residuals, the first-generation panel unit root test will be uset, but in case of availability, the second-generation panel unit root test will be used (Erkişi & Boğa, 2019). Null hypothesis (H_0) for this test is that no cross-section dependence (correlation) in residuals.The LM, CD and LM_{adj} tests were used respectively by (Breusch & Pagan, 1980), (Pesaran, 2004), and (Pesaran, Ullah, & Yamagata, 2008) to examine cross-sectional dependence in residuals. The results of the tests are shown in Table 4.

As shown in Table 4, the p-value is less than the 5% significance level, thus H_0 was rejected. This means that there is a cross-section dependence in residuals.

In order to verify the degree of Stationarity of the variables in case of the cross-sectional dependence in residuals and the heterogeneity of parameters, the panel unit root was tested using the first-generation tests through the LLC and IPS tests. See Table 5:

As Table 5 shows, it can be understood that the variables contain a unit root in their level except for the MCS variable, where the p-values of the test parameters is greater than the 5% level. As for the variables EGR and FTS, the test results are opposite. LLC test indicates that the two variables are stationary

Table 5. Panel unit root test results

Variables	LLC		IPS		Inferences
	Level	First dif.	Level	First dif.	
EGR	-3.10071**	.....	-0.42373	-2.54661**	I(1)
MCS	-3.35648**	.....	-3.31552**	.....	I(0)
FTS	-1.78964**	.....	-0.41093	-4.73879**	I(1)
INT	0.73402	-5.61663**	4.17175	-5.70883**	I(1)
Note(s): ** p < 0.01					

Table 6. Estimation results of PMG method

Dependent Variables	Coefficient	Std. Error	t-Statistic	Prob.*
	Long Run Equation			
MCS	0.115178	0.014233	8.092052	0.0000
FTS	-0.0566	0.091649	-0.61755	0.5395
INT	-0.20345	0.020211	-10.066	0.0000
	Short Run Equation			
ECT-1	-1.67515	0.429989	-3.89579	0.0003
D(MCS)	0.176607	0.239052	0.73878	0.4633
D(FTS)	-0.38083	1.238892	-0.3074	0.7597
D(INT)	-0.30317	0.184521	-1.64301	0.1063

in the level, while the IPS test indicates that they are not stationary in the same level. Hence, the two variables will be treated as non-stationary in the level meaning that they contain a unit root. Generally speaking, all of the variables do not contain the unit root at their first difference, because the p-values of the test parameters are less than the 5% level; so the null hypothesis is rejected. Therefore, each of the model variables is I (1), except the MCS variable is I (0).

The results of the unit root test indicates that the degrees of integration of the variables are I(0) and 1(1). Thus, the dynamic analysis of the panel model was carried out by estimating the ARDL model which allows estimating the long- and short-term effects separately (Pesaran, Shin, & Smith, 1999). The ARDL model has certain advantages. It allows estimation by a different combination of stationary variable I(0) or I(1). It also allows identifying both the short-run and long-run dynamics with the error correction limit (Yasmeen & Tufail, 2015). By analyzing the results of the dynamic model, it is possible to know the different influences and relationships between economic growth variables and ICT variables in the long and short-run.The short-run dynamic model shows how to make the different time series adaptive to rebalance the long-run. Regarding the relationship between the variables in the long-run, it is accessed by the error correction term (ECT), which shows the speed by which the system returns to equilibrium after the shock and when the error correction term is negative and significant. This supports the idea that there is a long-run relationship between the variables (Belloumi & Touati, 2022).

Table 6 shows the results of the impact of information and communication technology variables on economic growth in the long and short-run, and the adaptive parameter in the error correction term.

With reference to Table 6, in the long-run, it is found that the variable of mobile phone subscribers has a positive and significant effect on economic growth, while the variable of Internet users has a negative and significant effect. The variable of fixed telephone subscribers has a negative effect but insignificant. In the short run, it is noted that the three variables of ICT have no statistically significant effects on the economic growth of the countries of the Arabian Peninsula region. It is also understood that the error correction term parameter ECT is negative and significant, thus supporting the idea that there is a long-run relationship between the variables, and that any imbalance in the independent variables must affect the model and push it towards equilibrium in the long-run with an adaptive speed of 1.67%, indicating that it returns to complete equilibrium after 7 months.

In theory, ICT is expected to positively affect economic growth in the long and short-run. However, empirical results have indicated that the impact of ICT on economic growth can vary depending on the economic development status of a country and investment in ICT infrastructure (Kurniawati, 2022; Majeed & Ayub, 2018; Pradhan, Mallik, & Bagchi, 2018). In this context, the countries of the Arabian Peninsula region have witnessed a great development in the spread of ICT, and it occupies the lead among the Arab countries in the ICT index as well as in per capita income level, with the exception of Yemen which comes in the last ranks. These results are consistent with the study of (Al-Amin & Benlahabib, 2018). Moreover, the impact of ICT on economic growth varies between countries in line with our findings (Kurniawati, 2022; Bahrini & Qaffas, 2019). These results are also consistent with those of (Bouallaga & Kebir, 2021; Bahrini & Qaffas, 2019) as regards the impact of mobile phone and fixed telephone, and with (Benlahbib, 2018) regarding the impact of the Internet and fixed telephone. They also go in line with the findings of (Ben Yaba, Bouaziz, & Lnassari, 2021; Kacham & Kebir, 2020) regarding the impact of the internet and mobile phone, with those of (Kurniawati, 2022; Hussein, 2020) regarding the impact of the mobile phone, and with (Al-Amin & Benlahabib, 2018) regarding the impact of the fixed telephone. İnspite of agreement with most of the studies mentioned above, the results of this study are diffrent from the results of (Kurniawati, 2022; Hussein, 2020) as regards the impact of the internet and fixed telephone. They also differ from those findings of (Al-Amin & Benlahabib, 2018) regarding the impact of the internet and mobile phone, (Bouallaga & Kebir, 2021; Bahrini & Qaffas, 2019) regarding the impact of the Internet, and (Kacham & Kebir, 2020) regarding the impact of the fixed telephone. Personally speaking, the difference in the results of this study is due to the nature of the study sample, as the current study includes a partial sample of previous studies containing the majority of Arab countries. This is in addition to the nature of the dependent variable most of the previous studies used the average per capita income variable to express economic growth, while this study used the real economic growth rate variable. The difference may be also explained in terms of the method used for estimating the relationship between the variables, which is different from one study to another. With regard to the negative impact of the Internet indicator on economic growth, it is likely that this is due to the nature of the use of the Internet in the countries included in the study sample, where the Internet is used in social life, especially social media, more than economic activities. Internet users spend long hours on the internet without any real production, but only to pass the time, with the difficulty in dealing with the Internet for productive activities as a result of the scientific level of the citizens of the study sample countries, as well as the continuation of many institutions with traditional work within the framework of their internal and external environment (Benlahbib, 2018).

CONCLUSION

ICT has a positive impact on economic growth in the long and short-run. However, empirical results have indicated that the impact of ICT on economic growth can vary depending on the economic development status of countries and investment in ICT infrastructure. This study aims at examining the effects of ICT on the economic growth of the Arabian Peninsula countries during the period 1996-2020. This has been done through the two indicators of access (the number of subscribers to mobile phone and fixed telephone per 100 people) and the use of the internet (individuals using the internet, % of population). The study used static and dynamic analysis of panel models. Findings show that there is a difference in the impact of ICT on economic growth according to the indicator used and the time range. While there was a positive effect of the mobile phone and a negative impact of the Internet on economic growth in the long-run, there was no effect of any variable in the short-run, The error correction indicates that there is a long-run relationship between the variables and that any imbalance between them will return to equilibrium after 7 months. The difference in the results of this study with the results of previous studies is due to the nature of the study sample, the nature of the dependent variable, and the method used for estimating the relationship between the variables, which differs from one study to another. With regard to the negative impact of the Internet indicator on economic growth, it is likely that this is due to the nature of the use of the Internet in the study sample countries, where the internet is used more in social life than in economic activities, with the difficulty of using the internet for productive activities, as well as the continuation of many institutions with traditional work within the framework of their internal and external environment.

The present study recomments that the countries of the Arabian Peninsula region shall target investment that enhances the spread of ICT through increased investment in ICT infrastructure, and in helping easy access to the internet and its applications in economic life. Increasing investment in ICT should be a priority in government policies. The present study may suffer a small weakness. This can be explained in the idea that the investigation is limited to the countries of the Arabian Peninsula region, which vary in income levels and information technology indicators. Therefore, as a future research direction, we aim to expand the analysis with data for other Arab countries, while conducting the analysis according to the economic situation of the countries and the extent of their classification within the ICT index.

REFERENCES

Aghaei, M., & Rezagholizadeh, M. (2017). The impact of information and communication technology (ICT) on economic growth in the OIC countries. *Economic and Environmental Studies (E&ES), 17*(2), 257-278.

Al-Amin, B. M., & Benlahabib, T. (2018). *The impact of information and communication technology on economic growth (case study of Arab countries 2018). In Modern Technology and Quality of Life.* Hamma Lakhdar Eloued University.

Bahrini, R., & Qaffas, A. A. (2019). Impact of Information and Communication Technology on Economic Growth: Evidence from Developing Countries. *Economies, 7*(21), 1–13. doi:10.3390/economies7010021

Belloumi, M., & Touati, K. (2022). Do FDI Inflows and ICT Affect Economic Growth? An Evidence from Arab Countries. *Sustainability, 14*(6293), 1–21. doi:10.3390u14106293

Ben Yaba, M., Bouaziz, A., & Lnassari, A. (2021). A Standard Study of The Impact of Information and Communication Technology on Economic Growth in the Arab Countries During The Period (2004-2018). *Journal of International Economy and Globalization, 4*(1), 39–50.

Benlahbib, T. (2018). The impact of ICTs on economic growth in developing countries econometric study during 2005-2015. *Journal of Economic and Financial Research, 5*(1), 559–580.

Bouallaga, L., & Kebir, M. (2021). Measuring the impact of information and communication technology on economic growth in Algeria and some Arab countries Standard study accrediting Analysis - Dynamic Panel for the period: 2000-2018. *The Journal of Business Administration and Economic Studies*, *7*(1), 205–526.

Breusch, T. S., & Pagan, A. R. (1980). The Lagrange Multiplier Test and its Applications to Model Specification in Econometrics. *The Review of Economic Studies*, *47*(1), 239–253. doi:10.2307/2297111

Conole, G., & Dyke, M. (2004). What are the affordances of information and communication technologies? *ALT-J Research in Learning Technology*, *12*(2), 113–124. doi:10.3402/rlt.v12i2.11246

Erkişi, K., & Boğa, S. (2019). High-technology products export and economic growth: A panel data analysis for EU-15 countries. *Bingöl Üniversitesi Sosyal Bilimler Enstitüsü Dergisi*, *9*(18), 669–683.

Farhadi, M., Ismail, R., & Fooladi, M. (2012). Information and Communication Technology Use and Economic Growth. *PLoS One*, *7*(11), e48903. doi:10.1371/journal.pone.0048903 PMID:23152817

Hodrab, R., Maitah, M., & Luboš, S. (2016). The Effect of Information and Communication Technology on Economic Growth: Arab World Case. *International Journal of Economics and Financial Issues*, *6*(2), 765–775.

Hussein, E. F. (2020). The impact of information and communication technology on inclusive growth: An empirical study on developing and Arab countries. *Journal of the Faculty of Economics and Political Science*, *21*(3), 47–77.

IMF. (2022). *GDP per capita, current prices*. Retrieved from IMF DATAMAPPER: https://www.imf.org/external/datamapper/NGDPDPC@WEO/MEQ

IMF. (2022a). *IMF DATAMAPPER*. Retrieved from Real GDP growth: https://www.imf.org/external/datamapper/NGDP_RPCH@WEO/OEMDC/ADVEC/WEOWORLD

Ishida, H. (2015). The effect of ICT development on economic growth and energy consumption in Japan. *Telematics and Informatics*, *32*(1), 79–88. doi:10.1016/j.tele.2014.04.003

ITU. (2018). *Measuring the Information Society Report: Executive summary*. ITU Publications.

ITU. (2022). *The ICT Development Index*. Retrieved from Committed to connecting the world: https://www.itu.int/en/ITU-D/Statistics/Pages/IDI/default.aspx

ITU. (2022a). *The ICT Development Index (IDI): Conceptual framework and methodology*. Retrieved from Committed to connecting the world: https://www.itu.int/en/ITU-D/Statistics/Pages/publications/mis2017/methodology.aspx

ITU. (2022b). *Select an option IDI 2017 Rank*. Retrieved from ICT development Index ITUdata: https://www.itu.int/net4/ITU-D/idi/2017/index.html

Kacham, I., & Kebir, M. (2020). Information and Communication Technology and its Impact on Economic Growth in the Organization of Arab Petroleum Exporting Countries (OAPEC) A Standard Study for the period 2000-2018. *The Economic Science Management and Business Science*, *13*(1), 460–479.

Khireddine, B., Noureddine, B., & Bouzelifa, S. (2018). The impact of information and communication technology on economic activity. In *The Third National Forum on Consumers and the Digital Economy: The necessity of transition and protection challenges* (pp. 1-18). Mila: Abdelhafid Boussouf University.

Kurniawati, M. A. (2022). Analysis of the impact of information communication technology on economic growth: Empirical evidence from Asian countries. *Journal of Asian Business and Economic Studies*, *29*(1), 2–18. doi:10.1108/JABES-07-2020-0082

Majeed, M. T., & Ayub, T. (2018). Information and communication technology (ICT) and economic growth nexus: A comparative global analysis. *Pakistan Journal of Commerce and Social Sciences*, *12*(2), 443–476.

Maneejuk, P., & Yamaka, W. (2020). An analysis of the impacts of telecommunications technology and innovation on economic growth. *Telecommunications Policy*, *44*(10), 102038. doi:10.1016/j.telpol.2020.102038

Nabi, A. A., Tunio, F. H., Azhar, M., Syed, M. S., & Ullah, Z. (2022). Impact of Information and Communication Technology, Financial Development, and Trade on Economic Growth: Empirical Analysis on N11 Countries. *Journal of the Knowledge Economy*, 1–18. doi:10.100713132-022-00890-6

Nguyen, T. P., Dinh, T. T., Tran Ngoc, T., & Duong Thi Thuy, T. (2022). Impact of ICT diffusion on the interaction of growth and its volatility: Evidence from cross-country analysis. *Cogent Business & Management*, *9*(1), 1–27. doi:10.1080/23311975.2022.2054530

Niebel, T. (2018). ICT and economic growth – Comparing developing, emerging and developed countries. *World Development*, *104*(April), 197–211. doi:10.1016/j.worlddev.2017.11.024

OIC. (2022). *OIC Statistics Database (OICStat) - Query*. Retrieved from The Statistical, Economic and Social Research and Training Centre for Islamic Countries (SESRIC): https://www.sesric.org/query.php

Özkaya, M. H., & Alhuwesh, M. A. (2017). Determinants of the Republic of Yemen's Foreign Trade with the Gulf Cooperation Council Countries: Gravity Model. In *International Conference on Scientific Cooperation for the Future in the Economics and Administrative Sciences* (pp. 79-88). Thessaloniki: Usak University & Aristotle University of Thessaloniki.

Pesaran, M. H. (2004). *General Diagnostic Tests for Cross Section Dependence in Panels.* University of Cambridge, Discussion Paper No. 1240, 1-42.

Pesaran, M. H., Shin, Y., & Smith, R. P. (1999). Pooled Mean Group Estimation of Dynamic Heterogeneous Panels. *Journal of the American Statistical Association*, *94*(446), 621–634. doi:10.1080/01621459.1999.10474156

Pesaran, M. H., Ullah, A., & Yamagata, T. (2008). A bias-adjusted LM test of error cross-section independence. *The Econometrics Journal*, *11*(1), 105–127. doi:10.1111/j.1368-423X.2007.00227.x

Pradhan, R. P., Arvin, M. B., & Normancd, N. R. (2015). The dynamics of information and communications technologies infrastructure, economic growth, and financial development: Evidence from Asian countries. *Technology in Society*, *42*, 135–149. doi:10.1016/j.techsoc.2015.04.002

Pradhan, R. P., Mallik, G., & Bagchi, T. P. (2018). Information communication technology (ICT) infrastructure and economic growth: A causality evinced by cross-country panel data. *IIMB Management Review*, *30*(1), 91–103. doi:10.1016/j.iimb.2018.01.001

Sabbagh, K., Friedrich, R., El-Darwiche, B., Singh, M., & Koster, A. (2013). Digitization for Economic Growth and Job Creation: Regional and Industry Perspectives. In *The Global Information Technology Report 2013 Growth and Jobs in a Hyperconnected World* (pp. 35-42). World Economic Forum.

Saidi, K., Ben Hassen, L., & Hammami, M. S. (2015). Econometric Analysis of the Relationship Between ICT and Economic Growth in Tunisia. *Journal of the Knowledge Economy*, *6*(4), 1191–1206. doi:10.100713132-014-0204-9

Sart, G., Bayar, Y., Corpădean, A. G., & Gavriletea, M. D. (2022). Impact of ICT and Globalization on Educational Attainment: Evidence from the New EU Member States. *Sustainability*, *14*(3039), 1–13. doi:10.3390u14053039

Sepehrdoust, H., & Ghorbanseresht, M. (2019). Impact of information and communication technology and financial development on economic growth of OPEC developing economies. *Kasetsart Journal of Social Sciences*, *40*(3), 546–551.

Stanley, T. D., Doucouliagos, H., & Steel, P. (2018). Does ICT Generate Growth? A Meta-regression Analysis. *Journal of Economic Surveys*, *32*(3), 705–726. doi:10.1111/joes.12211

Usman, A., Ozturk, I., Hassan, A., Zafar, S. M., & Ullah, S. (2021). The effect of ICT on energy consumption and economic growth in South Asian economies: An empirical analysis. *Telematics and Informatics*, *58*(101537), 1–9. doi:10.1016/j.tele.2020.101537

Vu, K. M. (2011). ICT as a source of economic growth in the information age: Empirical evidence from the 1996–2005 period. *Telecommunications Policy*, *35*(4), 357–372. doi:10.1016/j.telpol.2011.02.008

WB. (2002). *Information and Communication Technologies: A World Bank Group Strategy*. The World Bank.

WB. (2012). *ICT for Greater Development Impact: World Bank Group Strategy for Information and Communication Technology, 2012-2015.* World Bank. Retrieved from https://openknowledge.worldbank.org/bitstream/handle/10986/27411/715400WP0WBG0I0sclosed0July02502012.pdf?sequence=1&isAllowed=y

WB. (2022). *Indicators*. Retrieved from World Bank Open Data: https://data.worldbank.org/indicator

Yasmeen, R., & Tufail, S. (2015). Impact of Internet Technology on Economic Growth in South Asia with Special Reference to Pakistan. *Pakistan Journal of Social Sciences*, *35*(2), 777–784.

Yousefi, A. (2011). The impact of information and communication technology on economic growth: Evidence from developed and developing countries. *Economics of Innovation and New Technology, 20*(6), 581–596. doi:10.1080/10438599.2010.544470

ENDNOTE

[1] The ICT Development Index (IDI) was designed in 2009, and in March 2017, a revised set was approved and included in IDI, bringing the number to 14 indicators. However, countries faced challenges in collecting new data, and since 2018, all attempts to publish or modify IDI have failed.

Chapter 5
Exploring the ICT Diffusion and Economic Growth Nexus:
Evidence From BRICS Nations

Oğuzhan Yelkesen
https://orcid.org/0000-0002-3314-5068
Bandirma Onyedi Eylul University, Turkey

ABSTRACT

In the last few decades, there has been a surge of interest in the effects of information and communication technologies (ICTs) on economic growth. Although ICT diffusion affects economic growth in many ways, the magnitude of ICT penetration is not time-invariant and depends on many factors. Using pooled mean group (PMG) panel ARDL model, this study aims to explore the nexus between ICT penetration and economic growth for BRICS nations over 2003-2020. The findings show that, in the short run, fixed broadband and mobile cellular subscriptions have a positive effect on economic growth, while both R&D expenditure and ICT goods import have a negative insignificant effect. In the long run, however, the effect of both R&D expenditure and ICT goods import turns out to be significantly positive, whereas fixed broadband and mobile cellular subscriptions have a negative effect on economic growth. The results have significant implications for policymakers and practitioners who are willing to harness ICTs to improve the economic and social conditions that would lead to higher economic growth.

INTRODUCTION

Economic growth is a fundamental determinant of a country's welfare and a significant indicator for social and institutional development. Conceptually, economic growth stands for higher value for goods and services produced in a country during a specific period and is generally proxied by Gross Domestic Product (GDP). From a broader perspective, however, the concept of economic growth can be extended as it is one of the most important components of economic development, which is a much more complex issue for economists and policy makers. With a greater ratio of economic growth, it is not only expected to obtain higher level of growth rate but also is expected to channel more resources to the areas that

DOI: 10.4018/978-1-6684-6620-9.ch005

would yield more benefit to future economic activities through encouraging innovation, providing a more effective health care system, and lowering transaction costs, which are highly crucial for economic activities. In this sense, this creates a virtuous cycle and paves the way for further economic development more generally.

Over the past century, however, there has been a dramatic increase in the use of Information and Communication Technologies (hereafter, ICT). In this vein, the impact of ICT on economic growth has been an object of research since the early years of ICT use in the 1990s. The use of ICTs well contributed to the process of establishing an institutional structure through lowering costs in communication and documentation as well as expanding storage capacity of countries and helping countries provide better public services (Ndung'u, 2019; Mano, 2021). Besides its contribution to institutional structure, ICT has been an important concept in the study of economic growth and development. Some scholars have included the technology as a significant explanatory factor in their growth models (see, for example, Solow, 1965; Romer, 1986, 1990; Lucas, 1988). Although academics treat technology differently in their models, their common ground was in part the same: investigating the role of technology in explaining economic growth.

In the aid literature, ICT diffusion appears to be positively related to economic growth (for example, see, Jin & Cho, 2015; Cheng & Chien, 2021). Existing studies are largely empirical in nature, mostly relying on econometric methods that analyze both the linear and nonlinear association between the variables. For instance, Vu (2011) clearly defines the channels through which ICT diffusion positively affects economic growth: firstly, it encourages innovation. Secondly, it enables firms and individuals to make better decisions, which is quite similar to the previous theoretical assumptions as it lowers transaction costs. Lastly, in addition to lowering transaction costs, production costs would decrease, and demand would increase, which in turn would increase the output level. In parallel to this, for instance, using a Generalized Method of Moments (GMM) over the period 1990-2007, Nasab and Aghaei (2009) investigated the impact of ICT on economic growth for OPEC member countries. The authors find positive and statistically significant results, indicating that a 10 percent increase in ICT investment leads to a 0.2 percent increase in the average rate of GDP per capita.

On the other side, even though there is no doubt that ICT diffusion helps countries reach higher economic growth and standard of living, the magnitude of the sprawling technology does not seem to be similar across countries due to the differences in the level of economic development (Porter et al., 2001; Mayer et al., 2020). In a study which set out to discuss ICT and economic growth nexus, Niebel (2018) well explained comparatively how ICTs play a key role to increase productivity growth in developing, emerging and developed countries. The author highlights the fact that the linkage between ICT investments and economic growth is rather weak due to the lack of both micro- and macro-level data. In addition, some others point out the paradox of productivity where an increase in ICT investment does not result in higher productivity and capacity (Solow, 1987; Carr, 2003). According to Haller and Lyons (2015), however, the key point in achieving higher productivity through ICT diffusion lies not in the adoption of technology but in the use of it.

At first blush, contrasting views across the literature might seem to be confusing. This is, however, a natural reflection of the dynamic interaction happening between the dynamic phenomena. That is to say, the extent to which ICT diffusion affects economic growth may change across countries due to cultural, social, and institutional factors. Referencing to the first channel affecting economic growth through ICT by Vu (2011) above, for ICT investment to create innovation, it requires a process that would take idiosyncratic factors into consideration (Suchman, 1987). In this sense, therefore, we aim at exploring

ICT diffusion and economic growth nexus for BRICS nations over the period 2003-2020. The main reason for choosing this country group is that it consists of both types of countries, the one producing and exporting technology and services like China and India, and the other one importing and utilizing technology like Brazil. It is aimed at treating these countries as one group and analyzing the relationship between ICT diffusion and economic growth by referencing the differences at play. Data availability was the determinant factor of choosing the relevant time period.

Using the pooled mean group (PMG) panel ARDL model, we investigated both short and long run interrelationship between ICT diffusion and economic growth for BRICS countries for the period 2003-2020. ICT diffusion is proxied by fixed broadband subscriptions (per 100 people), mobile cellular subscriptions (per 100 people), Research and Development (R&D) expenditure (% of GDP), and ICT goods imports (% of total import) while measuring economic growth by GDP per capita growth. The results show that R&D expenditure and ICT goods imports have a significant and positive effect, while fixed broadband and mobile cellular subscriptions have a negative, but only the former is significant, effect on economic growth. In general, therefore, it confirms the idea that some countries' absorptive capacity is relatively low compared to other countries in the group. Additionally, some countries, like China, have been producing and exporting a large amount of technology, generating more revenue for further innovation and investment in technology, and therefore resulting in the digital divide in the group. Another important but often neglected factor is that BRICS nations have been undergoing a process of development in terms of both economic and social.

Therefore, our results are in line with the characteristics of BRICS nations in the sense that R&D investment and utilization from imported technology would take time and necessary institutional and social infrastructure would be necessary for ICT diffusion to positively affect economic growth. On the other side, mobile cellular and fixed broadband subscriptions provide people with opportunities for utilizing and taking advantage of the existing technological groundwork, hence their short run effect is positive on economic growth. Their long run effect, however, would depend on investments in research and development and human capital that would create value added products for further technological breakthroughs. In this way, the contribution of this study has been to confirm the positive relationship between ICT diffusion and economic growth for BRICS nations. In addition, the insights gained from this study may be of assistance to policymakers in terms of implementing short run policies that would lead to further technological advancements in the long run. Finally, the findings of the present study will help researchers and policymakers design policy based on countries' social, cultural and institutional differences.

In the next section, conceptual framework and theoretical concepts regarding the use of technology in contemporary growth models will be discussed. Thereafter, ICT diffusion will be analyzed in BRICS nations. After discussing ICT penetration in BRICS nations, in the subsequent chapter, the data that have been used in our analysis will be defined and the model will be specified. Following this chapter, next section will present results. In the following last two sections, the results will be briefly discussed, and final remarks will be presented.

CONCEPTUAL FRAMEWORK

In the extant literature, some scholars have attempted to explain economic growth by using different growth models. Theoretically, the neoclassical growth model proposed by Solow (1956) sees technological change as an exogenous factor, that is, not included in the model itself, for economic growth, while endogenous growth model by Romer (1986, 1990) and Lucas (1988) considers technology one of the most important production factors and claims that technology promotes economic growth through increasing ability and skills of the labor force. Although both growth theories well explain the role of technology in economic growth, endogenous growth theory proposes that technology will play a contributing role in reviving economic growth as it leads to new production processes and business models for the economy (Czernich et al., 2011).

As stated above, the neoclassical growth model does not include technological change within the growth model and sees it as an exogenous factor. This causes a special notion to arise, which is called "Solow residual". This explains the portion of economic growth that cannot be explained by standard production factors of the model, like capital and labor. Neoclassical growth theory explains economic growth by either increasing capital and labor or using both efficiently and does not deny the role of technological change to be able to sustain economic growth. However, since the role of technological changes is crucial, even if it is exogenous, explaining output growth solely by capital and labor factors becomes prominently vague. Thus, Solow residual helps us explain the mechanism through which technology diffusion affects economic growth.

Considering Cobb-Douglas (1928) production function for *n* inputs will be as following:

$$Y = f\left(x_1, x_2, x_3, \ldots, x_n\right) = \mu \prod\nolimits_{i=1}^{n} x_i^{\varphi_i} \qquad (1)$$

where *Y* represents output, x_i's are inputs, respectively. Additionally, μ and φi $_c$an be considered changing parameters to gauge the level of output towards the impulses of inputs.

The output-elasticity, that is, φi $_p$arameters, is restricted to be φiÎ($_0$,1) and their sum to be equal to one. Thus, a standard Cobb-Douglas production function can be rewritten in the following way:

$$Y = \mu L^{\varphi_1} K^{1-\varphi_2} \qquad (2)$$

where φ1$_+$$\varphi$2=$_1$ and L *and* K *r*epresent labor and capital as production factors, sequentially. In Eq. (2), μ stands for total factor productivity, which has been discussed under "Solow residual" above. Moreover, μ explains the channels through which technology promotes economic growth, such as the dissemination of information in a more efficient and quick way, resulting in lower transaction costs and effective market structure (Biagi, 2013). If μit=f($_{Ti}$t*)*, $_{th}$en equation can be rewritten in its most popular form as follows:

$$\text{Yit} = {}_f(\text{Li}t, {}_{Ki}\text{t}, {}_{Ti}\text{t}) \qquad (3)$$

where Eq. (3) defines economic growth as the function of labor, capital and technology.

In this sense, some studies have defined technology within the growth models in order to better capture the relative effect of technology use in the output level. Nasab and Aghaei (2009), for instance, add technology as a significant input in the growth model, which is turning out to be a more efficient production process by which the technology upgrades labor skills and improves capital utilization. In a similar but not identical pattern, Jin and Cho (2015) set out an alternative growth model, which contains other socioeconomic factors like corruption, regulation and education. The authors refer to these factors as non-economic factors and well substantiate that these factors should also be integrated into the growth model as they significantly affect economic growth. With a similar logic, the Eq. (3) can be extended to a larger set of explanatory factors that would influence the level of output.

Against this backdrop, in analyzing the nexus between ICT diffusion and economic growth, we assume that the gradual diffusion of technology has significantly changed the way in which economies work, but the magnitude of this effect varies across countries due to differences in the level of developments and country-specific factors. Therefore, economic growth is treated as a process which is highly ICT diffusion is identified as a gradual and exogenous factor for BRICS nations. Following up Eq. (3) and benefiting from the Appiah-Otoo and Song's (2021) approach, we rearrange the model to obtain a panel data form and substitute T into the model. To show econometrically, each input will take a coefficient and an error term to be consistent with a stochastic model. Therefore, the equation will be:

$$y_{it} = \beta_0 + \beta_1 L_{it} + \beta_2 K_{it} + \beta_3 T_{it} + \varepsilon_{it} \tag{4}$$

Eq. (4) represents the most basic neoclassical production function. In the following sections, we will delve into the details of the data set that have been used for our analysis and substitute the right-hand side variables with that of our explanatory variables.

ICT DIFFUSION IN BRICS NATIONS

Recently, BRICS group consists of five major emerging market economies, namely Brazil, Russia, India, China, and South Africa. Those countries, however, have been paying a lot more attention to the diffusion and use of ICTs to reach their predetermined goals and benefit from the opportunities that technological development would deliver. In fact, in the 7th BRICS communications ministers' virtual meeting held on 22 October 2021, the community pointed out the importance of ICTs in preventing the crippling effects of the COVID-19 pandemic. More specifically, they considered ICTs one of the significant factors that would help countries provide necessary services like health and education and sustain other financial related activities in such disastrous times. In this sense, it is possible to say that BRICS nations already highlighted the importance of ICTs in helping countries achieve their goals and weather some challenges caused by both global and regional factors.

The stance of BRICS nations regarding the role of ICTs has long been discussed. It seems that each of the BRICS member countries is well aware of the fact that technology plays an important role in every sphere of social life. Therefore, they initiated some reforms and put more effort in action to keep up with technological development that is seemingly led by developed countries. In this parallel, China and India introduced a series of programs that would help them create an environment where the up-to-date technological developments will be followed. For instance, DigitalIndia has been implemented in India while a state program called "Information Society (2011-2020)" has been put into action in 2010 in Russia and a strategic plan called "Plan 863" has been initiated in China in 1986 (Revinova, 2016). Additionally, governments in other BRICS countries have supported ICT exporting countries, exempted them from taxes partly or completely, and provided some tax incentives for both ICT export-and-import.

To concretely measure the progress of BRICS in terms of producing, exporting and using technology, we benefit from World Bank indicators that have also been used in our empirical analysis. Figure 1 below

Figure 1. Individuals using the internet in BRICS (% of people, 2003-2020)
Source: Authors' calculations, World Development Indicator

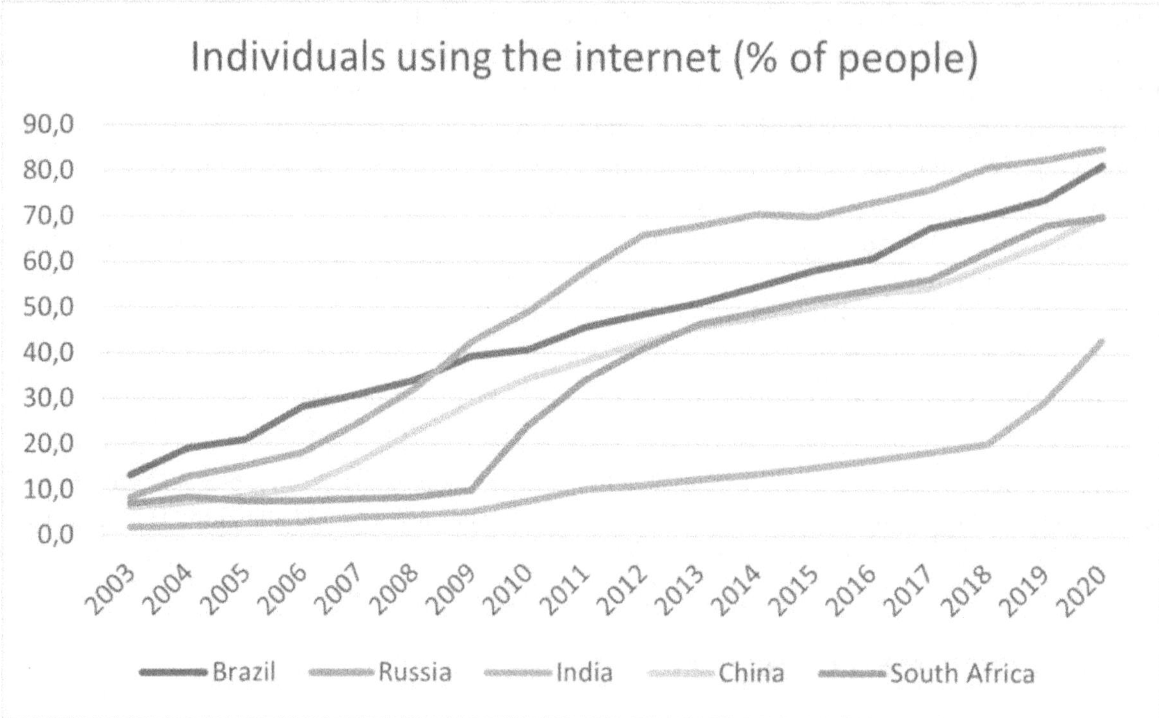

shows that there has been a steady rise in the number of people using the internet in Brazil, Russia and China. What is striking in this figure is the rapid increase in people using the internet in South Africa starting in 2009. In addition, what can be clearly seen in this figure is that Russia has the highest rate of internet use, approximately 85% in 2020 while India has the lowest ratio with 43%. Apparently, four members of BRICS follow the same pattern in internet use except India, which is rather diverse until 2018.

As another indicator for ICT capacity of BRICS, the ratio of exports over total export has been used. As can be seen in Figure 2, two salient factors come into prominence: Firstly, China's export ratio is far ahead of other BRICS countries. This clearly shows that China has been producing more technological breakthroughs, and thus generating more revenue that can be allocated to the factors that would increase capacity use. This would also be expected to serve China's economic development through higher investment in productive sectors like health, infrastructure and most importantly education, which would highly enhance the absorptive capacity of the country in terms of exploiting and producing technological goods and services. Secondly, other countries did not seem to have produced technology and gained much from its export compared to China.

Although almost all countries recognize the significance of ICT diffusion, it seems that each country has different characteristics in using and absorbing new technology. The difference lies primarily in their pre-investment stages as well as institutional and social structure. When we look at the level of digital development of BRICS nations using the OECD toolkit in measuring the level of digital economy, the differences across countries reveal much more prominently. According to Figure 1 and 3, India is the laggard compared to other BRICS nations in almost each category below. What stands out in these fig-

Figure 2. ICT goods exports in BRICS (% of total export, 2003-2020)
Source: Authors' calculations, World Development Indicator

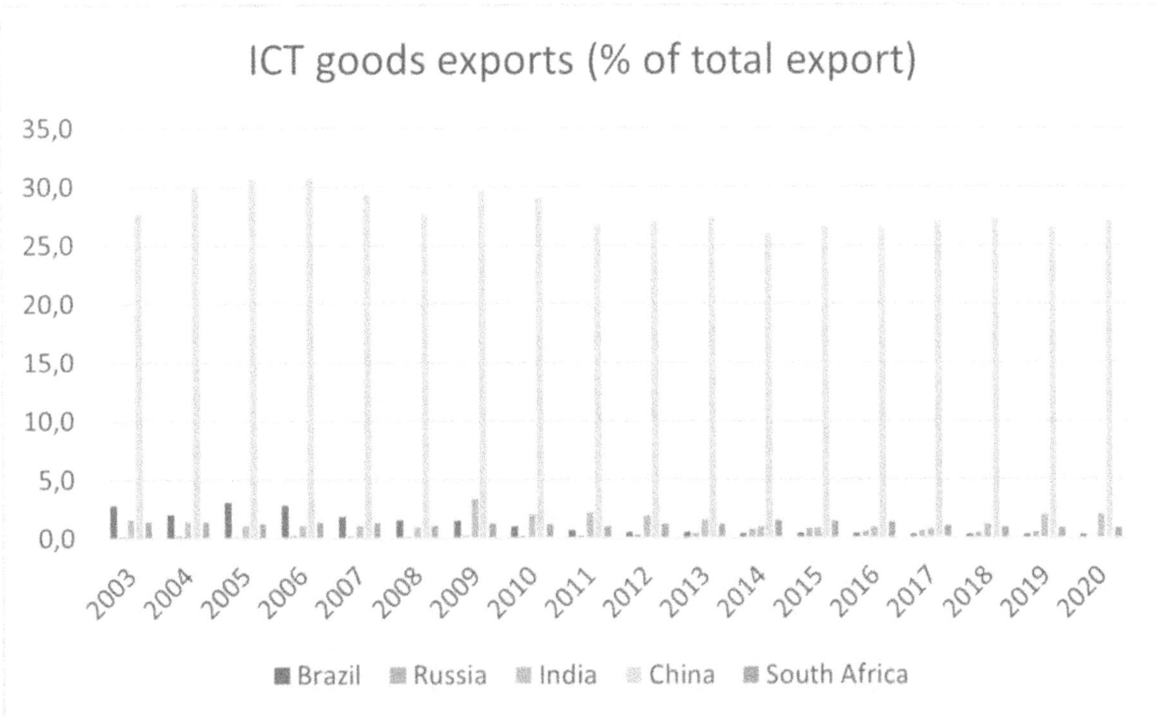

ures is the existence of an explicit chasm between countries in having required infrastructure as well as economic and social structure to harness technology to build up an environment where it better serves the economic and political goals of the BRICS association.

All in all, Figure 3 above well helps us to recognize the differences among BRICS countries in absorbing and harnessing technology diffusion. Underlying differences would cause the effect of ICT indicators on economic growth to vary over time, which is justifying the assessment of both short and long run coefficients of the variables to obtain robust and reliable results as well as to better analyze the effect of technology evolution for those countries.

DATA, MODEL SPECIFICATION, AND METHODOLOGY

Data and Model Specification

This chapter lays the foundations for the method and identifies the data that are used to empirically investigate the nexus between ICT diffusion and economic growth in BRICS nations over the period 2003-2020. The specific time period is chosen due to data availability for all countries. The variables that we have utilized for our estimation model are as follows: (i) GDP per capita growth as a proxy for economic growth and used as the dependent variable, for ICT variables; (ii) fixed broadband subscription (per 100 people), (iii) mobile cellular subscriptions (per 100 people), (iv) research and development

Figure 3. Digital development indicators for BRICS nations
Source: Compiled by Ignatov (2020) depending on G20 Argentina Presidency Report (2018)

Category	Indicators	China	Brazil	South Africa	India	Russia
Infrastructure	Fixed broadband subscriptions per 100 inhabitants	28.0	13.7	3.0	1.3	21.4
	Active mobile broadband subscriptions per 100 inhabitants	83.6	90.2	70.0	25.8	80.8
	Average speed, Mbps	7.6	6.8	6.7	6.5	11.8
	Fixed and mobile broadband prices, $ PPP	16.1	29.9	18.0	17.1	10.1
Empowering Society	Internet users	54.3	67.5	56.2	34.5	76.0
	Individuals who purchased online in the last 12 months	x	40.2	x	x	19.6
	Individuals using the Internet to interact with public authorities	x	24.0	x	x	x
	Tertiary graduates in the natural sciences, engineering and ICTs (NSE & ICT)	x	15.0	x	31.0	29.0
Innovation and Technology Adaptation	Enterprises using cloud computing services, with 10 or more persons employed	x	44.6	x	x	x
Jobs and Growth	ICT goods as a percentage of merchandise trade	26.5	0.39	1.4	0.95	0.55
	Telecommunications, computer and information services as a percentage of services trade	x	10.0	x	36.6	14.2

expenditure-to-GDP ratio, (v) the ratio of ICT goods import over total import are used. The data have been taken from the World Development Indicators by the World Bank.

According to Table 1 below, which shows the descriptive statistics of the variables, mobile cellular subscription has the highest volatility as well as maximum value compared to other variables. On the other hand, Table 2 displays the correlation matrix of the variables. As can be seen, except fixed broadband and mobile cellular subscriptions, research and development expenditure and the ratio of ICT import is positively correlated with GDP per capita growth. It seems that there is no evidence of high correlation among the regressors that would give rise to the multicollinearity problem.

Relying on Eq. (4), we may arrange our empirical equation to be appropriate for a panel data analysis. With the dependent and independent variables, the equation will be shown as follows:

$$gdpc_{it} = \beta_0 + \beta_1 fbs_{it} + \beta_2 mcs_{it} + \beta_3 imp_{it} + \beta_4 rde_{it} + \varepsilon_{it} \quad (5)$$

Table 1. Descriptive statistics

Variable	Obs	Mean	Std. Dev.	Min	Max
gdpc	90	3.468133	4.320451	-8.164894	13.63582
fbs	90	7.493673	8.389542	.0126279	33.5956
mcs	90	93.33179	46.1595	3.030976	165.661
imp	89	11.30839	5.904366	4.088295	26.06887
rde	79	1.101312	.3839587	.65282	2.14058

where *gdpc* stands for GDP per capita growth, *fbs* represents fixed broadband subscription, *mcs* denotes mobile cellular subscription, *imp* represents ICT goods import, and *rde* refers to research and development expenditure. In this study, therefore, short and long run coefficients of the explanatory variables will be estimated through deploying Pooled Mean Group (PMG) panel data model.

Note parenthetically that the impact of technology would vary over time. For those countries that have not well invested in hardware, software and telecommunication infrastructure in the previous period, the effect of ICT would be limited. Because they determine absorptive capacity of countries and are seen as main components of ICTs (Kaiser, 2004; Adeleye & Eboagu, 2019). In this sense, the impact of ICT diffusion may be seen with lag in some countries. As our study offers both short and long run interaction between ICT diffusion and economic growth, one would have the chance to see the impact of ICTs on economic growth in a 17-year time horizon.

Methodology

Panel Unit Root Tests

To avoid the spurious regression problem, unit root characteristics of the variables should be checked first. Running a model with non-stationary variables would lead to biased and unreliable results that would, in turn, raise a question mark over the validity of the empirical findings. Therefore, we employ Im-Pesaran-Shin (IPS) unit root test developed by Im et al. (2003). The rationale behind choosing this test specifically and not to prefer others like Livin-Lin-Chu (LLC) developed by Levin et al. (2002) is that there are some missing data in our dataset. Therefore, the best option was to choose IPS unit root test to check stationarity of the variables. The specification of the IPS unit root can be written as follows:

Table 2. Correlation matrix

	gdpc	fbs	mcs	imp	rde
gdpc	1.0000				
fbs	-0.1523	1.0000			
mcs	-0.5684	0.5432	1.0000		
imp	0.5604	0.3320	-0.2602	1.0000	
rde	0.2868	0.6891	0.0540	0.7433	1.0000

$$y_{it} = \rho_i y_{it-1} + \sum_{n=1}^{p_i} \partial_{in} \Delta y_{it-n} + \varnothing_{it} \delta + \varepsilon_{it} \quad (6)$$

According to the Eq. (6), null hypothesis is tested if H_0: $\rho i_= 1$, implying that all the series have unit root, and in an alternative case, that is, if $H0_: \rho i <_1$, stationarity is ensured.

Panel ARDL Method

In exploring the nexus between the variables, we deploy a pooled mean group (PMG) panel ARDL model, which is developed by Pesaran et al. (1999). This is the one that enables us to assess both short and long run relationships among variables that have different unit-root characteristics (i.e., I(0) and I(1)). Another advantage of this method is that it enables us to reconcile longitudinal data with a cross-section approach, thus allowing us to generalize our results and making findings more robust and reliable. The PMG estimator, however, allows the long run parameters to be constant, providing a basis for homogeneity while enabling the short run constant, the variance of the error terms and the coefficients of the variable to vary. Due to long run homogeneity, there is no need for running cointegration test. Therefore, it will not be specified in this section. The ARDL for the model can be specified as follows:

$$y_{it} = \sum_{j=1}^{z} \partial_{ij} y_{it-j} + \sum_{j=0}^{g} \varnothing_{ij} \gamma_{it-j} + \mu_i + \varepsilon_{it} \quad (7)$$

where γ is the vector of independent variables. Rearranging the Eq. (7) and turning it into the form which includes the coefficient of error correction term as well as short and long run coefficients of the variables, it can be rewritten in the following way:

$$\Delta y_{it} = \varphi_i \left(y_{it-1} - \beta_i \gamma_{it} \right) + \sum_{j=1}^{z-1} \partial_{ij}^{*} \Delta y_{it-j} + \sum_{j=0}^{g-1} \varnothing_{ij}^{*} \Delta \gamma_{it-j} + \mu_i + \varepsilon_{it} \quad (8)$$

where φ represents the coefficient of the error correction term while $\beta i'_s$ refer to the long run coefficients of the explanatory variables. The other parameters denote short run coefficients.

It should be noted that there is another method developed by Pesaran et al. (1999) referred to as Mean Group (MG) estimator. The main difference between these estimators is to put some constraints on parameters that have been used throughout the analysis. To determine which estimator would be deployed for a panel ARDL model, Hausman test, which is developed by Hausman (1978), is generally used. In this analysis, therefore, we also resorted to the Hausman test before running PMG estimator. The Hausman test result will be presented and discussed at length in the next section.

RESULTS

Prior to running the model, the stationarity of the variables should be checked due to a possible spurious regression problem. Since panel data use the standard OLS basis, nonstationary variables may create biased results. As can be seen in Table 3 below, we applied Im-Pesaran-Shin unit root test developed

Table 3. Unit root test results

Variable	Level (I(0))	First-Difference (I(1))
fbs	2.3507 (0.99)	-2.2357 (0.01)*
gdpc	1.1988 (0.88)	-3.5679 (0.00)*
mcs	-1.1752 (0.03)**	4.0769 (0.00)*
imp	-0.6891 (0.24)	-4.0769 (0.00)*
rde	-0.6822 (0.24)	-3.5255 (0.00)*

*; ** Denotes respectively 1% and 5% significance level.

by Im et al. (2003). According to the table, only mobile cellular subscription (mcs) is stationary at level *(I(0))*, but after taking the first difference of each variable, all become stationary at *I(1)*.

Before running PMG estimator for the ARDL model, we make sure that whether PMG or MG estimator will be deployed for the present analysis. In order to concretize the use of PMG for our analysis, we benefit from the Hausman test developed by Hausman (1978). As Table 5 below shows, the probability value is greater than 5% (p>5%), justifying that the PMG estimator is convenient in assessing the nexus between ICT diffusion and economic growth.

Following the PMG estimator results, we obtain both short and long run estimation coefficients regarding the nexus between ICT diffusion and economic growth for BRICS nations. Table 4 clearly shows the short run interaction between the variables of interest. According to Table 4, the coefficient of fixed broadband subscription (*fbs*) is positive and significant at the 5% level of statistical significance, indicating that fixed broadband subscription significantly and positively affects economic growth in BRICS nations in the short run. This result, however, backs the findings of Minges (2015), Bahrini and Qaffas (2019), Shodiev et al. (2021), Odhiambo (2022). Interestingly, as can be seen from Table 5 below, *fbs* has significant but negative effect on economic growth in the long run. This can be interpreted as people much benefit from fixed internet line in the short run, but investment in technical know-how as well as in both hardware and software infrastructure is needed for fixed line internet to be much more effective in the long run for BRICS nations.

The effect of mobile cellular subscription (*mcs*) on economic growth is positive but statistically insignificant in the short run. Conversely, the long run effect of *mcs* on economic growth is negative and not statistically significant. In the aid literature, studies have more generally found positive and statistically significant relationship between *mcs* and economic growth (Adeleye and Eboagu, 2019; Makun and Jayaraman, 2022; Nizami et al., 2022). However, the findings of Ejemeyovwi and Osabuohien (2018) set out the opposite effect of *mcs* on growth, which is in line with our long run estimation result. Mobile cellular subscription is expected to positively affect economic growth as it enables information to flow easily and quickly between people and reduces transaction costs, which are highly effective in creating a more efficient way for economic transactions that would lead to higher productivity. However, it is not solely taken as a representative of ICT diffusion but can be taken only as a part of it. So, its short and long run effect might naturally differ across countries and is dependent on other ICT components. Although short run coefficient of *mcs* is consistent with common knowledge of the ICT literature, the

Table 4. PMG short run estimation results

Coef.	Std.	Err.	z	P>lzl
D(fbs)	1.599514	.6291326	2.54	0.011**
D(mcs)	.0852412	.0953913	0.89	0.372
D(imp)	-.0699953	.3223459	-0.22	0.828
D(rde)	-9.330431	7.020184	-1.33	0.184
C	-4.672747	1.686496	-2.77	0.006*
ECT	-.9507896	.0768898	-12.37	0.000*

*; ** Denotes respectively 1% and 5% significance level.

negative long run coefficient can basically be explained by the lack of required base for information infrastructure (Wang, 1999).

As for ICT goods import as the ratio of total import (imp), the short run coefficient is negative and not statistically significant. In the short run, some ICT goods import like computers, mobile phones and other related technological devices may create a huge burden on the government's budget, and therefore, may retard economic growth. Our short run finding seems to be in line with that of Jin and Cho (2015) and Chatterjee (2020) who find that import of ICT technologies has a negative and insignificant effect for 128 ITU countries. In the long run, however, ICT import is expected to affect positively economic growth through enabling people to reach technology that would accelerate communication with lower cost in terms of both time and effort, therefore, would increase productivity. Our long run result reveals that *imp* significantly and positively affects economic growth in BRICS nations, which is backed by Sassi and Goaied (2013), Khalili et al. (2014), and Nguyen et al. (2020).

When it comes to the role of research and development (R&D) expenditures on economic growth, it seems that in the short run research and development expenditure over GDP ratio has a negative and statistically insignificant effect on economic growth. R&D expenditures would yield desired outcomes in the medium- and long run as it contributes to technological advancement and cumulative knowledge as well as encourages innovation and creates an environment that would improve human capital over some period. As can be clearly seen, our long run finding for *rde* is in line with the theory where R&D expenditures positively and significantly affect economic growth. Similarly, Hong (2017) also points out the existence of bidirectional causality between economic growth and R&D investment, which works from economic growth to R&D in the short run while it is from R&D to economic growth in the long run. For R&D, however, being regarded as the long run engine of economic growth is not new. Some other studies also confirm the long run positive impact of R&D on economic growth (see, for example, Romer (1990), Nair et al., (2020)).

As can be seen from Table 4, the error correction term is negative and statistically significant, implying that there is a convergence to the long run steady state equilibrium in the case of a short run shock. Alternatively saying, the negative and significant coefficient of error correction term reveals that the variables would convergence in the long run, again verifying the long run homogeneity assumption.

Table 5. PMG long run estimation results

Coef.	Std. Err.	z	P>|z|	Hausman Test	
fbs	-.4157148	.0822978	-5.05	0.000*	0.9201
mcs	-.0038168	.0076672	-0.50	0.619	
imp	.5620149	.1291995	4.35	0.000*	
rde	3.494704	1.642755	2.13	0.033**	

*; ** denotes respectively 1% and 5% significance level.

DISCUSSION

Based on the results presented above, the findings should be interpreted through the lens of country-specific differences that make each ICT component work differently for each BRICS nation. To what extent BRICS nations would benefit from ICT diffusion is dependent on the countries' investment in technological, economic, social areas that would enhance the absorptive capacity and led these countries to benefit from ICT diffusion effectively and efficiently. Moreover, the existence of the digital divide between BRICS nations, which has been previously discussed by Lazanyuk and Revinova (2019) and referred to as being a crucial factor in bringing the average of ICT utilization down, may determine the extent to which ICT diffusion associates with economic growth, or vice versa. It seems that, therefore, despite a variety of factors affecting the nexus between ICT penetration and economic growth, governments should also take necessary measures in bridging the gaps between the utilization level of BRICS nations, which is arising from country-specific economic, social, political, and institutional differences. To give a concrete example, there are notable differences in the level of development between BRICS nations. As Gorkey-Aydinoglu and Yalki-Berker (2016) pointed out, for instance, China is the clearest example of producing ICT products. In this way, its export rate in the ICT sector is the highest one compared to other countries in BRICS. It is clear that higher export would generate more revenue for China, which will, in turn, make the country more prosperous in economic terms.

The most obvious finding to emerge from the analysis is that technology import (*imp*) and R&D expenditure (*rde*) are the most effective factors that have a considerable impact on economic growth in the long run. As highlighted by Jin and Cho (2015), import of ICT devices will pave the way for development. This will create an environment where people can better adopt technology and produce innovation that would foster economic growth. On the other side, countries may devote money from the budget to importing required sources, labor, services and devices in an effort to promote innovation and technological advancement. In this parallel, it has been found that R&D investment enhances economic activity and stimulates economic growth (Cronin et al., 1991; Hong, 2017). The strong association of higher GDP per capita with technology import and R&D expenditure is interesting, but not surprising. As Romer (1990) highlighted the importance of R&D in determining long run productivity of developed countries, R&D will also help countries accumulate intangible capital that would foster innovation and enhance competitiveness through higher standards for further research.

The results also indicate that although fixed broadband subscription (*fbs*) and mobile cellular subscription (*mcs*) positively affect GDP per capita growth in the short run, their impact on growth turns out to be negative in the long run. These findings are somewhat surprising given the fact that other research shows a positive association (see, for example, Sassi and Goaied, 2013; Khalili et al., 2014; Adeleye and Eboagu, 2019; Nguyen et al., 2020; Appiah-Otoo and Song, 2021; Makun and Jayaraman, 2022; Nizami et al., 2022). However, the observed relationship between *mcs* and GDP per capita growth in this study is not significant while the interaction with *fbs* is significant. This rather intriguing result can be explained by several factors. First, there are political, demographic, economic, social differences, causing Technology Readiness Level (TRL) to vary across BRICS nations. As Lucas (1988) well underlined the role of educated and high-skilled workers in bolstering economic growth, some others (Kumar and Sharma, 2015) stipulate the necessity of some conditions that would improve the use, utilization and efficiency of ICT devices to reach a high level of economic growth. To cut the long story short, in order for *fbs* and *mcs* to positively affect economic growth, differences in the TRL across the countries should be minimized. For instance, Lazanyuk and Revinova (2019) well substantiate that it will take some time for both organizations and institutions to adapt themselves into new technologies if they are at a level somewhat below the required threshold of adequately utilizing ICT devices. Parallel to this, the authors underline the fact that Russia has the highest TRL while others like China, South Africa and India have a long way to go before adapting technology to obtain higher economic growth. In other terms, there will be a need for investment in both hardware and software as well as in human capital as to how to use ICT devices for some of BRICS nations that would make them utilize ICT devices at maximum capacity.

Secondly, other factors like trade openness, economic growth and financial development of countries should also be taken into consideration when analyzing the impact of ICT diffusion on economic growth. It is clear that China is one of the most open countries to international trade compared to other BRICS nations. This fairly distinguishes the country from other BRICS nations in terms of generating more revenue, which would later be invested in R&D and other areas that would induce positively GDP per capita. Moreover, this leads some of BRICS nations not to access ICT devices simultaneously, which would cause a digital divide. In general, therefore, it seems that utilization from the network like fixed broadband and mobile cellular subscriptions would deliver inconsistent results for BRICS nations.

Lastly, cultural, institutional and political factors should not be overlooked. Political atmosphere and institutional structure would determine the way how resources for technology and economic opportunities are distributed. On the other hand, an even dissemination of ICT devices and the integration of ICT diffusion into the natural evolution of social, cultural and economic development of countries are rather important (for more information, see, Morales-Gómez and Melesse, 1998). In this sense, it does not seem possible for some countries, especially for underdeveloped and developing ones, to leapfrog to further processes of producing and obtaining higher amounts of product to close the gap with developed countries. The absence of such steps in enhancing the utilization rate of ICT devices and making them more useful for GDP per capita growth would be expected to result in ICT penetration to be highly unlikely to deliver intended outcomes for economic growth and provide a basis for overall development.

CONCLUDING REMARKS

In the present study, we analyze the nexus between ICT diffusion and economic growth for BRICS nations over the period 2003-2020 by using a pooled mean group (PMG) panel ARDL model. When surveying the related nexus, we consider not only the role of ICTs but also the possible impact of other factors like research and development and technology import over the short and long run period on economic growth.

Although economic growth can be measured in numerical terms, economic development seems to be a more complex issue that requires not only economic advancements but also social, institutional and political arrangements to achieve higher standards of production processes and of living. In the case of BRICS nations, we explicitly see how economic, demographic and country-specific differences across countries may significantly affect the utilization of technological breakthroughs in general. That is to say, differences in the countries' level of development and its determinant factors like economic, cultural and demographic structure seem to be significantly determining the extent to which countries might benefit from ICT penetration and create an environment where those ICTs can be effectively used. In this sense, this study offers a wider perspective as to steps those countries should take as a part of their development process.

Overall, our analysis reveals that fixed broadband and mobile cellular subscriptions have a positive, but only the former significant, impact on economic growth in the short run. As they enable people, organizations and institutions to benefit from the IT infrastructure that would create a positive externality in the first place, which is expected and in line with theory. However, it seems that their impact on economic growth turns out to be negative, with only the fixed broadband subscription being significant. It reveals that countries should design strategies and implement policies that would provide a basis for a long run positive return on investment made in both hardware and software networks. Moreover, efforts to promote technology readiness at very early stages of development should be backed by governments in order to minimize TRL differences, which would, in turn, help diminish the digital divide, and to integrate people to the process of technology explosion. This would also prevent those countries from leapfrogging to the next sections of technology use without adequately completing the early stages of development.

The results also show that although the effect of ICT imports and R&D expenditures is negative and insignificant in the short run, their long run effect on GDP per capita growth is significant and positive. Consistent with the theory and extant literature, higher import rates and R&D expenditures will scale government expenditures up, which will be rather cumbersome for the government budget to sustain short and medium run growth dynamics. This would inevitably have a negative impact on economic growth in the short run. On the other hand, the effect of ICT imports and R&D expenditures is expected to positively affect economic growth through creating an environment where innovation is spurred, and technology readiness is promoted. Parallel to this result, economic growth should be backed up by not only tangible factors like ICT devices and their installations, but also by the infrastructure that improves human capital and creates an institutional structure that would help governments to ensure economic growth.

Taken together, these results suggest that BRICS nations should prioritize actions and perform policies that would support short run network efforts with long run investments in both ICT imports and R&D expenditures. It seems that differences in the TRL between BRICS nations should be minimized to reach a higher absorptive capacity, that would, in turn, increase GDP per capita growth in these countries. Although country-specific factors play a significant role in determining the utilization rate for ICT devices, necessary steps and reforms may turn out to benefit each country and therefore BRICS nations in general a lot from ICT diffusion and technological breakthrough.

REFERENCES

G20 Argentina Presidency. (2018). *Toolkit for Measuring the Digital Economy*. https://www.oecd.org/g20/summits/buenos-aires/g20-detf-toolkit.pdf

Adeleye, N., & Eboagu, C. (2019). Evaluation of ICT development and economic growth in Africa. *NETNOMICS: Economic Research and Electronic Networking*, *20*(1), 31–53. doi:10.100711066-019-09131-6

Appiah-Otoo, I., & Song, N. (2021). The impact of ICT on economic growth-Comparing rich and poor countries. *Telecommunications Policy*, *45*(2), 102082. doi:10.1016/j.telpol.2020.102082

Bahrini, R., & Qaffas, A. A. (2019). Impact of information and communication technology on economic growth: Evidence from developing countries. *Economies*, *7*(1), 21. doi:10.3390/economies7010021

Biagi, F. (2013). *ICT and productivity: A review of the literature (JRC-IPTS working papers on digital economy No. 2013–09)*. Institute of Prospective Technological Studies, Joint Research Centre.

Carr, N. G. (2003). IT doesn't matter. *EDUCAUSE Review*, *38*, 24–38. PMID:12747161

Chatterjee, A. (2020). Financial inclusion, information and communication technology diffusion, and economic growth: A panel data analysis. *Information Technology for Development*, *26*(3), 607–635. doi:10.1080/02681102.2020.1734770

Cheng, C. Y., Chien, M. S., & Lee, C. C. (2021). ICT diffusion, financial development, and economic growth: An international cross-country analysis. *Economic Modelling*, *94*, 662–671. doi:10.1016/j.econmod.2020.02.008

Cobb, C. W., & Douglas, P. H. (1928). A Theory of Production. *The American Economic Review*, *18*(1), 139–165. https://www.jstor.org/stable/1811556

Cronin, F. J., Parker, E. B., Colleran, E. K., & Gold, M. A. (1991). Telecommunications infrastructure and economic growth: An analysis of causality. *Telecommunications Policy*, *15*(6), 529–535. doi:10.1016/0308-5961(91)90007-X

Czernich, N., Falck, O., Kretschmer, T., & Woessmann, L. (2011). Broadband infrastructure and economic growth. *Economic Journal (London)*, *121*(552), 505–532. doi:10.1111/j.1468-0297.2011.02420.x

Ejemeyovwi, J. O., & Osabuohien, E. S. (2018). Investigating the relevance of mobile technology adoption on inclusive growth in West Africa. *Contemporary Social Science*.

Gorkey-Aydinoglu, S., & Yalki-Berker, Ġ. (2016). The Comparative Development of ICT in BRICS: A cluster analysis. *The Journal of Knowledge Economy and Knowledge Management*, *11*(1), 17–28.

Haller, S. A., & Lyons, S. (2015). Broadband adoption and firm productivity: Evidence from Irish manufacturing firms. *Telecommunications Policy*, *39*(1), 1–13. doi:10.1016/j.telpol.2014.10.003

Hausman, J. A. (1978). Specification tests in econometrics. *Econometrica*, *46*(6), 1251–1271. doi:10.2307/1913827

Hong, J. P. (2017). Causal relationship between ICT R&D investment and economic growth in Korea. *Technological Forecasting and Social Change, 116*, 70–75. doi:10.1016/j.techfore.2016.11.005

Ignatov, A. (2020). The digital economy of BRICS: Prospects for multilateral cooperation. *International Organisations Research Journal, 15*(1), 31–62. doi:10.17323/1996-7845-2020-01-02

Im, K. S., Pesaran, M. H., & Shin, Y. (2003). Testing for unit roots in heterogeneous panels. *Journal of Econometrics, 115*(1), 53–74. doi:10.1016/S0304-4076(03)00092-7

Jin, S., & Cho, C. M. (2015). Is ICT a new essential for national economic growth in an information society? *Government Information Quarterly, 32*(3), 253–260. doi:10.1016/j.giq.2015.04.007

Kaiser, U. (2003). *Strategic complementarities between different types of ICT-expenditures* (No. 03-46). ZEW Discussion Papers.

Khalili, F., Cheong, K. C., & Lau, W. Y. (2014). ICT as a source of economic growth in the information age: Empirical evidence from ICT leading countries. *Research Journal of Economics, Business and ICT, 9*(1).

Kumar, S., & Sharma, D. (2015). Factors Influencing ICT Development in BRICS Countries. *International Journal of Social and Organizational Dynamics in IT, 4*(2), 30–40. doi:10.4018/IJSODIT.2015070103

Lazanyuk, I., & Revinova, S. (2019). Digital economy in the BRICS countries: myth or reality? In *International scientific and practical conference on digital economy (ISCDE 2019)* (pp. 227-231). Atlantis Press. 10.2991/iscde-19.2019.97

Levin, A., Lin, C. F., & Chu, C. S. J. (2002). Unit root tests in panel data: Asymptotic and finite-sample properties. *Journal of Econometrics, 108*(1), 1–24. doi:10.1016/S0304-4076(01)00098-7

Lucas, R. E. Jr. (1988). On the mechanics of economic development. *Journal of Monetary Economics, 22*(1), 3–42. doi:10.1016/0304-3932(88)90168-7

Makun, K., & Jayaraman, T. K. (2022). *Tourism and economic growth in South Asian countries: Asymmetric analysis and lessons for mitigating the adverse effects of Covid-19 pandemic*. Academic Press.

Mano, R. (2021). The institutionalization of ICT and civic participation: Evidence from eight European nations. *Technology in Society, 64*, 101518. doi:10.1016/j.techsoc.2020.101518

Mayer, W., Madden, G., & Wu, C. (2020). Broadband and economic growth: A reassessment. *Information Technology for Development, 26*(1), 128–145. doi:10.1080/02681102.2019.1586631

Minges, M. (2015). *Exploring the relationship between broadband and economic growth*. Academic Press.

Morales-Gómez, D., & Melesse, M. (1998). Utilising information and communication technologies for development: The social dimensions. *Information Technology for Development, 8*(1), 3–13. doi:10.1080/02681102.1998.9525287

Nair, M., Pradhan, R. P., & Arvin, M. B. (2020). Endogenous dynamics between R&D, ICT and economic growth: Empirical evidence from the OECD countries. *Technology in Society, 62*, 101315. doi:10.1016/j.techsoc.2020.101315

Nasab, E. H., & Aghaei, M. (2009). The effect of ICT on economic growth: Further evidence. *International Bulletin of Business Administration*, *5*(2), 46–56.

Ndung'u, N. (2019). Digital technology and state capacity in Kenya. Academic Press.

Nguyen, T. T., Pham, T. A. T., & Tram, H. T. X. (2020). Role of information and communication technologies and innovation in driving carbon emissions and economic growth in selected G-20 countries. *Journal of Environmental Management*, *261*, 110162. doi:10.1016/j.jenvman.2020.110162 PMID:32148259

Niebel, T. (2018). ICT and economic growth–Comparing developing, emerging and developed countries. *World Development*, *104*, 197–211. doi:10.1016/j.worlddev.2017.11.024

Nizami, U., Niazi, M. F. K., & Kadri, A. S. (2022). Impact of ICT Investment and Diffusion on Economic Growth: A Case Study of Pakistan. *Advances in World Economics*, *1*(1), 12–20.

Odhiambo, N. M. (2022). Information technology, income inequality and economic growth in sub-Saharan African countries. *Telecommunications Policy*, *46*(6), 102309. doi:10.1016/j.telpol.2022.102309

Pesaran, M. H., Shin, Y., & Smith, R. P. (1999). Pooled mean group estimation of dynamic heterogeneous panels. *Journal of the American Statistical Association*, *94*(446), 621–634. doi:10.1080/01621459.1999.10474156

Porter, M., Sachs, J., & McArthur, J. (2001). *Executive summary: Competitiveness and stages of economic development*. Academic Press.

Revinova, S. (2016). Information technology development in the BRICS countries-comparative analysis. In *3rd International Multidisciplinary Scientific Conference on Social Sciences and Arts SGEM* (*Vol. 5*, p. 353). Academic Press.

Romer, P. M. (1986). Increasing returns and long-run growth. *Journal of Political Economy*, *94*(5), 1002–1037. doi:10.1086/261420

Romer, P. M. (1990). Endogenous technological change. *Journal of Political Economy*, *98*(5, Part 2), S71–S102. doi:10.1086/261725

Sassi, S., & Goaied, M. (2013). Financial development, ICT diffusion and economic growth: Lessons from MENA region. *Telecommunications Policy*, *37*(4-5), 252–261. doi:10.1016/j.telpol.2012.12.004

Shodiev, T., Turayey, B., & Shodiyev, K. (2021). ICT and Economic Growth Nexus: Case of Central Asian Countries. *Procedia of Social Sciences and Humanities*, *1*, 155–167. doi:10.21070/pssh.v1i.37

Solow, R. (1987). We'd better watch out. *New York Times Book Review, 36.*

Solow, R. M. (1956). A contribution to the theory of economic growth. *The Quarterly Journal of Economics*, *70*(1), 65–94. doi:10.2307/1884513

Suchman, L. A. (1987). *Plans and situated actions: The problem of human-machine communication*. Cambridge University Press.

Vu, K. M. (2011). ICT as a source of economic growth in the information age: Empirical evidence from the 1996–2005 period. *Telecommunications Policy*, *35*(4), 357–372. doi:10.1016/j.telpol.2011.02.008

Wang, E. H. H. (1999). ICT and economic development in Taiwan: Analysis of the evidence. *Telecommunications Policy*, *23*(3-4), 235–243. doi:10.1016/S0308-5961(99)00005-1

ADDITIONAL READING

Colecchia, A., & Schreyer, P. (2002). ICT investment and economic growth in the 1990s: is the United States a unique case? A comparative study of nine OECD countries. *Review of Economic Dynamics*, *5*(2), 408–442. doi:10.1006/redy.2002.0170

Edquist, H., & Henrekson, M. (2017). Swedish lessons: How important are ICT and R&D to economic growth? *Structural Change and Economic Dynamics*, *42*, 1–12. doi:10.1016/j.strueco.2017.05.004

Ishida, H. (2015). The effect of ICT development on economic growth and energy consumption in Japan. *Telematics and Informatics*, *32*(1), 79–88. doi:10.1016/j.tele.2014.04.003

Jalava, J., & Pohjola, M. (2008). The roles of electricity and ICT in economic growth: Case Finland. *Explorations in Economic History*, *45*(3), 270–287. doi:10.1016/j.eeh.2007.11.001

Jorgenson, D. W., & Vu, K. M. (2016). The ICT revolution, world economic growth, and policy issues. *Telecommunications Policy*, *40*(5), 383–397. doi:10.1016/j.telpol.2016.01.002

Kallal, R., Haddaji, A., & Ftiti, Z. (2021). ICT diffusion and economic growth: Evidence from the sectorial analysis of a periphery country. *Technological Forecasting and Social Change*, *162*, 120403. doi:10.1016/j.techfore.2020.120403

Sarangi, A. K., & Pradhan, R. P. (2020). ICT infrastructure and economic growth: A critical assessment and some policy implications. *Decision (Washington, D.C.)*, *47*(4), 363–383.

Vu, K., Hanafizadeh, P., & Bohlin, E. (2020). ICT as a driver of economic growth: A survey of the literature and directions for future research. *Telecommunications Policy*, *44*(2), 101922. doi:10.1016/j.telpol.2020.101922

Vu, K. M. (2013). Information and communication technology (ICT) and Singapore's economic growth. *Information Economics and Policy*, *25*(4), 284–300. doi:10.1016/j.infoecopol.2013.08.002

KEY TERMS AND DEFINITIONS

ICT Penetration: The adoption, integration and use of information and communication technologies in general.

Solow Residual: It is one of measurements of total factor productivity. It therefore represents the part of economic growth that cannot be explained by altering both capital and labor.

Chapter 6
ICT Penetration and Entrepreneurial Activity:
An Empirical Analysis for High Income Countries

Lina Karabetyan
Independent Researcher, Turkey

ABSTRACT

Entrepreneurial activity is a critical factor for performance of firms and economies. Therefore, firms and governments try to create an entrepreneurial environment given positive economic and social gains of entrepreneurship. Furthermore, governments stimulate the entrepreneurial activities through financial and legal incentives. This study investigates the mutual interaction between ICT penetration and entrepreneurial activity in 13 high income countries for the 2002-2018 period by a panel causality test which considers the cross-sectional dependence and heterogeneity. The results of panel level causality analysis reveal a bidirectional causal interaction between ICT penetration and entrepreneurial activity. However, the results of country level causality analysis uncover that interaction between ICT penetration and entrepreneurial activity differs among higher income countries depending on country specific characteristics.

INTRODUCTION

The economic and social benefits of the entrepreneurship make the entrepreneurship a vital factor for firms and the whole economies. Entrepreneurs can foster the growth of whole economy and firms and provide new employment opportunities by developing the innovation, products, and services and increase the productivity of the economies and firms, encourage the firms to be more competitive by the increased competition, and accelerate the structural change in the economy (Kritikos, 2014, Franco, 2020). However, the innovation and increased competition resulting from entrepreneurship can lead some firms to

DOI: 10.4018/978-1-6684-6620-9.ch006

be closed and in turn increase the unemployed and the economic costs of unsuccessful entrepreneurs are sometimes borne by taxpayers (Kritikos, 2014, Neumann, 2021).

The influence of entrepreneurship on economic growth and employment has been extensively researched, but the influence of entrepreneurship on innovation, competitiveness, and social welfare has been explored by relatively few researchers (Neumann, 2021). The related empirical literature has suggested that the findings about influence of entrepreneurship on economic growth and employment for different countries and regions are consistent with the wave-pattern theory to a great extent (Matejovsky et al., 2014; Delfmann and Koster; 2016, Dvouletý, 2017). However, the relatively few empirical studies about the influence of entrepreneurship on innovation and competitiveness has remained inconclusive (Amorós et al., 2012, Albulescu & Draghici, 2016).

Similarly, social and environmental effects of entrepreneurship have been empirically explored by few scholars. In this context, Lee and Rodríguez-Pose (2021) revealed that entrepreneurship in tradeables decreased the poverty and raised the incomes for non-entrepreneurs in the US cities. Dzingirai (2021) found that entrepreneurship made a contribution to poverty reduction in agricultural communities of Lower Gweru in Zimbabwe. Sall (2022) also uncovered that entrepreneurship decreased the poverty in Africa. On the other hand, Ben Youssef et al. (2018) and Dhahri & Omri (2018) discovered a negative environmental effect of entrepreneurship.

The entrepreneurial orientation also can positively affect the firm performance through innovative, proactive, risk-taking, autonomy, and competitive aggressiveness (Putniņš & Sauka, 2020). In this context, Ardhi et al. (2021) analyzed the impact of entrepreneurial orientation dimensions on 80 business performance in micro, small, and medium enterprises from Special Region of Yogyakarta and uncovered that dimensions of proactive, autonomy, and competitive aggressiveness positively influenced the business performance. Hina et al. (2021) also reached the conclusion that five dimensions of entrepreneurial orientation had a significant influence on firm performance through a sample of 50 respondents from software houses of Pakistan.

Many researchers have researched the critical factors underlying entrepreneurship considering the extensive social and economic gains of the entrepreneurship. In this regard, many demographic and individual factors, economic factors, cultural factors, social and psychological factors, technological factors, legal factors, and political factors have been suggested as the determinants of entrepreneurship (Kastrati, 2015; Cepel et al., 2019; Ahadi & Kasraie, 2020; Martínez-Rodriguez et al., 2020; Kurjono et al., 2020; Méndez-Picazo et al., 2021). However, the interaction between ICTs (information and communications technologies) and entrepreneurship has not been explored sufficiently although ICTs and entrepreneurship have potential to influence each other. Therefore, this paper centers on the interaction between ICT penetration and entrepreneurial activity.

ICTs comprise "hardware, software, networks, and media for collection, storage, processing, transmission, and presentation of information. (World Bank, 2002)". ICTs can foster the entrepreneurial activity through the following direct and indirect channels:

- ICTs ease the entrepreneurs to access the knowledge, new business ideas, funding opportunities, and new markets (Giudice & Straub, 2011; Niles & Hanson, 2003)
- ICTs enables the entrepreneurs to sell their products and services through online markets and reach the potential customers (Chen et al., 2015).
- ICTs enable entrepreneurs to participate to the training and collaboration activities and technology transfer at national and global levels (Viju, 2010)

- ICTs can foster the entrepreneurial activity through education, economic growth and development.

On the other hand, improvements in entrepreneurial activities can enhance the ICT penetration, because the entrepreneurs can easily access to the new customers and markets through ICTs. Therefore, a mutual interaction between ICT and entrepreneurial activities is theoretically expected.

This study investigates the interaction between ICT penetration and entrepreneurial activities in selected high-income countries for 2002-2018 period through causality test with cross-sectional dependence and heterogeneity. The selected high-income countries made a considerable improvement in ICT penetration as seen in Table 1. The countries with low ICT penetration in 2002 such as Chile, Croatia, Italy, and Spain made a relatively higher improvement in ICT penetration during the 2002-2018 duration.

The related empirical literature has generally analyzed the influence of ICTs on entrepreneurial activities and discovered a positive influence of ICTs on entrepreneurial activities (e.g. see Alderete (2014), Ukommi and Okeagu (2015), Asongu and Nwachukwu (2018), Gomes and Lopes (2022), and Afawubo and Noglo (2022)). However, this paper intends to contribute to the relevant empirical literature by analyzing the mutual interaction between ICT penetration and entrepreneurial activities, because ICTs also have many other economic, social, and environmental effects. The next part of the paper introduces the limited empirical literature, and data and methods are briefly described. The econometric analyses are done and the results of the econometric tests are argued. The study is concluded with Conclusions.

LITERATURE REVIEW

Entrepreneurship has become a key factor for success of firms and economies. Therefore, factors underlying entrepreneurship have been widely explored in the relate literature. However, few researchers have researched the influence of ICTs on entrepreneurship and discovered a positive influence of ICTs on entrepreneurship (Alderete, 2014; Ukommi & Okeagu, 2015; Asongu & Nwachukwu, 2018; Gomes & Lopes, 2022; Afawubo & Noglo, 2022). However, the effect of entrepreneurship on ICT penetration has not been explored sufficiently, although ICTs have many economic, social and environmental effects for firms and economies. Therefore, the mutual interaction between ICT penetration and entrepreneurial activity is analyzed considering the research gap in the related empirical literature.

In the limited empirical literature, Alderete (2014) investigated the influence of ICT development on entrepreneurial activity in 59 countries for the 2007-2011 period and discovered a positive influence of ICT development on entrepreneurial activity. On the other hand, Ukommi & Okeagu (2015) explored the effect of ICT education on youth entrepreneurship in Uyo, Akwa Ibom State of Nigeria with the data obtained from 1,890 participants by questionnaires and ICT education raised the youth entrepreneurship

Wamuyu (2015) analyzed the influence of ICT adoption and penetration on technology entrepreneurship in Kenya and discovered that unpredictable ICT trends and slow ICT adoption influenced the ICT-based enterprises negatively. Cuevas-Vargas et al. (2016) also explored the influence of ICTs on innovation in industrial micro, small and medium sized enterprises in Mexico via structural equation modeling and disclosed that ICTs fostered the innovation.

Alderete (2017) examined the influence of mobile broadband on entrepreneurship in 58 countries for 2007-2012 period with regression analysis and disclosed a positive effect of mobile broadband on the entrepreneurial activity. Asongu and Nwachukwu (2018) also explored the influence of ICTs represented by mobile phone penetration and internet access on entrepreneurship in 49 sub-Saharan African

Table 1. Entrepreneurial activity and ICT development in selected high-income countries

Countries	Year	Total early-stage entrepreneurial activity*	ICT index**
Chile	2002	15.68	7.49301
	2018	25.06	15.09722
Croatia	2002	3.62	9.283607
	2018	9.61	18.67002
France	2002	3.13	13.31639
	2018	6.13	20.44134
Germany	2002	5.16	14.61531
	2018	4.97	27.65142
Greece	2002	4.12	11.404
	2018	6.35	20.88104
Ireland	2002	9.14	11.625
	2018	9.64	21.29693
Italy	2002	5.74	8.736272
	2018	4.18	16.60966
Netherlands	2002	4.59	14.46153
	2018	12.29	31.03546
Slovenia	2002	4.63	11.2093
	2018	6.37	21.02446
Spain	2002	4.59	10.79822
	2018	6.39	20.91808
Sweden	2002	3.9	17.56389
	2018	6.82	25.99413
United Kingdom	2002	5.36	14.75572
	2018	8.24	26.79893
United States	2002	10.62	16.10887
	2018	15.59	24.42884

* A percentage of a nascent entrepreneur or owner-manager of a new business to 18-64 population.
** It is calculated based on internet accessibility, server security, fixed line and mobile phones users.
Source: Global Entrepreneurship Monitor (2022), United Nations Conference on Trade and Development (2022)

economies for the 2000-2012 term through regression analysis and found that ICT indicators improved the entrepreneurship. Barnett et al. (2018) examined the impact of ICT on entrepreneurship in China using the data obtained from China Family Panel Studies (CFPS) survey and reached the conclusion that ICT proxied by internet use and cell phone ownership positively influenced the entrepreneurship.

Ejemeyovwi et al. (2019) also investigated the interaction between ICT adoption and sustainable entrepreneurial development in ECOWAS countries by regression analysis and discovered a positive influence of ICT adoption on entrepreneurial sustainability. Shrivastava et al. (2021) analyzed the effect of ICT infrastructure, skills, and use on entrepreneurship in 90 countries for the 2012-2016 term through regression analysis and uncovered that ICT was a significant positive factor for entrepreneurial

Table 2. Descriptive characteristics of the variables

Characteristics	TEA	ICT
Mean	7.463575	18.68524
Median	6.190000	18.67002
Maximum	26.83000	31.03546
Minimum	1.630000	7.493010
Std. Dev.	4.366557	5.097537
Skewness	2.370140	0.114128
Kurtosis	9.506512	2.499076

development. Taiba et al. (2021) examined the influence of ICT on entrepreneurial self-efficacy, social networking in sample of 398 microentrepreneurs from Pakistan during COVID-19 pandemic through structural equation modelling and reached the conclusion that ICT had a significant impact on entrepreneurial self-efficacy and social networking of microentrepreneurs.

Gomes and Lopes (2022) investigated the effect of ICTs proxied by mobile-cellular telephone subscriptions, fixed-broadband subscriptions, and individuals using the internet on entrepreneurial activity in 37 OECD members for the 2000-2019 period via regression method and discovered a positive effect of three ICT indicators on entrepreneurial activity. However, mobile-cellular telephones had the largest impact on entrepreneurial impact of entrepreneurial activity, and fixed-broadband subscriptions and internet access respectively followed the mobile-cellular telephones in terms of influence size. Lastly, Afawubo and Noglo (2022) analyzed the influence of ICTs proxied by ICT capital services on total entrepreneurial activity in developed, emerging and developing economies for the 2006-2016 period and uncovered a positive influence of ICTs on entrepreneurial activity in all countries.

DATA AND ECONOMETRIC METHODOLOGY

In this study, the interaction between ICT penetration and entrepreneurial activity in 13 high income countries for the 2002-2018 period is analyzed via Emirmahmutoglu and Kose (2011) causality test. The entrepreneurial activity is represented by total early-stage entrepreneurial activity (TEA) as a percentage of a nascent entrepreneur or owner-manager of a new business to 18-64 population by Global Entrepreneurship Monitor (2022). On the other hand, ICT penetration is proxied by ICT index of United Nations Conference on Trade and Development (2022) and the index is calculated by considering internet accessibility, server security, fixed line and mobile phones users.

The sample of study consists of 13 high income countries (Chile, Croatia, France, Germany, Greece, Ireland, Italy, Netherlands, Slovenia, Spain, Sweden, the United Kingdom, and the United States) based on income classification by World Bank (2022). The study period is 2002-2018, because the entrepreneurship data exists as of 2002 and ICT index ends in 2018. The econometric analyses are conducted via EViews 11.0 and Stata 15.0 statistical packages.

The mean of total early-stage entrepreneurial activity and ICT penetration in the sample is respectively 7.46% and 18.68, but both variables display a significant variation among the selected high-income countries. Table 2 shows the descriptive characteristics of the variables.

The interaction between ICT penetration and entrepreneurial activity is examined with Emirmahmutoglu and Kose (2011) causality test in view of cross-sectional and heterogeneity's presence among the variables. Emirmahmutoglu and Kose (2011) causality test makes the causality analysis for each cross-section by applying the bootstrap method to Fisher statistic. The stationarity of the variables $\left(d_{max_i}\right)$ and optimal lag length (p_i) are specified before the causality analysis. Then the error terms for each cross-section are obtained by following regression:

$$TEA_{i,t} = \alpha_{i,t} + \sum_{j=1}^{p_i+d_{max_i}} \beta_{ij} TEA_{i,t-j} + \sum_{j=1}^{p_i+d_{max_i}} \gamma_{ij} ICT_{i,t-j} + \varepsilon_{it} \tag{1}$$

$$ICT_{i,t} = \alpha_{i,t} + \sum_{j=1}^{p_i+d_{max_i}} \beta_{ij} ICT_{i,t-j} + \sum_{j=1}^{p_i+d_{max_i}} \gamma_{ij} TEA_{i,t-j} + \varepsilon_{it} \tag{2}$$

The null hypothesis of the test suggests the absence of the causality between two variables.

EMPIRICAL ANALYSIS

In the econometric analysis, cross-sectional dependence and heterogeneity characteristics of the series are firstly examined by cross-sectional dependence and heterogeneity tests. In this context, Breusch and Pagan (1980) LM test, Pesaran (2004) LM CD test, and Pesaran et al. (2008) LM adj. test are utilized to control the subsistence of cross-sectional dependence and the tests' findings are reported in Table 3. The null hypothesis of cross-sectional independence is declined at 1% significance level and the subsistence of cross-sectional dependence between two variables is concluded. The subsistence of homogeneity is investigated by delta tilde tests of Pesaran and Yamagata (2008) and the delta tests' results are reported in Table 3. The null hypothesis of homogeneity is declined at 1% significance level and the subsistence of heterogeneity is concluded. As a result, the subsistence of cross-sectional dependence and heterogeneity is revealed for the variables and in turn use of a causality tests which takes heterogeneity and cross-sectional dependence into consideration leads us to obtain relatively more robust results.

The stationarity of TEA and ICT is examined with Pesaran (2007) CIPS (Cross-sectionally augmented IPS (Im- Pesaran-Shin, 2003) unit root test in view of cross-sectional dependence between two variables and the results of the unit root test are reported in Table 4. The unit root disclosed that TEA and ICT was I(1).

The interaction between ICT penetration and entrepreneurial activity is analyzed by Emirmahmutoglu and Kose (2011) causality test in view of the subsistence of cross-sectional dependence and heterogeneity and the results of causality analysis are reported in Table 5. The panel level causality analysis reveals a bidirectional causality between ICT penetration and entrepreneurial activity. On the other hand, country level causality analysis uncovers a bidirectional causality between ICT penetration and entrepreneurial activity in Italy, a unidirectional causality from ICT penetration to entrepreneurial activity in Germany, Slovenia, and the United Kingdom and a unidirectional causality from entrepreneurial activity to ICT penetration in France.

Table 3. Cross-sectional dependence and heterogeneity tests' results

Test	Test Statistic	Prob.
LM	167.6	0.0000
LM adj*	13.96	0.0000
LM CD*	4.633	0.0000
$\tilde{\Delta}_{adj}$	9.588	0.000
$\tilde{\Delta}$	8.701	0.000

*Two-sided test

Table 4. Results of CIPS panel unit root test

Variables	Constant	Constant + Trend
TEA	-1.046	1.127
D(TEA)	-3.170***	-2.870***
ICT	0.010	0.454
D(ICT)	-6.143***	-4.707**

*** Denotes that it is significant at 1%.

Table 5. Causality analysis between ICT and TEA

Countries	ICT ½ TEA		TEA½ ICT	
	Test statistic	P value	Test statistic	P value
Chile	0.015	0.904	0.273	0.602
Croatia	0.687	0.407	0.389	0.533
France	1.663	0.197	**8.018**	**0.005**
Germany	**6.701**	**0.010**	0.271	0.602
Greece	0.256	0.613	0.575	0.448
Ireland	0.425	0.514	2.311	0.128
Italy	**3.232**	**0.072**	**3.226**	**0.072**
Netherlands	0.353	0.552	0.230	0.631
Slovenia	**3.774**	**0.052**	2.468	0.116
Spain	0.403	0.526	0.515	0.473
Sweden	0.192	0.662	1.255	0.263
United Kingdom	**2.863**	**0.091**	1.857	0.173
United States	0.218	0.994	0.558	0.455
Panel	**36.123**	**0.089**	**39.476**	**0.044**

A bidirectional causality between ICT penetration and entrepreneurial activity is expected at theoretical terms. However, the researchers have generally investigated the influence of ICTs on entrepreneurship in various countries and regions and verified that ICTs positively influenced the entrepreneurship (Alderete, 2014; Ukommi and Okeagu, 2015; Alderete, 2017; Asongu and Nwachukwu, 2018; Gomes and Lopes, 2022; Afawubo and Noglo, 2022). The results of panel level causality analysis are revealed to be consistent with theoretical expectation and empirical findings. However, the results of country level causality point out a bidirectional causality between two series only for Italy, a unidirectional causality from ICT penetration to entrepreneurial activity for Germany, Slovenia, and the United Kingdom, and a unidirectional causality from entrepreneurial activity to ICT penetration for France.

CONCLUSION

The economic and social gains of the entrepreneurship have led the entrepreneurship to be a crucial factor for success of firms and economies. The entrepreneurship can influence the economic growth and development, firm performance, employment, innovation, competitiveness, poverty, income inequality, and environment, but the size and direction of entrepreneurship's impact on the economic, social and environmental variables can vary depending on country specific characteristics. Therefore, determinants of entrepreneurship have become critical factors to encourage the entrepreneurial activities. In this context, the influence of demographic, individual, economic, and social factors on entrepreneurship have been extensively researched by many researchers until today. But the effect of ICTs, a leading and driving factor for globalization, on the entrepreneurship has been rarely explored. Furthermore, the effect of entrepreneurship on ICTs also have not been sufficiently researched in the related empirical literature yet.

In this paper, the mutual interaction between ICT penetration and entrepreneurial activity is investigated in sample of high-income countries by a causality test which takes cross-sectional dependence and heterogeneity into consideration. The results of causality analysis uncover a bidirectional causality between ICT penetration and entrepreneurial activity for the whole panel. But, country level causality analysis uncovers a bidirectional causality between ICT penetration and entrepreneurial activity for Italy, a unidirectional causality from ICT penetration to entrepreneurial activity in Germany, Slovenia, and the United Kingdom and a unidirectional causality from entrepreneurial activity to ICT penetration in France. The findings of causality analysis indicate that on one hand ICTs can be employed as an effective instrument to encourage the entrepreneurial activity and on the other hand entrepreneurship can be used to improve the ICT penetration.

REFERENCES

Afawubo, K., & Noglo, Y. A. (2022). ICT and entrepreneurship: A comparative analysis of developing, emerging and developed countries. *Technological Forecasting and Social Change*, *175*, 121312. doi:10.1016/j.techfore.2021.121312

Ahadi, S., & Kasraie, S. (2020). Contextual factors of entrepreneurship intention in manufacturing SMEs: The case study of Iran. *Journal of Small Business and Enterprise Development*, *27*(4), 633–657. doi:10.1108/JSBED-02-2019-0074

Albulescu, C. T., & Draghici, A. (2016). Entrepreneurial activity and national innovative capacity in selected European countries. *International Journal of Entrepreneurship and Innovation Management*, *17*(3), 155–172. doi:10.1177/1465750316655902

Alderete, M. V. (2014). ICT incidence on the entrepreneurial activity at country level. *International Journal of Entrepreneurship and Small Business*, *21*(2), 183–201. doi:10.1504/IJESB.2014.059472

Alderete, M. V. (2017). Mobile broadband: A key enabling technology for entrepreneurship? *Journal of Small Business Management*, *55*(2), 254–269. doi:10.1111/jsbm.12314

Amorós, J. E., Fernández, C., & Tapia, J. (2012). Quantifying the relationship between entrepreneurship and competitiveness development stages in Latin America. *The International Entrepreneurship and Management Journal*, *8*(3), 249–270. doi:10.100711365-010-0165-9

Ardhi, M.K., Mulyo, J.H., & Irham. (2021). How does entrepreneurial orientation affect the business performance of coffee shop MSMEs in Indonesia? *E3S Web of Conferences, 306*, 03011. doi:10.1051/e3sconf/202130603011

Asongu, S. A., & Nwachukwu, J. C. (2018). Openness, ICT and entrepreneurship in sub-Saharan Africa. *Information Technology & People*, *31*(1), 278–303. doi:10.1108/ITP-02-2017-0033

Barnett, W. A., Hu, M., & Wang, X. (2018). *Does the utilization of information communication technology promote entrepreneurship: Evidence from rural China.* http://www2.ku.edu/~kuwpaper/2018Papers/201802.pdf

Ben Youssef, A., Boubaker, S., & Omri, A. (2018). Entrepreneurship and sustainability: The need for innovative and institutional solutions. *Technological Forecasting and Social Change*, *129*, 232–241. doi:10.1016/j.techfore.2017.11.003

Breusch, T. S., & Pagan, A. R. (1980). The lagrange multiplier test and its applications to model specification in econometrics. *The Review of Economic Studies*, *47*(1), 239–253. doi:10.2307/2297111

Cepel, M., Belas, J., & Dvorský, J. (2019). The impact of significant political factors on starting a new business. *14th European Conference on Innovation and Entrepreneurship ECIE 2019*, 220-228. 10.34190/ECIE.19.117

Chen, Y., Wang, Y., Nevo, S., Benitez-Amado, J., & Kou, G. (2015). IT capabilities and product innovation performance: The roles of corporate entrepreneurship and competitive intensity. *Information & Management*, *52*(6), 643–657. doi:10.1016/j.im.2015.05.003

Cuevas-Vargas, H., Estrada, S., & Larios-Gómez, E. (2016). The effects of ICTs as innovation facilitators for a greater business performance. Evidence from Mexico. *Procedia Computer Science*, *91*, 47–56. doi:10.1016/j.procs.2016.07.040

Delfmann, H., & Koster, S. (2016). The effect of new business creation on employment growth in regions facing population decline. *The Annals of Regional Science*, *56*(1), 33–54. doi:10.100700168-015-0738-1

Dhahri, S., & Omri, A. (2018). Entrepreneurship contribution to the three pillars of sustainable development: What does the evidence really say? *World Development*, *106*, 64–77. doi:10.1016/j.worlddev.2018.01.008

Dvouletý, O. (2017). Can policy makers count with positive impact of entrepreneurship on economic development of the Czech regions? *Journal of Entrepreneurship in Emerging Economies*, *9*(3), 286–299. doi:10.1108/JEEE-11-2016-0052

Dzingirai, M. (2021). The role of entrepreneurship in reducing poverty in agricultural communities. *Journal of Enterprising Communities: People and Places in the Global Economy*, *15*(5), 665–683. doi:10.1108/JEC-01-2021-0016

Ejemeyovwi, J. O., Osabuohien, E. S., Bowale, E. K., Abuh, O. O., Adedoyin, J. P., & Ayanda, B. (2019, December 1). Information and communication technology adoption and innovation for sustainable entrepreneurship. *Journal of Physics: Conference Series*, *1378*(2), 022085. doi:10.1088/1742-6596/1378/2/022085

Emirmahmutoglu, F., & Kose, N. (2011). Testing for granger causality in heterogeneous mixed panels. *Economic Modelling*, *28*(3), 870–876. doi:10.1016/j.econmod.2010.10.018

Franco, M. (2020). Entrepreneurship, competitiveness and innovation. *Sustainability*, *12*(16), 6576. doi:10.3390u12166576

Giudice, M., & Straub, D. (2011). Editor's comments: IT and entrepreneurism: An on-again, off-again love affair or a marriage? *Management Information Systems Quarterly*, *35*(4), 3–8. doi:10.2307/41409961

Global Entrepreneurship Monitor. (2022). *Entrepreneurial Behaviour and Attitudes*. https://www.gem-consortium.org/data

Gomes, S., & Lopes, J. M. (2022). ICT access and entrepreneurship in the open innovation dynamic context: Evidence from OECD countries. *Journal of Open Innovation*, *8*(2), 102. doi:10.3390/joitmc8020102

Hina, S. M., Hassan, G., Parveen, M., & Arooj, S. (2021). Impact of entrepreneurial orientation on firm performance through organizational learning: The moderating role of environmental turbulence. *Performance Improvement Quarterly*, *34*(1), 77–104. doi:10.1002/piq.21343

Im, K. S., Pesaran, M. H., & Shin, Y. (2003). Testing for unit roots in heterogeneous panels. *Journal of Econometrics*, *115*(1), 53–74. doi:10.1016/S0304-4076(03)00092-7

Kastrati, V. (2015). Technological innovation of small and medium enterprises in Kosovo: Challenges and barriers. *European Journal of Sustainable Development*, *4*(3), 145–150.

KOF Swiss Economic Institute. (2022). *KOF Globalisation Index*. https://kof.ethz.ch/en/forecasts-and-indicators/indicators/kof-globalisation-index.html

Kritikos, A. S. (2014). Entrepreneurs and their impact on jobs and economic growth. *IZA World of Labor*, *8*. Advance online publication. doi:10.15185/izawol.8

Kurjono, K., Mulyani, H., & Murtadlo, Y. (2020). The Effect of Psychological Factors on Entrepreneurial Intention. *Advances in Economics, Business and Management Research*, *117*, 150–152. doi:10.2991/aebmr.k.200131.032

Lee, N., & Rodríguez-Pose, A. (2021). Entrepreneurship and the fight against poverty in US cities. *Environment and Planning A. Economy and Space*, *53*(1), 31–52. doi:10.1177/0308518X20924422

Martínez-Rodriguez, I., Callejas-Albiñana, F., & Callejas-Albiñana, A. (2020). Economic and socio-cultural drivers of necessity and opportunity entrepreneurship depending on the business cycle phase. *Journal of Business Economics and Management*, *21*(2), 373–394. doi:10.3846/jbem.2020.11848

Matejovsky, L., Mohapatra, S., & Steiner, B. (2014). The dynamic effects of entrepreneurship on regional economic growth: Evidence from Canada. *Growth and Change*, *45*(4), 611–639. doi:10.1111/grow.12055

Méndez-Picazo, M., Galindo-Martín, M., & Castaño-Martínez, M. (2021). Effects of sociocultural and economic factors on social entrepreneurship and sustainable development. *Journal of Innovation & Knowledge*, *6*(2), 69–77. doi:10.1016/j.jik.2020.06.001

Neumann, T. (2021). The impact of entrepreneurship on economic, social and environmental welfare and its determinants: A systematic review. *Management Review Quarterly*, *71*(3), 553–584. doi:10.100711301-020-00193-7

Niles, S., & Hanson, S. (2003). A new era of accessibility. *URISA J.*, *15*, 35–41.

Pesaran, M. H. (2004). *General diagnostic tests for cross section dependence in panels*. CESifo Working Paper No. 1229.

Pesaran, M. H. (2007). A simple panel unit root test in the presence of cross-section dependence. *Journal of Applied Econometrics*, *22*(2), 265–312. doi:10.1002/jae.951

Pesaran, M. H., Ullah, A., & Yamagata, T. (2008). A bias-adjusted LM test of error cross-section independence. *The Econometrics Journal*, *11*(1), 105–127. doi:10.1111/j.1368-423X.2007.00227.x

Pesaran, M. H., & Yamagata, T. (2008). Testing slope homogeneity in large panels. *Journal of Econometrics*, *142*(1), 50–93. doi:10.1016/j.jeconom.2007.05.010

Putniņš, T. J., & Sauka, A. (2020). Why does entrepreneurial orientation affect company performance? *Strategic Entrepreneurship Journal*, *14*(4), 711–735. doi:10.1002ej.1325

Sall, M. A. (2022). Entrepreneurship and poverty alleviation in Africa. *Industrial Policy*, *2*(1), 33–37.

Shrivastava, U., Ofstein, L., & Golhar, D. (2021). Direct and indirect effects of ICT infrastructure, skills, and use on entrepreneurship: A cross-country empirical investigation. *Journal of Global Information Management*, *29*(6), 1–25. doi:10.4018/JGIM.20211101.oa48

Taiba, S., Zeng, J., Muhammad, B., & Nausheen, S. (2021). Impact of ICT on entrepreneurial self-efficacy in emerging economy: Sustaining lock-down during COVID-19 pandemic. *Human Systems Management*, *40*(2), 299–314. doi:10.3233/HSM-201066

Ukommi, A.S., Okeagu, B.N. (2015). Information and communication technology (ICT) education and youth entrepreneurship in Uyo, Akwa Ibom State of Nigeria. *IOSR Journal of Humanities and Social Science*, *20*(7), 01-14. doi:10.9790/0837-20740114

United Nations Conference on Trade and Development. (2022). *Data Center.* https://unctadstat.unctad.org/wds/ReportFolders/reportFolders.aspx?sCS_ChosenLang=en

Viju, M. (2010). Women entrepreneurship in Middle East: Understanding barriers and use of ICT for entrepreneurship development. *The International Entrepreneurship and Management Journal, 6*(2), 163–181. doi:10.100711365-010-0144-1

Wamuyu, P. K. (2015). The impact of information and communication technology adoption and diffusion on technology entrepreneurship in developing countries: The case of Kenya. *Information Technology for Development, 21*(2), 253–280. doi:10.1080/02681102.2014.948372

World Bank. (2002). *Information and Communication Technologies: A World Bank Group Strategy.* World Bank. https://openknowledge.worldbank.org/handle/10986/15243

World Bank. (2022). *World Bank Country and Lending Groups.* https://datahelpdesk.worldbank.org/knowledgebase/articles/906519-world-bank-country-and-lending-groups

ADDITIONAL READING

Afawubo, K., & Noglo, Y. A. (2022). ICT and entrepreneurship: A comparative analysis of developing, emerging and developed countries. *Technological Forecasting and Social Change, 175*, 121312. doi:10.1016/j.techfore.2021.121312

Gomes, S., & Lopes, J. M. (2022). ICT access and entrepreneurship in the open innovation dynamic context: Evidence from OECD countries. *Journal of Open Innovation, 8*(2), 102. doi:10.3390/joitmc8020102

Chapter 7
A Detailed Analysis of the Digital Divide and Its Impact on the Development of Countries

Özlem Yorulmaz
https://orcid.org/0000-0003-1029-5749
Istanbul University, Turkey

ABSTRACT

Information and communication technologies (ICT) lead to significant social and economic changes in the lives of individuals. In parallel with this change, new criteria are needed when evaluating nations' development levels. In this study, the Digital Economy and Society Index (DESI), the Women in Digital Index (WID), and Rural Digital Index (RDI) proposed by the European Commission were considered as the proxies of ICT. The effects of these indexes on human development and gender inequality were studied. Findings showed that there is a digital divide among EU countries. The effect of DESI on HDI is higher in developed countries than in developing countries. In addition, it was observed that the increase in the WID value affected countries with high gender inequality more, and the RDI has an impact on the development of countries.

INTRODUCTION

Today, Information and Communication Technologies (ICT) have become indispensable to life in many areas, from education and health to work and entertainment. ICT provides countless opportunities for personal fulfillment, professional development, and value creation (ITU, 2021). The digital age we are in causes a transformation in many areas, particularly economic, social and cultural. ICT makes an essential contribution to improving the flow of information; with access to ICT, individuals access more information in a much shorter time, and as a result, a competitive advantage is obtained by making more optimal decisions in many areas. The problems experienced in accessing ICT, on the other hand, lead to a slowdown in sustainable development (Tjoa & Tjoa, 2016).

DOI: 10.4018/978-1-6684-6620-9.ch007

It has become a necessity to use ICT to work, learn and stay in touch, especially during the intense effects of the Covid-19 pandemic. As a matter of fact, 54% of the world's population was using the internet in 2019, while this rate increased to 63% with the pandemic (ITU, 2021). If ICT had been insufficient in this recent pandemic, more serious consequences could have been endured. For example, the health sector may have needed more time to provide effective services, information flow may not have been provided, education, production, and distribution processes may have been disrupted, and practical decisions may not have been made on many issues.

The impact of ICT on countries is not the same, and these effects occur at different rates. For this reason, countries are also affected by the benefits of ICT at varying levels. While internet usage in developed countries is 90% in 2021, it is 57% in developing countries and 2% in less developed countries. (ITU, 2021). The growth rate in internet usage is slower in developed countries, so the gap between developed and less developed countries is also narrowing; While the difference was 66% in 2017, it became 63% in 2021.

In terms of ICT access and Internet use, the difference between individuals, households, businesses, and geographic regions is defined as the digital divide (OECD, 2001). The digital divide is unequal access to and use of ICT (Hamburg & Lütgen,2019). Not only in terms of access to digital but also in terms of individuals' ability to benefit from technological applications, the digital divide emerges depending on age, gender, and geographical area. Different social and economic conditions of countries inevitably prepare the ground for the formation of the digital divide. The disadvantaged group cannot benefit from digital technologies regarding economic inclusion, communication, and access to improved public or private services (Mariscal et al., 2019). If the gap between those with access to ICTs and those disadvantaged increases in this regard, the welfare gap will widen, and inequality will increase worldwide (Mariscal et al., 2019).

When the statistics of 2021 were examined, it was seen that internet usage was 90% in developed countries, 57% in developing countries, and 2% in less developed countries (ITU, 2021). As a result of both the ICT policy reforms followed and the falling prices in accessing mobile services, internet usage is increasing year by year, especially in developing and underdeveloped countries (Byrne & Corrado, 2017). In terms of internet usage, the growth rate is slower in developed countries, so the gap between developed and less developed countries is also closing; while the difference was 66% in 2017, it was 63% in 2021 (ITU, 2021).

Although there has been increased access to digital worldwide recently, there are still points to overcome regarding women's participation in this process (Mariscal et al., 2019). Since digital technology affects all areas of life, if the existing gender digital divide is not dealt with, this case can cause gender inequality problems in labor markets and other areas. This will result in women being unable to take advantage of better job opportunities, find fair-paying jobs, and ultimately a wider gender pay gap (Mariscal et al., 2019). The most critical social inequality in technology is the lack of women in areas that impact sustainable development, such as university research and new technologies (Kerras et al., 2020). BIT plays an essential role in ensuring gender equality and empowering women, as they allow women to have a say in social and economic development by allowing access to information. Women and Men are common actors and beneficiaries of development, so women's use of ICT, which constitutes half of the world's population, should be considered in terms of justice, human rights, and economic contribution (UNDP, 2015). Gender equality is an essential human right necessary for a prosperous, peaceful, and sustainable world; gender equality is among the sustainable development goals of the United Nations (UN, 2021). Gender-based inequality of interests and skills will be a problem for society as many areas of life become more dependent on ICT (Cooper, 2006). The inequalities between men and women in terms of participation in basic internet use and the frequency of internet connection and computer use are defined as the "gender digital divide" (Castano, Martin, & Martinez, 2011). Digital Gender Divide is considered "both the result and the cause of human rights violations" (OHCHR, 2017). In developing and underdeveloped countries, women's access to and use digital technology is lower (Hilbert, 2011; Alozie & Akpan-Obong, 2017). Generally, women are less confident in using ICT than men (OECD, 2018). While internet usage rates for women and men in developed countries are 88% and 89%, respectively, these rates are 50% and 57% for developing countries; for underdeveloped countries, these rates are 19% and 31% (ITU, 2021). In terms of internet use, the gender digital divide increases as the level of development of countries decreases. Women are less likely than men to own digital devices and are technology-related (UNICEF, 2021). The fact that the proportion of women in the ICT sector in the European Union is approximately 17% explains this less interest (Acilar, Sæbø, 2021).

ICT has direct and indirect effects on the quality of life of individuals. Namely, ICT improves people's quality of life by creating new job opportunities and facilitating access to information, education, and services (Karaman et al., 2021). It also helps to increase human freedoms and the productivity of various human activities (Sen, 2010). Since ICT is effective in many fields, such as education, health, career, environment, and regional development, it causes to increase in the productivity of humans and machines (Azadnia et al., 2017; Kerras et al., 2020). Although many studies deal with the relationship between ICT and economic growth, the number of studies on the relationship between human development and ICT, which is influential on development, is fewer. The most prominent among these studies is Kim et al. (2008) 's study. This study investigated the impact of ICT investments in 51 countries on economic development. They discussed ICT investments in terms of hardware, software, and telecommunication. They evaluated the impact of ICT investments on GDP per capita and education (literacy rates, school enrollments), and health (life expectancy at birth). Kim et al. (2008) used the components of the human development index by considering the countries at different economic levels.

In this study, the impacts of technological development on both human development and gender inequality were evaluated separately. Several proxy measures of ICT were used in the study. The first of these proxies is the Digital Economy and Society Index (DESI), which the European Commission proposed in 2014. The other two are the Digital Women's Index (WID) and the Rural Digital Index (RDI) proposed by the European Commission in its 2022 report. As summarized above, the impact of ICT on countries varies depending on the level of development of the countries. Therefore, unlike previous studies, the effect of DESI on HDI was examined with quantile regression, which considers the different development levels of countries. In the following two stages, the relationship of WID with the Gender Inequality Index (GII), the composite metric of gender inequality, and then the relationship between RDI and HDI were evaluated. These WID and RDI indexes, which were developed based on gender and geographical area, which are the parameters of the digital divide, made it possible to make a detailed evaluation in this study.

Background

Digital Divide was first used in the 1990s to describe the social inequalities that emerged with the widespread use of ICT. The digital divide was defined shortly as the difference between individuals using the Internet and those with relatively disadvantaged Internet use (Rogers, 2001) or the gap between those who can afford to buy the computer software and hardware needed to benefit from the global information network and those who do not (Dragulanescu, 2002). With a broader definition, the digital divide does not only correspond to the use and access of ICT; it includes digital skills (Kerras et al., 2020).

The digital divide can be considered in two dimensions: measuring the gap between countries and between groups within the country (Cuervo & Menendez, 2006). The Local Digital Divide shows whether there is a division between factors such as gender, age, education level, and income level in the same society (Cuervo & Menendez, 2006).

At a macro level, ICT provides benefits at a lower rate for less developed countries compared to developed countries (Chinn & Fairlie, 2007). ICT offers opportunities for educated and resourced individuals at a micro level, but disadvantaged groups are less likely to benefit; this situation inevitably widens the social gap (Singh, 2003). This social gap can be considered in terms of developed and underdeveloped countries and between men and women. The participation of women in every field where information and communication technologies are used is essential both for correcting the gap in human capital and for developing a knowledge-based society (Fountain, 2000). Many studies indicate that women are less involved in ICT-related fields (Kerras et al., 2020). Some of the professions in the IT sector, computer design, database maintenance, and computer security programming, are especially preferred by men; these professions are seen as more masculine (Kerras et al. al, 2020).

When the studies examining the digital divide in European countries from the perspective of geographical area, gender, age, income, and education level are examined, it is seen that a digital divide exists in the EU. Some of these studies can be summarized as follows. Çilan, Bolat, and Coşkun (2009) investigated the digital divide among EU countries, new members, and candidate countries in their studies. According to the results of this study, in which they used Discriminant Analysis and MANOVA, they concluded that there is a digital divide between the mentioned country groups. Kılıç (2011) evaluated the local digital divide in Turkey in his study and concluded that inequalities between low- and high-income families, education level, and geographical region cause significant differences. Cruz-Jesus et al. (2012) used clustering analysis to investigate the digital divide in the EU. Based on the common ICT indicators, five clusters were obtained of the countries in these clusters. The clusters were named as follows: Digital followers, digital laggards, digital leaders, individual-side focused, firm-side, and low access costs focused. The study concluded that there is a statistically significant difference between these clusters, that is, the digital divide. Tsetsi & Rains (2017) evaluated the digital divide considering smartphone users and stated that education level, age, race, gender, and income level stand out in determining the difference between groups. Ragnedda and Kreitem (2018) investigated the existence of digital disparities in digital access and skills among European countries. Accordingly, Internet access is increasing rapidly all over Europe, but there are differences in Internet access between Northern and Eastern EU countries and even within Eastern Europe itself; Romania and Bulgaria are at the bottom. The study also considered inequalities in the level of digital skills, revealing differences in access, use, and benefits across Europe, particularly in Eastern EU countries. Serrano-Cinca et al. (2018) examined Internet use, including social networks, e-government services, and e-commerce, from the perspective of the digital divide, from a survey of 2304 people in Spain; they concluded that

- The gender digital divide affects middle-aged people but not the oldest and youngest groups;
- The business group most affected by the digital divide is housewives;
- In rural environments, internet use is low in the elderly groups.
- Age, education, and income are the most important factors explaining the digital gap.

Perez-Morote et al. (2020) examined the use of e-government services in 27 EU countries between 2010-2018. They determined that trust in governments and the income-related digital divide play a role in using this service. Elena-Bucea et al. (2021) assessed the digital divide between and within the European countries. Their analysis consists of four socio-demographic contexts: age, education, gender, and income. Two important dimensions, namely, e-Services and Social Networks, were explored by factor analysis. According to their findings, e-Services adoption is affected essentially by the education level of individuals, whereas Social Networks adoption is more affected by individuals' age.

Besides economic growth, researchers have recently examined the effects of education, the health system, human development, and firm profitability on ICT (Karaman et al., 2021). ICT causes the development of SMEs and increases employment. Ngwenyama et al. (2006) examined the influence of ICT investments on human development regarding education and health and stated that this effect was positive. Gholami et al. (2010), the impact of ICT on the human development index is more remarkable in underdeveloped countries than in developed countries. Bankole et al. (2011, 2013) emphasized that ICT has a different effect on human development components, and this effect differs according to the economic development level of countries. De la Hoz-Rosales et al. (2019), on the other hand, concluded that ICT has a positive effect on human development regardless of the level of development of the countries. Perez-Castro et al. (2021) evaluated the impact of technological development on human development in Mediterranean countries, taking into account the period from 2010 to 2015. According to their results, technological and human development have an average growth rate of 23% and 2.5%, respectively. Differences between countries also narrowed by 5.2% for ICT and 7.4% for HDI. It has been observed that the digital divide exists and decreases over time. According to their findings, there is a digital divide between northern and southern Mediterranean countries. Technological development has an impact on human development. Overall, the empirical literature agrees that ICT positively affects human development (Karaman et al., 2021).

DESI has been used as a proxy of ICT in many studies. In 2021, the calculation of DESI was updated due to two new policies, the Recovery, and Resilience Facility (RRF) and the Digital Decade Compass, affecting the EU's digital transformation (Kovacs et al., 2022). The DESI index was calculated considering OECD guidelines and recommendations. The data was obtained from the Member States' competent authorities and various survey studies carried out by the Commission (EC, 2020). Some of the recent studies employing DESI can be listed as follows. Stoica and Bogoslov (2017), using the DESI index between 2014 and 2017, compared Romania and the EU and concluded that Romania made progress in this process. Kontolaimou and Skintzi (2018) reached similar findings for Greece in their study. Scupola (2018) studied Denmark at the top of the DESI ranking and examined the e-government experience in detail. In their study, Nikolov and Krumova (2019) examined the e-Government issue for all EU countries. Russo (2020) made a local assessment of the Abruzzo region of Italy, adhering to the guidelines of the DESI index. Banhidi, Dobos, and Nemeslaki (2020) classified all EU countries based on the DESI index components.

EMPIRICAL ANALYSIS

This section provides information about the statistical methods used in the analysis and the dataset, and the findings obtained from the research are shared.

Methodology

In the first stage, HDI and i-DESI indexes were evaluated separately; each index's variation rate was calculated to interpret the evaluation process between 2015 -2018, considering 44 countries. In the next step, clustering analysis was used to group the countries by considering the index values in question. Clustering analysis includes two methods in the form of hierarchical and non-hierarchical. This study adopted a hierarchical clustering approach based on Minkowsky and squared Euclidean distances with centroid and ward methods. The number of clusters was determined by taking into account the dendrograms. Then, a non-hierarchical clustering approach based on the k-means algorithm was used. Using both methods in clustering analysis allows for better results (Sharma, 1995). After the countries were divided into clusters, whether the indices (WID, RDI, HDI) under consideration presented statistically different values in each cluster was evaluated by the Kruskal-Wallis test. The Kruskal-Wallis test is a non-parametric version of ANOVA that evaluates whether there is a difference between the group averages. After the Kruskal-Wallis test, the differences between the cluster averages were evaluated by applying the "post-hoc test" (pairwise comparison).

To evaluate the impact of ICT on the development of countries, three different regression models were used to test this relationship in this study. The first regression model investigates the relationship between i-DESI and HDI. The second model emphasizes Gender in Digital and gender inequality. This model is based on Women in Digital and its impact on Gender Inequality. The last regression model evaluates the relationship from the perspective of rural in digital and human development. For this reason, RDI and HDI indexes were employed. The first two regression models were estimated by the Quantile regression approach. Quantile regression was developed by Koenker and Bassett (1978) and is estimated based on the following objective function.

$$\arg\min_{\beta} = \sum_{e_i \geq 0} q|e_i| + \sum_{e_i < 0} (1-q)|e_i|$$

(q (0<q<1) is to show the quantile value and e is to show the residues obtained from the model)

Linear programming is used to solve these estimators. Unlike the ordinary least squares approach, quantile regression provides the whole picture, as it provides information for different quantile levels of the dependent variable. The use of quantile regression analysis is suitable for situations where conditional quantiles vary, especially in studies of income and wage distribution inequalities. With the usage of quantile regression, the impact of digitalization was investigated from a broad perspective. Quantile regression estimates the dependent variable's conditional median (and other quantiles). For instance, the estimation results for the 10th, 50th, and 90th quantiles of HDI and GII indexes were considered.

Dataset

In this study, Digital Economy and Society Index (DESI), Women in Digital (WID), and Rural in Digital (RDI) indexes were used to evaluate the digital divide issue in detail. In addition, Human Development Index (HDI) and Gender Inequality Index (GII), which are development indicators, were used to examine the relationship between Digitalization and Development from the perspective of the digital divide. The dimensions that make up these indexes can be summarized as follows.

The Digital Economy and Society Index (DESI) was first presented in 2014 by the European Commission in an annual report tracking the progress of EU Member States on digital. When DESI was first developed, it consisted of five main dimensions: Connectivity, Human Capital, Internet Usage, Integration of Digital Technology, and Digital Public Services. However, in the current calculations of this index, the internet usage dimension is not included in the calculations, and the remaining four main dimensions are taken into account with equal weight. The sub-dimensions and weights of the dimensions are as follows (European Commison, 2021):

A) Human capital -25%
Internet user skills (50%) and advanced digital skills (50%)
B) Connectivity- 25%
Fixed broadband take-up (25%), fixed broadband coverage (25%), mobile broadband (40%), and broadband prices (10%)
C) Integration of digital technology- 25%
Digital intensity (15%), digital technologies for businesses (70%), and e-commerce (15%)
D) Digital public services -25%
e-Government (100%)

The European Commission has published a DESI index for 44 countries, considering various countries outside the EU. This index, called I-DESI and calculated based on five main dimensions, was calculated only for 2015-2018. The latest I-DESI was published in 2020 and is calculated based on 2018 data. The European Commission presented the index of women in digital (WID) in its 2022 report. This index calculated for 26 EU countries is based on EUROSTAT survey data. It consists of three main dimensions of equal weight: Internet use, Internet user skills, and Specialization skills-employment. These three main dimensions contain 12 sub-dimensions in total. (WDI, 2022). The Rural Digital Index presented in the European Commission 2022 report is calculated as the weighted average of three dimensions: Internet Usage, Human Capital and Connectedness. These dimensions contain 16 sub-dimensions in total.

The Gender Inequality Index (GII) is an index that was first presented in 2010 by the United Nations Development Program. It is calculated as the weighted average of three main dimensions of the GII: reproductive health, empowerment, and labor market. GII ranges from 0 to 1. Higher GII values mean higher inequalities between women and men, interpreted as a loss in human development (HDR, 2020). The Human Development Index (HDI) is calculated based on the dimensions of a long and healthy life, knowledge, and good living standards. It is measured by expected schooling times for children of incipient age. The standard of living dimension is measured by gross national income per capita (HDR, 2020).

While the I-DESI and HDI indices used in the study belong to 44 countries between 2015 and 2018, the WID, RDI, and GII indices include 2022 data of 26 EU members. Since the DESI index started to be calculated in 2014 and the WID and RDI indexes were first calculated in 2022, it was studied with a limited time dimension.

Findings obtained through descriptive statistics, quantile regression analysis, the Kruskal-Wallis test, and cluster analysis were presented in this section. As mentioned above, the relationship between i-DESI and HDI was evaluated in the first stage. Table 1 shows the HDI and DESI scores of 44 countries for 2015-2018, including the variations for each index. As can be seen from the table, according to HDI, the countries with the lowest scores are "China, Brazil, Mexico, Serbia, and Bulgaria," whereas the ones having the highest scores are "Denmark, Germany, Iceland, Switzerland, and Norway". The gap between the least and most developed countries decreased (-6.85%) during the period, from 0.22 to 0.20. The highest growth rates in HDI belong to Turkey and China (2.7% and 3%, respectively. The distribution of DESI scores exhibits differences from the HDI. Contrary to HDI, the gap between the DESI scores of the least and the most developed countries increased (4.23%) in this period.

It can be explicitly seen that the DESI scores of highly developed countries are higher than developing countries. However, China has the highest growth rate in DESI (58.84%), and Israel follows it (27.28%). Regarding the DESI variation rate, some of the countries shrank, contrary to most of the countries. The countries that shrank can be ranked as follows: Croatia, Chile, New Zealand, Brazil, Portugal, and Greece. When the rankings of 2018 i-DESI and HDI scores in the table are examined together, it is seen that the countries with the highest scores are similar and are in the high-income class. However, the countries in the lower ranks of these indices differ; the lower ranks of i-DESI include upper-middle-income countries and European countries such as Italy, Greece, and Poland.

The second stage of the analysis investigates the impact of information and technology on the development of countries. Firstly, the MMS test (Machado-Santos Silva, 2000), which examines the heteroscedasticity problem, was applied to choose the proper regression estimation technique. Since the findings of the MMS test indicate the heteroscedasticity problem ($p< 0.01$), the quantile regression approach was preferred here. Due to the data studied being panel data with a relatively limited time dimension (2015-2018), the nonadditive fixed effect estimator developed by Powell (2016) was used in the analysis.

According to the findings of the quantile regression in Table 2, the effect of DESI decreases for countries with higher HDI values. In other words, the impact of information and technology on development is more remarkable in less developed countries. The low value of the slope coefficient at the 0.9 quartile level may occur since developed countries have reached the peak in ICT, and cognitive technologies are being used.

As mentioned in the previous section, i-DESI represents information about 44 countries, whereas DESI, WID, and RDI indexes give information about 26 European countries. To have a closer look at gender digital divide and rural digital divide, WID and RDI indexes were investigated here. As can be seen from Figure 1, in countries where gender inequality is low, it is observed that "women in digital" and "rural digital" values are high. This situation is clearly observed in countries such as Finland and Denmark. Romania and Bulgaria are the countries with the highest gender inequality and low WID and RDI scores to that extent. However, it can be seen that scores of WID and RDI are lower than expected compared to the gender inequality values of Germany, Italy, and Portugal.

The correlation coefficients between these indexes are shown in Table 3. The correlation between GII and HDI, the development indexes endorsed by UNDP, is strong as expected (-0.812). In addition, WID and RDI and the correlation between the development indexes are also strong.

In the third stage of the analysis, European countries were classified using the scores of WID, RDI, and HDI. The evaluation of the classification was done through hierarchical and k-means clustering approaches. According to the findings, the optimal number of clusters was defined as three. Figure 2 presents the averages of these clusters with bar charts. The classification of countries and the average scores for each cluster can be seen in Table 4.

The bar chart where the cluster averages were drawn by taking into account can be seen in Figure 2. Although the HDI scores of the three clusters are close to each other, the average RDI and WID scores are quite different. The Kruskal-Wallis test applied by considering the HDI, WID, and RDI variables showed that these three variables were statistically significant in determining these clusters. The Kruskal-Wallis p-values based on all three variables are all $p<0.05$. However, the pairwise comparison (post

Table 1. Human development index (HDI) & International digital economy and society index (i-DESI)

Countries	HDI_{2015}	HDI_{2016}	HDI_{2017}	HDI_{2018}	$HDI^{a}_{\%}$	$DESI_{2015}$	$DESI_{2016}$	$DESI_{2017}$	$DESI_{2018}$	$DESI^{a}_{\%}$
Australia	0,93	0,93	0,94	0,94	0,64	0,55	0,58	0,52	0,59	8,23
Austria	0,91	0,91	0,92	0,92	0,77	0,50	0,50	0,53	0,52	2,98
Belgium	0,92	0,92	0,93	0,93	1,09	0,46	0,50	0,51	0,50	7,84
Brazil	0,75	0,76	0,76	0,76	1,46	0,35	0,34	0,31	0,34	-4,24
Bulgaria	0,81	0,81	0,81	0,81	0,00	0,36	0,36	0,41	0,39	8,43
Canada	0,93	0,93	0,93	0,93	0,65	0,50	0,53	0,55	0,54	7,41
Chile	0,84	0,85	0,85	0,85	1,43	0,37	0,42	0,43	0,33	-12,03
China	0,73	0,74	0,75	0,76	3,00	0,28	0,33	0,42	0,44	58,84
Croatia	0,84	0,84	0,85	0,85	1,31	0,39	0,45	0,41	0,32	-18,16
Cyprus	0,87	0,87	0,88	0,89	2,54	0,42	0,43	0,47	0,43	1,65
Czechia	0,89	0,89	0,90	0,90	1,13	0,41	0,39	0,45	0,49	18,25
Denmark	0,93	0,94	0,94	0,94	1,29	0,61	0,64	0,64	0,68	12,50
Estonia	0,88	0,88	0,89	0,89	0,91	0,55	0,55	0,53	0,56	2,56
Finland	0,93	0,93	0,93	0,93	0,76	0,63	0,63	0,64	0,68	7,12
France	0,89	0,89	0,90	0,90	0,67	0,49	0,52	0,51	0,58	18,13
Germany	0,94	0,94	0,94	0,94	0,75	0,51	0,50	0,52	0,58	14,00
Greece	0,88	0,88	0,88	0,88	0,11	0,38	0,39	0,41	0,37	-1,32
Hungary	0,84	0,84	0,84	0,85	0,84	0,36	0,41	0,45	0,39	6,63
Iceland	0,94	0,95	0,95	0,95	1,71	0,61	0,62	0,68	0,62	0,98
Ireland	0,91	0,93	0,93	0,93	2,19	0,52	0,51	0,54	0,60	16,12
Israel	0,91	0,91	0,91	0,92	0,99	0,47	0,42	0,41	0,60	27,29
Italy	0,88	0,88	0,89	0,89	0,57	0,34	0,38	0,37	0,37	8,26
Japan	0,91	0,92	0,92	0,92	0,88	0,50	0,48	0,49	0,57	12,77
Korea	0,90	0,90	0,91	0,91	1,34	0,46	0,51	0,52	0,54	15,52
Latvia	0,85	0,85	0,86	0,86	1,65	0,42	0,40	0,41	0,42	1,44
Lithuania	0,86	0,86	0,87	0,88	1,74	0,44	0,44	0,47	0,44	0,23
Luxemburg	0,93	0,92	0,92	0,92	-0,76	0,58	0,63	0,66	0,62	7,40
Malta	0,88	0,89	0,89	0,90	2,27	0,49	0,50	0,51	0,49	-0,82
Mexico	0,77	0,77	0,78	0,78	1,17	0,32	0,33	0,35	0,33	4,42
Netherlands	0,93	0,93	0,93	0,94	0,64	0,56	0,59	0,64	0,68	21,33
NewZealand	0,93	0,93	0,93	0,93	0,54	0,59	0,51	0,55	0,55	-6,50
Norway	0,95	0,95	0,96	0,96	0,74	0,62	0,59	0,63	0,64	4,21
Poland	0,87	0,87	0,87	0,88	1,16	0,32	0,36	0,35	0,36	14,60
Portugal	0,85	0,85	0,85	0,86	1,30	0,41	0,37	0,43	0,40	-2,95
Romania	0,81	0,81	0,82	0,82	1,48	0,38	0,39	0,45	0,38	2,13
Russia	0,82	0,82	0,82	0,83	1,35	0,36	0,41	0,43	0,42	17,37
Serbia	0,79	0,79	0,80	0,80	1,78	0,33	0,37	0,34	0,36	7,53
Slovakia	0,85	0,85	0,85	0,86	0,82	0,37	0,37	0,41	0,39	7,38
Slovenia	0,90	0,90	0,91	0,91	1,22	0,42	0,42	0,43	0,46	8,57
Spain	0,88	0,89	0,90	0,90	1,47	0,44	0,50	0,47	0,45	2,52
Sweden	0,94	0,94	0,94	0,94	0,64	0,60	0,60	0,65	0,66	9,95
Switzerland	0,95	0,95	0,96	0,96	0,53	0,56	0,57	0,61	0,67	19,18
Turkey	0,81	0,82	0,82	0,83	2,97	0,29	0,27	0,33	0,31	7,17
UK	0,92	0,92	0,92	0,93	0,54	0,55	0,61	0,60	0,60	9,64
Difference[b]	0,22	0,21	0,21	0,20	-6,85	0,36	0,37	0,37	0,37	4,23
Mean	0,88	0,88	0,88	0,89	1,14	0,46	0,47	0,49	0,49	6,52

a Variation Rate period 2015–2018.
b Difference Rank 1 – Rank 44.
Sources: EU Database & UNDP Database

Table 2. The relationship between HDI and I-DESI

HDI	Coef. (.1 quantile)	Coef. (.5 quantile)	Coef.(.9quantile)
I-DESI	0.506***	0.392***	0.284***
CONS	0.603***	0.699***	0.783***

R-squared = .68
***: p<0.01

Figure 1. Distribution of WID, RDI, and GII index scores for European countries based on 2021

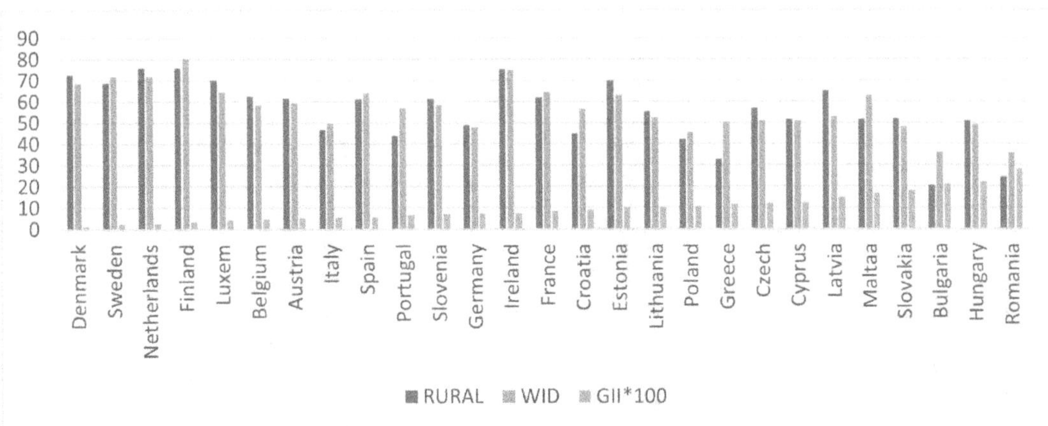

hoc) results (Table 5) indicate that it is necessary to be careful about supporting the claim that clusters 1 and 3 are statistically different from each other (p=0.046). Considering that this situation may arise due to similar HDI scores of clusters 1 and 3, the process was repeated, taking into account only WID and RDI variables. Although the results of cluster analysis and Kruskal-Wallis results were the same as before, it was concluded that the averages of Clusters 1 and 3 were statistically different (p=0.02).

At the last stage of the analysis, the relationships between the values of the GII and WID, besides the relationship between RDI and HDI, were evaluated separately by regression analysis. The findings of MMS tests indicated that the quantile regression approach is suitable for explaining the relationship between GII and WID due to the heteroscedasticity problem (p=0.04).

Table 3. Correlation coefficients between indexes

Variables	(1)	(2)	(3)	(4)
(1) WID	1.000			
(2) GII	-0.738	1.000		
(3) RDI	0.872	-0.691	1.000	
(4) HDI	0.785	-0.812	0.775	1.000

Figure 2. Classification of countries according to WID, RDI and HDI

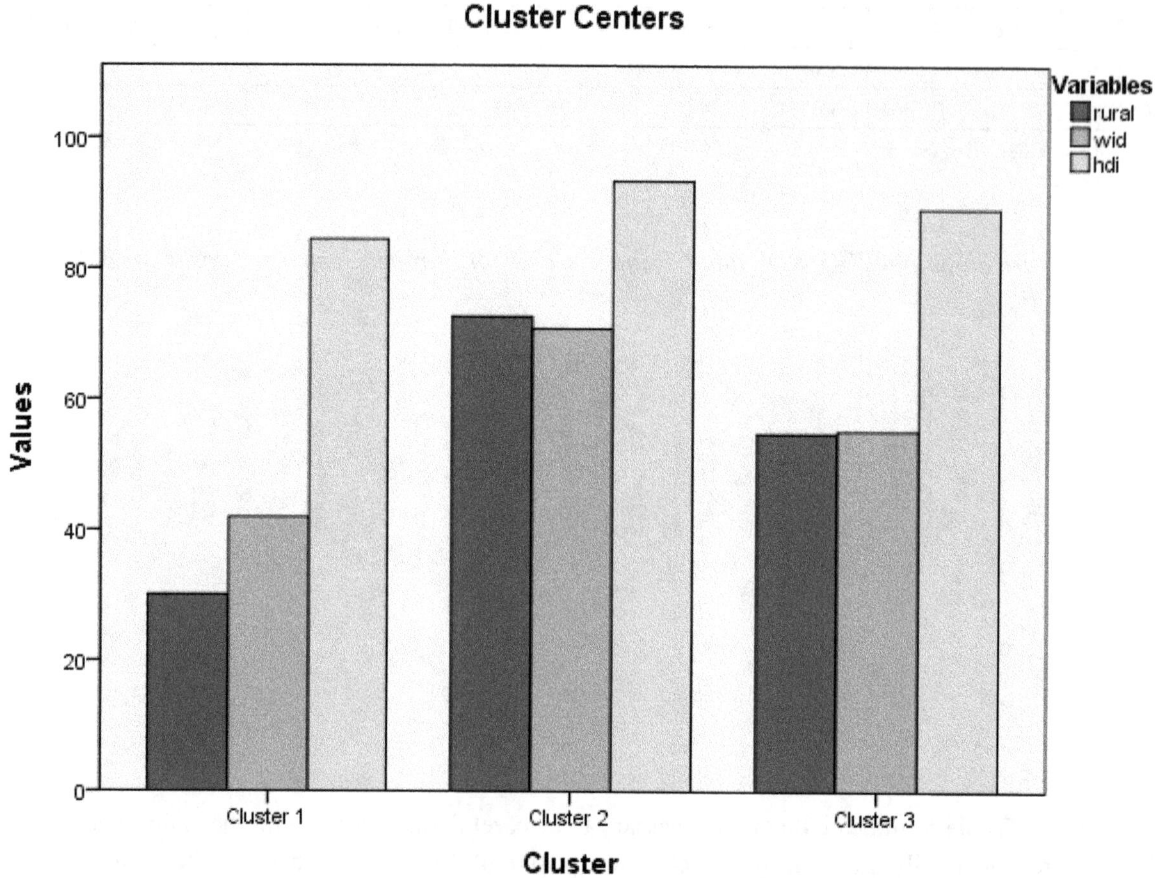

Table 4. List of countries in each cluster

Average-Score	Cluster I	Cluster II	Cluster III
(RDI- WID-HDI)	(30- 42- 84)	(73- 71- 93)	(55-55-89)
	Poland,Greece, Romania, Bulgaria	Netherlands, Finland, Ireland, Denmark, Estonia, Sweden,Luxemborg	Belgium, France, Austria, Spain, Czech, Germany, Hungary, Portugal, Crotia, Italy, Latvia, Slovenia, Slovakia,Malta, Lithuania, Cyprus

Table 5. Post-hoc analysis

	Pairwise Comparisons p-value (RDI&WID&HDI)	Pairwise Comparisions p-value (RDI&WID)
Cluster 1- Cluster 2	0.001	0.000
Cluster 1- Cluster 3	0.046	0.02
Cluster 2- Cluster 3	0.02	0.01

Table 6. The impact of WID on GII

GII	Coef. (.1 quantile)	Coef. (.5 quantile)	Coef.(.9quantile)
WID	-0.24***	-0.45***	-0.56***
cons	18.6***	34.9***	48.3***

(***, $p < .01$, calculations of standard errors based on 1000 bootstrap replications).

According to the findings of the quantile regression presented in Table 6, the effect of WID decreases for countries with higher GII values. As "women in digital" increases, countries with more inequivalent gender are more likely to be affected.

Since the homoscedasticity assumption was provided, the ordinary least squares estimator was applied to evaluate the relationship between HDI and RDI, as shown in Table 7. As the RDI index increases, the HDI score increases. WID, RDI, and DESI range from 0 to 100, whereas GII and HDI range from 0-1. Before the regression analysis, two scales were homogenized to provide a comparison. The proportion to which HDI varies in function of the RDI is 21%. Pointing out that if RDI increased by one unit, the HDI would increase by 0.21 units.

Table 7. The impact of RDI on HDI

HDI	Coef.	Sig
RDI	0.21	***
Constant	77.6	***
R-squared	0.601	
*** *p<.01,*		

The constant in Table 7 indicates that even if RDI were zero, the HDI score would be 0.77, meaning that besides the impact of technology and information, other factors influence HDI.

SOLUTIONS AND RECOMMENDATIONS

The study showed that the effect of ICT on HDI was different in developed and developing countries. Policymakers should make arrangements to remove the barriers that prevent especially developing countries from benefiting from ICT. Similarly, adopting policies to reduce the digital gender divide, especially for developing countries, contributes to gender equality. Policymakers should direct digital education, especially to disadvantaged groups. Training programs can be opened for women to develop digital skills, and scholarships can be given to encourage their education in technology.

FUTURE RESEARCH DIRECTIONS

One of the study's main limitations is that the DESI calculation was updated in 2021, and i-DESI was only collected until 2019. In future studies, the relationships between the sub-dimensions can be investigated in detail by considering the sub-dimensions of the WID and GII. In addition, HDI and DESI scores can be calculated at the regional level.

CONCLUSION

It is widely accepted in the literature that technological developments have positive effects on human development, and this effect differs according to the development level of countries, just like the rate of economic growth. Another common view is that the gender digital divide is higher in underdeveloped countries than in developed countries.

ICT innovations are effective in developing new criteria that can be used to compare countries. Here, besides DESI, the new indexes presented by European Commission, WID, and RDI were used as a proxy of ICT. First of all, the effect of DESI on HDI was examined by using quantile regression, which allows the heterogeneous structures of countries to be taken into account, unlike previous studies. In the next step, the relationship between WID and GII was evaluated for EU member countries with quantile regression. As far as we know, the relationship between this index, which measures women's place in ICT, and gender inequality has not been studied before. In the last stage, the relationship between RDI, which evaluates the use of ICT in rural areas, and HDI was investigated. In addition, in this study, EU countries were classified by cluster analysis considering WID, DESI, and HDI scores. The differences between the obtained clusters were tested with the Kruskal-Wallis test. The findings show a digital divide for EU countries, and technological development is effective in human development. However, this effect is more significant when the developed levels of countries, namely quantile levels, decrease. The effect of WID value is more remarkable in countries with high gender inequality, and the increase in RDI has shown that it will lead to an increase in human development.

As a recommendation policy, women should be encouraged to use information and communication technologies, take part in this sector, and develop their digital skills to achieve gender equality, which is an indicator of development.

REFERENCES

Acilar, A., & Sæbø, Ø. (2021). *Towards understanding the gender digital divide: A systematic literature review*. Global Knowledge, Memory and Communication.

Alozie, N. O., & Akpan-Obong, P. (2017). The digital gender divide: Confronting obstacles to women's development in Africa. *Development Policy Review*, *35*(2), 137–160. doi:10.1111/dpr.12204

Azadnia, M., Zahedi, S., Majjedin, A., & Reza, M. (2017). Analysis of the impact of ICT on sustainable development using sustainability indicators. *International Journal of Computers and Applications*, *169*(6), 13–24. doi:10.5120/ijca2017914766

Bánhidi, Z., Dobos, I., & Nemeslaki, A. (2020). What the overall Digital Economy and Society Index reveals: A statistical analysis of the DESI EU28 dimensions. *Regional Statistics*, *10*(2), 42–62. doi:10.15196/RS100209

Bankole, F. O., Osei-Bryson, K.-M., & Brown, I. (2013). The Impact of ICT Investments on Human Development: A Regression Splines Analysis. *Journal of Global Information Technology Management*, *16*(2), 59–85. doi:10.1080/1097198X.2013.10845636

Bankole, F. O., Shirazi, F., & Brown, I. (2011). Investigating the Impact of ICT Investments on Human Development. *The Electronic Journal on Information Systems in Developing Countries*, *48*(1), 1–19. doi:10.1002/j.1681-4835.2011.tb00344.x

Byrne, D. M., & Corrado, C. A. (2017). *ICT Services and their Prices: What do they tell us about Productivity and Technology?* Available at: https://ssrn.com/abstract=2920451

Castaño, C., Martín, J., & Luis Martínez, J. (2011). La brecha digital de género en España y Europa: Medición con indicadores compuestos. *Revista Espanola de Investigaciones Sociologicas*, *136*(1), 127–140. doi:10.5477/cis/reis.136.127

Chinn, M. D., & Fairlie, R. W. (2007). The determinants of the global digital divide: A cross-country analysis of computer and internet penetration. *Oxford Economic Papers*, *59*(1), 16–44. doi:10.1093/oep/gpl024

Cilan, Ç. A., Bolat, B. A., & Coşkun, E. (2009). Analyzing digital divide within and between member and candidate countries of European Union. *Government Information Quarterly*, *26*(1), 98–105. doi:10.1016/j.giq.2007.11.002

Cooper, J. (2006). The digital divide: The special case of gender. *Journal of Computer Assisted Learning*, *22*(5), 320–334. doi:10.1111/j.1365-2729.2006.00185.x

Cruz-Jesus, F., Oliveira, T., & Bacao, F. (2012). Digital divide across the European Union. *Information & Management*, *49*(6), 278–291. doi:10.1016/j.im.2012.09.003

Cuervo, M. R. V., & Menéndez, A. J. L. (2006). A multivariate framework for the analysis of the digital divide: Evidence for the European Union-15. *Information & Management*, *43*(6), 756–766. doi:10.1016/j.im.2006.05.001

De la Hoz-Rosales, B., Camacho-Ballesta, J. A., Tamayo-Torres, I., & Buelvas-Ferreira, K. (2019). Effects of information and communication technology usage by individuals, businesses, and government on human development: An international analysis'. *IEEE Access: Practical Innovations, Open Solutions*, *7*, 129225–129243. doi:10.1109/ACCESS.2019.2939404

Dewan, S., Ganley, D., & Kraemer, K. L. (2010). Complementarities in the diffusion of personal computers and the Internet: Implications for the global digital divide. *Information Systems Research*, *21*(4), 925–940. doi:10.1287/isre.1080.0219

Dragulanescu, N. G. (2002). Social impact of the "Digital Divide" in a Central–Eastern European country. *The International Information & Library Review*, *34*(2), 139–151. doi:10.1080/10572317.2002.10762570

Elena-Bucea, A., Cruz-Jesus, F., Oliveira, T., & Coelho, P. S. (2021). Assessing the Role of Age, Education, Gender and Income on the Digital Divide: Evidence for the European Union. *Information Systems Frontiers*, *23*(4), 1007–1021. doi:10.100710796-020-10012-9

European Commission. (2020). *Digital Economy and Society Index 2021 Methodological Note*. European Commission.

Fountain, J. E. (2000). Constructing the information society: Women, information technology, and design. *Technology in Society*, *22*(1), 45–62. doi:10.1016/S0160-791X(99)00036-6

Gholami, R., Higón, D. A., Hanafizadeh, P., & Emrouznejad, A. (2010). Is ICT The Key to Development? *Journal of Global Information Management*, *18*(1), 66–83. doi:10.4018/jgim.2010091104

Hamburg, I., & Lütgen, G. (2019). Digital divide, digital inclusion and inclusive education. *Adv. Soc. Sci. Res. J.*, *6*, 193–206. doi:10.14738/assrj.64.6457

Hilbert, M. (2011). Digital gender divide or technologically empowered women in developing countries? A typical case of lies, damned lies, and statistics. *Women's Studies International Forum*, *34*(6), 479–489. doi:10.1016/j.wsif.2011.07.001

ITU. (2019). *ITU ICT Facts and Figures 2017*. https://www.itu.int/en/ITUD/Statistics/Documents/facts/ICTFactsFigures2017.pdf

Karaman Aksentijević, N., Ježić, Z., & Zaninović, P. A. (2021). The Effects of Information and Communication Technology (ICT) Use on Human Development—A Macroeconomic Approach. *Economies*, *9*(3), 128. doi:10.3390/economies9030128

Kerras, H., Sánchez-Navarro, J. L., López-Becerra, E. I., & de-Miguel Gomez, M. D. (2020). The impact of the gender digital divide on sustainable development: Comparative analysis between the European Union and the Maghreb. *Sustainability*, *12*(8), 3347. doi:10.3390u12083347

Kiliç, Ç. (2011). Küreselleşen Dünyada Dijital Bölünme Sorunu. *Erzincan Üniversitesi Eğitim Fakültesi Dergisi*, *13*(1), 81–91.

Kim, Y. J., Kang, H., Sanders, G. L., & Lee, S. T. (2008). Differential Effects of IT Investments: Complementarity and the Effect of GDP Level. *International Journal of Information Management*, *28*(8), 508–516. doi:10.1016/j.ijinfomgt.2008.01.003

Koenker, R., & Bassett, G. Jr. (1978). Regression quantiles. *Econometrica*, *46*(1), 33–50. doi:10.2307/1913643

Kontolaimou, A., & Skintzi, G. (2018). 4.2. digitisation patterns of the Greek economy and society. *Greek Economic Outlook*, *2018*(37), 41–48.

Kovács, T. Z., Bittner, B., Huzsvai, L., & Nábrádi, A. (2022). Convergence and the Matthew Effect in the European Union Based on the DESI Index. *Mathematics*, *10*(4), 613. doi:10.3390/math10040613

Machado, J. A., & Silva, J. M. C. S. (2013). *Quantile regression and heteroskedasticity* [Unpublished manuscript]. Department of Economics, University of Essex.

Mariscal, J., Mayne, G., Aneja, U., & Sorgner, A. (2019). Bridging the gender digital gap. *Economics, 13*(1).

Ngwenyama, O., Andoh-Baidoo, F. K., Bollou, F., & Morawczynski, O. (2006). Is There A Relationship Between ICT, Health, Education and Development? An Empirical Analysis of five West African Countries from 1997–2003. *Journal of Information Systems for Developing Country, 23*(1), 1–11. doi:10.1002/j.1681-4835.2006.tb00150.x

Nikolov, H. S., & Krumova, M. Y. (2019). Hofstede's model in the context of e-government and open government in EU countries: Countries clustering based on similarities and differences *Smart Cities and Regional Development (SCRD). Journal, 3*(1), 29–46.

OECD. (2001). *Information and Communication Technologies and Rural Development.* doi:10.1787/9789264193079-en

OECD. (2018). *Bridging the digital gender divide: Include, upskill, innovate.* Available at: www.oecd.org/digital/bridging-the-digital-gender-divide.pdf

OHCHR. (2017). *Report on ways to bridge the gender digital divide from a human rights perspective.* Available at: www.ohchr.org/en/issues/women/wrgs/pages/waystobridgethegenderdigital.aspx

Pérez-Castro, M. Á., Mohamed-Maslouhi, M., & Montero-Alonso, M. Á. (2021). The digital divide and its impact on the development of Mediterranean countries. *Technology in Society, 64*, 101452. doi:10.1016/j.techsoc.2020.101452

Pérez-Morote, R., Pontones-Rosa, C., & Núñez-Chicharro, M. (2020). The effects of e-government evaluation, trust and the digital divide in the levels of e-government use in European countries. *Technological Forecasting and Social Change, 154*, 119973. doi:10.1016/j.techfore.2020.119973

Powell. (2016). *Quantile regression with nonadditive fixed effects.* Available at: https://works.bepress.com/david_powell/1/

Ragnedda, M., & Kreitem, H. (2018). The three levels of digital divide in East EU countries. *World of Media. Journal of Russian Media and Journalism Studies, 1*(4), 5–26.

Rogers, E. M. (2001). The digital divide. *Convergence, 7*(4), 96–111. doi:10.1177/135485650100700406

Russo, V. (2020). Digital Economy and Society Index (DESI). European guidelines and empirical applications on the territory. In *Qualitative and Quantitative Models in Socio-Economic Systems and Social Work* (pp. 427–442). Springer. doi:10.1007/978-3-030-18593-0_31

Scupola, A. (2018). Digital transformation of public administration services in Denmark: A process tracing case study. *Nordic and Baltic Journal of Information and Communications Technologies, 2018*(1), 261–284. doi:10.13052/nbjict1902-097X.2018.014

Sen, A. (2010). The Mobile and the World. *Information Technologies and International Development, 6*, 1–3.

Serrano-Cinca, C., Muñoz-Soro, J. F., & Brusca, I. (2018). A multivariate study of internet use and the digital divide. *Social Science Quarterly*, *99*(4), 1409–1425. doi:10.1111squ.12504

Sharma, S. (1995). *Applied Multivariate Techniques*. Wiley.

Singh, A. (2003). *Information and Communication Technologies (ICT) and Sustainable Development*. Development Alternatives.

Stoica, E. A., & Bogoslov, I. A. (2017, December). A comprehensive analysis regarding DESI country progress for Romania relative to the European average trend. In *Balkan Region Conference on Engineering and Business Education* (Vol. 2, No. 1, pp. 258-266). 10.1515/cplbu-2017-0034

Tjoa, A. M., & Tjoa, S. (2016, September). The role of ICT to achieve the UN Sustainable Development Goals (SDG). In *IFIP world information technology forum* (pp. 3–13). Springer. doi:10.1007/978-3-319-44447-5_1

Tsetsi, E., & Rains, S. A. (2017). Smartphone Internet access and use: Extending the digital divide and usage gap. *Mobile Media & Communication*, *5*(3), 239–255. doi:10.1177/2050157917708329

UN. (2021). *Gender equality*. Available at: www.un.org/en/global-issues/gender-equality

UNDP. (2015). *Human Development Report*. available at: https://hdr.undp.org/content/human-development-report-2015

UNICEF. (2021). *International day of the girl 2021 digital generation. Our generation*. Available at: www.unicef.org/gender-equality/international-day-girl-2021

ADDITIONAL READING

Antonio, A., & Tuffley, D. (2014). The gender digital divide in developing countries. *Future Internet*, *6*(4), 673–687. doi:10.3390/fi6040673

Gray, T. J., Gainous, J., & Wagner, K. M. (2017). Gender and the digital divide in Latin America. *Social Science Quarterly*, *98*(1), 326–340. doi:10.1111squ.12270

Huyer, S., & Sikoska, T. (2003). *Overcoming the gender digital divide: understanding ICTs and their potential for the empowerment of women*. INSTRAW.

Mardikyan, S., Yildiz, E. A., Ordu, M. D., & Simsek, B. (2015). Examining the global digital divide: A cross-country analysis. *Communications of the IBIMA*, *2015*, 1–10. doi:10.5171/2015.592253

Parra, J., Pérez-Pons, M. E., & González, J. (2020, June). Study Based on the Incidence of the Index of Economy and Digital Society (DESI) in the GDP of the Eurozone Economies. In *International Symposium on Distributed Computing and Artificial Intelligence* (pp. 164-168). Springer.

Sharma, R. S., Samuel, E. M., & Ng, E. W. (2009). Beyond the digital divide: Policy analysis for knowledge societies. *Journal of Knowledge Management*, *13*(5), 373–386. doi:10.1108/13673270910988178

Van Dijk, J. (2020). *The digital divide*. John Wiley & Sons.

Van Dijk, J. A. (2012). The evolution of the digital divide-the digital divide turns to inequality of skills and usage. In *Digital enlightenment yearbook 2012* (pp. 57–75). IOS Press.

KEY TERMS AND DEFINITIONS

Connectivity: Capacity for the interconnection of platforms, systems, and applications.
Developed Country: A country that has a mature economy and high quality of life.
Developing Country: A country that is seeking to become more advanced economically and socially.
Human Capital: The skills, knowledge, and experience an individual possesses.
Macro Level: At a level that is large in scale.
Micro Level: At a level that is small in scale.
Variation Rate: The difference between two numbers divided by the first number.

Chapter 8
The Impact of Digitalization on Innovation:
Evidence From Eurozone Countries

Funda Hatice Sezgin
Istanbul University-Cerrahpasa, Turkey

ABSTRACT

Today, the relationship between science and technology is one of mutual interaction, with technology developing on the basis of science, and developing technology opening up new fields of science, which in turn play a driving role in new technologies and innovative activities. The decisive role of science and technology in the process of economic development increases day by day and it is seen that all countries are in an increasingly intense search for science and technology policy. In the rapidly globalizing world economy, countries' conscious efforts and policy-making searches are intensifying on scientific and technological advances, which are among the main factors determining the competitive advantage of nations. In this study, the impact of digitalization on innovation for Euro Area countries is determined using panel regression. As a result of the analysis, it was determined that digitalization indicators have a statistically significant positive relationship on innovation.

INTRODUCTION

Changing world conditions with globalization have caused countries to focus on various marketing, innovation and digitalization efforts to sustain their competitive position, while the current competitive race has reached a more challenging level with the development of technology.

The introduction of the Internet into our lives and its use in every field from communication to trade has abstracted the concept of borders from the concept of borders and transformed the restricted market areas into a global marketplace (Gerguri-Rashiti et al., 2017). This has led to increased diversity and alternatives for both suppliers and customers. On the other hand, competition conditions have become tougher and customer demands have become more variable and diverse. Under all these conditions, businesses found the solution in sustainable and innovative approaches in order to meet the rapidly changing

DOI: 10.4018/978-1-6684-6620-9.ch008

demands of customers and not to fall behind in the competitive race in the market by developing existing competencies and resources without losing their advantages (Giesen et al., 2010:18).

The concept of innovation should be considered not only as a physical or qualitative change, development or improvement of a product, but also as a means of process improvement (Rogers, 2008). The improvements we have made in the way we do business in order to facilitate business processes by establishing a new and different working system in the flow in which we carry out our business processes are examples of minimal innovation efforts carried out on an individual basis. On a larger scale, streamlining processes is not the only benefit (Chandrashekar & Bala Subrahmanya, 2017). In addition to this benefit; creating added value for the customer provides significant benefits for businesses in terms of efficiency, effectiveness, performance improvement and profitability.

Therefore, having an innovative approach will provide businesses with vision, competitive advantage and customer satisfaction and loyalty in the long term. In the Industry 4.0 era, digitalization and innovation resources are integrated, creating new needs for customers and new sources of inspiration and production for businesses (Huang et al., 2015). The concept of digitalization has added a virtual innovation dimension to the understanding of innovation as tangible revisions to products, services and processes (Loukis, 2017).

Today, in addition to the development and widespread use of the Internet, concepts such as cloud technologies, artificial intelligence, Internet of Things, augmented and virtual reality are emerging along with technological developments. This process has placed computers and automation even more at the heart of production processes, ushering in a new era in manufacturing, also called Industry 4.0. In recent years, a holistic transformation process has begun. This process is called digital transformation. Digital transformation, which is brought about by the changes resulting from the application of digital technologies in all areas of society, is actually the transformation of the business world in the context of renewing business strategy or digital strategy, models, processes, products, marketing approach and goals by adopting digital technologies. The transformation phase means that digital uses enable new kinds of innovation and creativity in a given field, rather than simply enhancing and supporting traditional methods. The aim of this study is to determine the impact of selected digital transformation indicators on innovation for 15 Eurozone countries using panel regression analysis.

DIGITAL TRANSFORMATION

Digitalization incorporates technological resources into business processes and facilitates business processes by making these data and records available to everyone using digitization tools (Galindo-Martín et al., 2019). Digitalization is effective not only in production processes but also in the planning of value-adding activities in new product design and marketing (Iqbal, 2018).

Today, although not all businesses are experiencing a digital transformation, digitalization efforts are being carried out in different dimensions in the business processes of most companies. With digitalization, businesses can offer more opportunities to their customers (Morabito,2016). Digitalization has enabled businesses to create different innovative business and process structures by integrating their innovative ideas into their existing business strategies, thus providing flexibility to both businesses and employees in terms of space, time, time and resource use (Vial, 2019).

Innovation in the digital sense refers to the improvement and development of existing products, processes or business models by using technological resources as a tool or target within or between businesses. It is also the adoption and implementation of these innovations by the entire organizational structure (Ghalandari, 2013). The digital innovation process leads to outputs such as new products, applications, services, customized customer experience (Nambisan, et al., 2017).

Digital transformation practices involve the large-scale transition of an organization's routine activities and processes to data-driven technology-based systems (Díaz-Chao et al., 2015). They are practices that effectively streamline processes that can support sustainable activities, increase efficiency and achieve growth, and provide value gains for the organization in terms of key stakeholders and customers (Denner et al., 2018). Yoo et al., (2012) define digital transformation as the deep and rapid transformation of business activities, processes, competencies and models in order to strategically take full advantage of the changes and opportunities brought about by digital technologies and their impact on society.

The increase in the use of computer technologies since the 1990s and the widespread use of the internet have led to significant social and economic changes. Today, with the developments and innovations in technology, it is seen that the use of newly acquired concepts such as artificial intelligence, cloud technologies, internet of things, augmented and virtual reality is becoming increasingly widespread (Eze et al., 2018). This process, which includes a holistic transformation initiated by the new production era called Industry 4.0, which has placed computers and automation at the very heart of production processes, is defined as digital transformation. It is noteworthy that the concepts of "information" and "informatics", which are widely used in the definition of the great transformation process, and the technologies used in different fields, which seem to be very different from each other in essence, are gathered around a common point (Brunswicker, 2015). The fact that the basic inputs and outputs of the production process have always been "information", as well as the coding of information in digital format, is stated as a feature of the new era (Nambisan et al., 2017).

In simple terms, digitalization involves the increasing use of digital technologies and their integration into the firm's products and operations. Today, digitalization is driving fundamental changes, especially in how firms create and capture value. In simple terms, digitalization can be defined as the increased generation, analysis and use of data to improve the internal efficiency of the firm on the one hand, and to grow by adding value to customers on the other (Björkdahl, 2020; Kretschmer and Khashabi, 2020). In the information age, digital transformation and the resulting business models have fundamentally changed consumers' expectations and behavior, putting pressure on traditional firms and radically altering the structure of many markets (Verhoef et al., 2021). Therefore, digital transformation, which started with the emergence of computer technologies and accelerated with the discovery of the internet, is a mandatory process to gain competitive advantage and meet customer needs. Companies that are aware of this transformation are implementing digital transformation at different levels. Mostly this process takes place in phases and according to a plan (Nylén and Holmström, 2015).

Digital transformation has three phases: digitization, digitalization and digital transformation. Digitization is the conversion of analogue information into a digital format (i.e. zeros and ones) that computers can process and transmit (Vial, 2019). Digitalization phase focuses on how information technologies can be used to change existing business processes. An example of digitalization could be the creation of new online or mobile communication channels that allow all customers to easily connect with firms (Pan et al., 2022). Digital transformation is the most recent and widespread phase. It signals a company-wide change that leads to the development of new business models for the business or industry. The digital transformation phase is the implementation of a new business approach to create value (Usai et al., 2021).

THE RELATIONSHIP BETWEEN DIGITALIZATION AND INNOVATION

While digitalization is an innovation that can change organizations for the better, it is not the only method of change. Innovation, which is often mentioned when talking about the Digital Revolution, is a broad concept that includes digitalization, and they feed and complement each other in many ways. Although innovation is a form of innovation, it should not be confused with invention. While invention is the creation of a product or service from scratch, innovation consists of innovation and improvement to an existing concept (Urbinati, 2018). While digitalization is more likely to fall under the category of innovation, the main element of digitalization, computers or electronic devices in general, have been invented.

Innovation is a multi-stage process in which organizations develop and transform ideas into new/improved products, services or processes in order to successfully advance, compete and differentiate in their markets (Fichman et al., 2014). Although innovation has been a process going on throughout history, it has been one of the innovations brought by Industry 3.0 as a strategy method. Innovation is a concept that can be defined in many different ways, but in essence it can be summed up in two words: Improved innovation (Chesbrough et al., 2006). Digital Innovation not only changes the outline of an organization, but also has an impact at the micro level. Therefore, a better understanding of the process can be achieved by adapting the previously identified types of innovation to digital innovation. In addition, Henderson and Clark (1990) outline innovation in four categories.

Radical Innovation: Radical innovation creates a new dominant design and therefore a new set of fundamental design concepts embodied in components that are interconnected in a new architecture. Since this category of innovation implies fundamental change, it can be said to be the category into which digital innovation also enters.

Incremental Innovation: It improves and extends a built-in design. Individual components are improved, but the core design concepts and the connections between them remain the same. This type of innovation is used for changes made after digital innovation is completed, such as augmented reality and automation improvements.

Architectural Innovation: Only innovations that improve the relationship between the elements that make up the outline of the institution fall into this category. Innovations such as the strengthening and renewal of the relationship between departments as a result of digital innovation and the need to constantly interact can be examined in this category.

Modular Innovation: This is a type of innovation where only the main elements that make up the institutions undergo change and improvement. Examples include the radical change of the Information Department as both an employee and a system with the effect of digitalization, and the importance that customer satisfaction brings to departments such as human resources and marketing. Including the pioneers of this new change in the management table mentioned in the digitalization stages, and changes in the executive board in this direction are examples of changes in the main lines of institutions.

Digitalization is in essence a type of innovation for all markets, but it is not limited to this. While digitization can be seen as a method of innovation, digitalization is more than that. Digitalization refers to a system that is self-sustaining and capable of innovating and renewing itself (Giudice & Peruta, 2016). However, the concept of digitalization also includes a process known as digital innovation. Digital innovation is the use of digital technology and applications to improve existing business processes and workforce productivity, enhance customer experience, and launch new products or business models (Parida et al., 2019). These digital technologies have the potential to boost more inclusive and sustain-

Table 1. Dataset description

Variables	Variables Description	Data Source
R&D	R&D Expenditures /GDP	Eurostat databases
CIEM	Use of computers and the internet by employees	Eurostat databases
SM	Social media use by type, internet advertising	Eurostat databases
IOGS	Individuals using the internet for ordering goods or services	Eurostat databases
ICP	Percentage of the ICT sector on GDP	Eurostat databases
IPA	Individuals using the internet for interaction with public authorities	Eurostat databases
IB	Individuals using the internet for internet banking	Eurostat databases
LIN	Level of internet access - households	Eurostat databases

able growth by stimulating innovation, generating efficiencies and improving services, to name a few of the key benefits expected from innovation.

DATA AND ECONOMETRIC METHODOLOGY

The main purpose of the study is to examine the relationship between digitalization and innovation with the help of panel regression analysis, through annual data for the 2010-2020 period in emerging economy countries. In the study, digitalization was measured using seven different indicators and innovation variables was represented by R&D expenditures/GDP. All the series are annual and obtained from Eurostat databases. Table 1 shows the descriptions.

The study sample includes 15 *Eurozone countries* (Germany, France, England, Ireland, Netherlands, Belgium, Austria, Italy, Denmark, Finland, Portugal, Poland, Hungary, Romania, Greece).

Before carrying out panel regression, we first tested the existence of cross-sectional dependence between variables employing Breusch-Pagan (1980) (Lagrange Multiplier-LM) and Pesaran (2004) (Cross-section Dependence-CD) and Pesaran, Ullah and Yamagata (2008) (Bias-Adjusted Cross Sectionally Dependence Lagrange Multiplier- CDLM) tests. And then, based on the results of the cross-sectional dependence tests, we performed second generation panel unit root tests of CIPS test. In addition, the homogeneity of delta tilde tests of Pesaran and Yamagata (2008), and the test results were presented in Table 2.

Panel data estimation methods are performed with pooled, fixed and random effects. Pre-tests are required to determine which method to choose (Baltagi, 2005). Chow and Breush-Pagan's (BP) tests were applied for the panel regression estimation process in the first step, and Ho was rejected. In the other step, the Fixed Effects Model (FEM) was decided by applying the Haussmann test. Also, when this model was analyzed, different solution algorithms were tried, and the "Cross section SUR algorithm", which yields the smallest total squared error value, was employed.

EMPIRICAL ANALYSIS

The cross-sectional dependency between the series was analyzed with the LM CD test developed by Pesaran (2004) and the LM adj. test whose deviation was corrected by Pesaran et al. (2008) and the test

Table 2. Cross-sectional dependence tests and homogeneity test results

Cross-sectional Dependency Test (H_0: There is no cross-sectional dependency)		
Test	**Test Statistics**	**p**
LM (Breusch and Pagan, 1980)	30.741	0.012
LM_{adj} (Pesaran et. al, 2008)	31.097	0.005
LM CD (Pesaran, 2004)	31.784	0.000
Homogeneity test (H_0: Slope coefficients are homogeneous)		
Test	**Test Statistics**	**p**
Delta_tilde	18.473	0.000
Delta_tilde_adj	19.102	0.000

results are presented in Table 2. Since the probability values of the test results were below 1% and 5%, null hypothesis (there is no cross-sectional dependency) was rejected and it was determined that there was a cross-sectional dependency between the series. In addition, the homogeneity of the co-integration coefficients was tested using the delta tilde and corrected delta tilde tests of Pesaran and Yamagata (2008), and the test results were presented in Table 2. Since the probability values of the test results were below 1% and 5%, null hypothesis (slope coefficients are homogeneous) was rejected and it was determined that the cointegration coefficients were heterogeneous.

In this study, second-generation unit root tests should be used as cross-sectional dependency is determined. The second-generation unit root test of CIPS was used in this study. The results of the CIPS test are given in Table 3.

In CIPS tests, the maximum lag length was taken as 1 and the optimal lag length was determined according to the Schwarz information criterion. It is seen that hypothesis zero is rejected at 1% and 5% significance levels. Unit root test results show that the series are not stationary at the level, in other words, the variables are stationary at level I(1).

Regression analysis was made with first-order differences. Chow and Breush-Pagan's (BP) tests were applied in the first step for the panel regression estimation process. For the Chow test, the H_0 hypothesis

Table 3. Results of CIPS panel unit root test

Variables	**Level**		**First Difference**	
	Constant	**Constant + Trend**	**Constant**	**Constant + Trend**
R&D	-1.214	-1.299	-8.483*	-8.913*
CIEM	-1.087	-1.145	-7.445*	-8.114*
SM	-1.164	-1.207	-6.129*	-6.910*
IOGS	-0.987	-1.103	-9.357*	-10.043*
ICP	-1.277	-1.314	-8.021*	-9.258*
IPA	-1.023	-1.118	-7.485*	-8.246*
IB	-0.894	-0.992	-5.338*	-6.128*
LIN	-1.314	-1.426	-9.612*	-10.167*

*indicated that it is significant at 5% significance level.

Tablo 4. Panel regression model selection test results

Test	Probability (p)	Decision
Chow (F test)	0.015	H_0 rejected
BP (X^2 test)	0.004	H_0 rejected

Table 5. Haussmann test result

Test Summary	Chi-Square Statistics	Chi-Square s.d.	Probability (p)
Cross-Section Random	894.267	2	0.000
Period Random	917.316	2	0.000
Cross-Section and Period Random	995.428	2	0.000

was the pooled regression, the H_1 hypothesis was the fixed effects model (FEM), and the H_0 hypothesis was pooled regression for the BP test, and the H_1 random-effects model (REM).

As shown in Table 4, above, the H_0 hypothesis was rejected for both tests. For this reason, it was necessary to make a choice between REM and FEM models. In this case, selection will be made for two models with the help of the Haussmann test. The hypotheses of this test are given in Table 5, and the test results are seen in Table 6. H_0: Random effect (REM), H_1: No random effect (FEM).

According to the results of the Haussmann test, the H_1 hypothesis was accepted, and the FEM model was decided. Also, when this model was analyzed, different solution algorithms were tried, and the "Cross section SUR algorithm", that yielded the smallest total squared error value, was used.

According to the FEM results, significant relationships were obtained on R&D. As seen in the regression outputs, the independent variables had 72.3% explanatory power for the R&D variable. CIEM has

Table 6. Panel regression estimation results

Dependent Variable: D(R&D)				
Method: Panel EGLS				
Sampling: 2010.Q2 - 2020.Q4				
Variables	**Coefficient**	**Std. Error**	**t-Statistics**	**Probability**
D(CIEM)	0.307635	0.036203	8.497555	0.0000*
D(SM)	0.026372	0.004450	5.926141	0.0000*
D(IOGS)	0.039390	0.007752	5.081069	0.0000*
D(ICP)	0.465820	0.039823	11.69739	0.0000*
D(IPA)	0.018115	0.008394	2.158154	0.0319*
D(IB)	0.135807	0.062613	2.168982	0.0313*
D(LIN)	0.142751	0.035157	4.060345	0.0001*
Constant	1.686573	0.074882	22.52321	0.0000*
R^2 = 0.723; **F_{Ist}** = 38.592; **F(p)** = 0.000; **DW** = 2.10; **Wooldridge (p)** = 0.126; **Greene Heteroskedasticity Test (p)** = 0.328				

*Significant variable at 0.05 level, "D" notation shows 1st level difference

a statistically significant positive effect on R&D (β=0.307; $p<0.01$). SM has a statistically significant positive effect on R&D (β= 0.026; $p<0.01$). IOGS has a statistically significant positive effect on PER (β= 0.039; $p<0.01$). ICP has a statistically significant positive effect on R&D (β= 0.465; $p<0.01$). IPA has a statistically significant positive effect on R&D (β= 0.018; $p<0.05$). IB has a statistically significant positive effect on R&D (β= 0.135; $p<0.05$). LIN has a statistically significant positive effect on R&D (β= 0.142; $p<0.01$). According to the coefficient sizes, the ICP, CIEM and LIN variable is most effective on the R&D variable.

To test the assumptions of the model, the Wooldridge (2002) autocorrelation test was applied, and it was determined that there were no autocorrelations. As a result of the Greene Heteroscedasticity Test, the presence of homogeneity was determined, and it was found that the model resulted in which the assumptions were provided were interpretable.

CONCLUSION

Technological developments have transformed every aspect of life in various ways throughout history. The fourth industrial revolution is also affecting all areas of social life. These recent effects are also referred to as digital transformation. Digitalization offers huge economic and social potential that will not only shape the present, but also determine our future. Digital transformation is an economic necessity. It can also be said that not only the last industrial revolution but all industrial revolutions (technological developments) were carried out with the expectation of gaining economic advantages. Expectations such as reducing production costs, increasing production, reducing unemployment, increasing export revenues, reducing the need for imports, and the desire of countries to expand their superiority and become more competitive have carried these developments to advanced processes. The dynamism and speed of digital transformation is dizzying. So much so that while Industry 4.0 discussions are still fresh for many countries, companies, institutions and individuals, ideas of Industry 5.0 and Society 5.0 are rising from various parts of the world.

The aim of this study is to determine the impact of selected digital transformation indicators on innovation for 15 Eurozone countries using panel regression analysis. According to the FEM results, significant positive relationships were obtained on R&D variable. According to the coefficient sizes, the ICP, CIEM and LIN variable is most effective on the R&D variable. Thus, it has been seen that digitalization has a positive effect on innovation.

Developing technologies are an important factor in the emergence of inventions and innovations. Technological innovations lead to increased profitability and economic growth. In the process of digital transformation, rapid communication and close contact between the innovation activities of different countries has facilitated the innovation process and the spread of new ideas. Rapid changes in technology have played a role in increasing integration in the world trade system through digital transformation. Therefore, interest in efficiency and technology has naturally increased. Today, every product, production method or service that is related to innovation has been and continues to be linked to science and technology. As a result, the innovation process has become increasingly linked to science and technology, and new ideas in science and technology are being generated with the help of the digitalization process. When innovation means transforming science and technology into an economic or social benefit, the concept of technological innovation has gained importance.

Digitized data will be more easily accessible, interpretable and shareable. As a result of digitalization making people's lives easier, countries need to develop new business models from this data. The biggest contribution of digitalization is increased productivity. Since having knowledge, interpreting it correctly, and using technological tools will trigger more productive work and creativity, it is recommended that countries develop policies to ensure that they do not lag behind the digital process. Innovation centers need to be expanded in order to transform innovative ideas that will gain momentum as a result of digitalization into products. Thus, R&D and innovation projects will be supported and accelerated. It is necessary to create and expand national and regional networks, facilitate access to information on innovation and new technologies, and create the necessary structure for the development of technology exchange and transfer systems. It is recommended to coordinate industrial policies with innovation and technology policies, and to establish a system to overcome the disadvantages of digital transformation and turn them into advantages. Finally, in developing countries with growth targets, it is necessary to provide the necessary infrastructure to increase the productivity of R&D activities, to inform individuals and institutions about patenting, and to increase government incentives for patents, utility models and trademarks.

REFERENCES

Björkdahl, J. (2020). Strategies for Digitalization in Manufacturing Firms. *California Management Review, 62*(4), 17–36. doi:10.1177/0008125620920349

Brunswicker, S., Bertino, E., & Matei, S. (2015). Big data for open digital innovation-a research roadmap. *Big Data Research, 2*(2), 53–58. doi:10.1016/j.bdr.2015.01.008

Chandrashekar, D., & Bala Subrahmanya, M. H. (2017). Absorptive capacity as a determinant of innovation in SMEs: A study of Bengaluru high-tech manufacturing cluster. *Small Enterprise Research, 24*(3), 290–315. doi:10.1080/13215906.2017.1396491

Chesbrough, H., Vanhaverbeke, W., & West, J. (2006). *Open Innovation: Researching a New Paradigm.* Oxford University Press.

Del Giudice, M., & Della Peruta, M. R. (2016). The impact of IT-based knowledge management systems on internal venturing and innovation: A structural equation modeling approach to corporate performance. *Journal of Knowledge Management, 20*(3), 484–498. doi:10.1108/JKM-07-2015-0257

Denner, M.-S., Püschel, L. C., & Maximilian, R. (2018). How to Exploit the Digitalization Potential of Business Processes. *Business & Information Systems Engineering, 60*(4), 331–349. doi:10.100712599-017-0509-x

Díaz-Chao, Á., Sainz-González, J., & Torrent-Sellens, J. (2015). ICT, Innovation, and Firm Productivity: New Evidence from Small Local Firms. *Journal of Business Research, 68*(7), 1439–1444. doi:10.1016/j.jbusres.2015.01.030

EUROSTAT. (2022). *Eurostat data.* Available at: https://ec.europa.eu/eurostat/metadata/

Eze, S. C., Chinedu-Eze, V. C., & Bello, A. O. (2018). Actors and emerging information, communications and technology (EICT) adoption: A study of UK small and medium services enterprises. *Cogent Business & Management*, *5*(1), 1–19. doi:10.1080/23311975.2018.1480188

Fichman, R. G., Dos Santos, B. L., & Zheng, Z. E. (2014). Digital innovation as a fundamental and powerful concept in the information systems curriculum. *Management Information Systems Quarterly*, *38*(2), 329–353. doi:10.25300/MISQ/2014/38.2.01

Galindo-Martín, M. Á., Castaño-Martínez, M. S., & Méndez-Picazo, M. T. (2019). Digital transformation, digital dividends and entrepreneurship: A quantitative analysis. *Journal of Business Research*, *101*, 522–527. doi:10.1016/j.jbusres.2018.12.014

Gerguri-Rashiti, S., Ramadani, V., Abazi-Alili, H., Dana, L., & Ratten, V. (2017). ICT, innovation and firm performance: The transition economies context. *Thunderbird International Business Review*, *59*(1), 93–102. doi:10.1002/tie.21772

Ghalandari, K. (2013). The Effect of Information and Communication Technology on Export Performance of Iranian Firms. *Research Journal of Applied Sciences, Engineering and Technology*, *5*(17), 4367–4372. doi:10.19026/rjaset.5.4431

Giesen, E., Riddleberger, E., Christner, R., & Bell, R. (2010). When and how to innovate your business model. *Strategy and Leadership*, *38*(4), 17–26. doi:10.1108/10878571011059700

Henderson, R., & Clark, K. (1990). Architectural Innovation: The Reconfiguration of Existing Product Technologies and the Failure of Established Firms. *Administrative Science Quarterly*, *35*(1), 9–30. doi:10.2307/2393549

Huang, K. F., Lin, K. H., Wu, L. Y., & Yu, P. H. (2015). Absorptive capacity and autonomous R&D climate roles in firm innovation. *Journal of Business Research*, *68*(1), 87–94. doi:10.1016/j.jbusres.2014.05.002

Iqbal, R., Doctor, F., More, B., Mahmud, S., & Yousuf, U. (2018). Big data analytics: Computational intelligence techniques and application areas. *Technological Forecasting and Social Change*, *153*(1), 1–16.

Kretschmer, T., & Khashabi, P. (2020). Digital Transformation and Organization Design:An Integrated Approach. *California Management Review*, *62*(4), 86–104. doi:10.1177/0008125620940296

Loukis, E., Kyriakou, N., Pazalos, K., & Popa, S. (2017). Inter-organizational innovation and cloud computing. *Electronic Commerce Research*, *17*(3), 379–401. doi:10.100710660-016-9239-2

Morabito, V. (2016). *The Future of Digital Business Innovation*. Springer International Publishing. doi:10.1007/978-3-319-26874-3

Nambisan, S., Lyytinen, K., Majchrzak, A., & Song, M. (2017). Digital innovation management: Reinventing innovation management research in a digital world. *Management Information Systems Quarterly*, *41*(1), 223–238. doi:10.25300/MISQ/2017/41:1.03

Nylén, D., & Holmström, J. (2015). Digital innovation strategy: A framework for diagnosing and improving digital product and service innovation. *Business Horizons*, *58*(1), 57–67. doi:10.1016/j.bushor.2014.09.001

Pan, W., Xie, T., Wang, Z., & Ma, L. (2022). Digital economy: An innovation driver for total factor productivity. *Journal of Business Research, 139*, 303–311. doi:10.1016/j.jbusres.2021.09.061

Parida, V., Sjödin, D. & Reim, W. (2019). *Reviewing literature on digitalization, business model innovation, and sustainable industry: Past achievements and future promises*. Academic Press.

Rogers, E. M. (2008). Diffusion of innovations. In D. Stacks & M. Salwen (Eds.), *An integrated approach to communication theory and research* (pp. 409–420). Routledge.

Urbinati, A., Chiaroni, D., Chiesa, V., & Frattini, F. (2018). The role of digital technologies in open innovation processes: An exploratory multiple case study analysis. *R & D Management, 50*(1), 136–160. doi:10.1111/radm.12313

Usai, A., Fiano, F., Petruzzelli, A. M., Paoloni, P., Briamonte, M. F., & Orlando, B. (2021). Unveiling the impact of the adoption of digital technologies on firms; innovation performance. *Journal of Business Research, 133*, 327–336. doi:10.1016/j.jbusres.2021.04.035

Verhoef, P. C., Broekhuizen, T., Bart, Y., Bhattacharya, A., Qi Dong, J., Fabian, N., & Haenlein, M. (2021). Digital transformation: A multidisciplinary reflection and research agenda. *Journal of Business Research, 122*(1), 889–901. doi:10.1016/j.jbusres.2019.09.022

Vial, G. (2019). Understanding digital transformation: A review and a research agenda. *The Journal of Strategic Information Systems, 28*(2), 118–144. doi:10.1016/j.jsis.2019.01.003

Yoo, Y., Boland, R. J. Jr, Lyytinen, K., & Majchrzak, A. (2012). Organizing for innovation in the digitized world. *Organization Science, 23*(5), 1398–1408. doi:10.1287/orsc.1120.0771

ADDITIONAL READING

Benitez, J., Chen, Y., Teo, T. S. H., & Ajamieh, A. (2018). Evolution of the impact of e-business technology on operational competence and firm profitability: A panel data investigation. *Information & Management, 55*(1), 120–130. doi:10.1016/j.im.2017.08.002

Kotarba, M. (2018). Digital transformation of business models. *Found. Manag., 10*(1), 123–142. doi:10.2478/fman-2018-0011

Lee, M. X., Lee, Y. C., & Chou, C. J. (2017). Essential implications of the digital transformation in industry 4.0. *Journal of Scientific and Industrial Research, 76*, 465–467.

Preindl, R., Nikolopoulos, K., & Litsiou, K. (2020). Transformation strategies for the supply chain: The impact of industry 4.0 and digital transformation. Supply Chain. *Forum Int. J., 21*(1), 26–34. doi:10.1080/16258312.2020.1716633

Vendrell-Herrero, F., Bustinza, O. F., Parry, G., & Georgantzis, N. (2017). Servitization, digitization and supply chain interdependency. *Industrial Marketing Management, 60*, 69–81. doi:10.1016/j.indmarman.2016.06.013

KEY TERMS AND DEFINITIONS

Digitalization: Digitalization is the name given to the process of transferring accessible information to digital media in a way that can be read by any computer.

Innovation: Innovation is the adaptation and application of new creative ideas or inventions to economic fields – in other words, it is the process of creative solutions presented for the solution of obvious needs.

Panel Data Analysis: Panel data analysis is a method that combines time series and cross-section analysis and tests suitable models.

Chapter 9
The Journey of the Tourism Industry From Digital Marketing to Metaverse Network

Betül Garda
https://orcid.org/0000-0002-2406-6448
Selcuk University Social Science Vocational School, Turkey

ABSTRACT

Metaverse virtual tourism market, which is frequently mentioned as a result of this change, will appear as an alternative tourism type that allows people living in different countries to travel to the same country or different countries in the virtual world. Consequently, it is possible that the Metaverse virtual tourism market will become widespread as an alternative niche market to traditional tourism in the near future. For this reason, it is anticipated that the conceptual framework presented by the department about digital tourism marketing and metaverse network will contribute to the sector and future academic studies.

INTRODUCTION

The development of science and technology, health and economic crisis environments that affect the world are changing the functioning of every sector in a fast and dynamic way. On the other hand, as people adapt to the developing technology and digital world, their needs and wants are also changing (Khmiadashvili, 2019). In the digital age, digital transformation has become a necessity for businesses in all industries and for all humanity.

In 1990's digital marketing which first came to the fore with banner advertising and web crawler (spider) has opened new channels for the marketing and sales of all kinds of products, with applications such as social media marketing (SMM) and search engine optimization (SEO) (Kingsnorth, 2017). Therefore, digital marketing offers the opportunity to gain a flexibility that surpasses the brand power by providing the opportunity to reach online or mobile target audiences in the tourism industry, which has a wide range of alternative products and where there is intense competition (Luck & Lancester, 2003). In this context, tourism businesses have also focused on increasing their profitability by making all their

DOI: 10.4018/978-1-6684-6620-9.ch009

marketing processes online. Virtual fairs, online meetings, online forums, online virtual museums, and online tour services are some of the developments that help tourism move into the digital world. Metaverse, which is the reflection of the physical world in the virtual universe, use of technologies such as reality (VR), augmented reality (AR), and mixed reality (MR). In this virtual universe, which will be created with the new consumption types and cultures that emerged as a result of this contain new opportunities within the tourism industry (Cilesiz & Aydın, 2022).

Social networking sites such as LinkedIn, Facebook, Twitter, Pinterest, Instagram, YouTube have become a part of our lives as popular environments used for business networking, targeted marketing, and entertainment, sharing information of users, and enabling them to interact with each other on global. The possibility of sharing personal information, photos, and posts, chatting, and commenting on these sites will create various alternatives for tourism businesses to quickly obtain information about their potential customers and create market niches. While digital platforms will enable us to appeal to large market audiences, they will also allow us to significantly reduce marketing costs. In addition, social media advertisements, where users share their experiences, are more reliable than mass media such as radio, television and newspapers, and appeal to the tourism industry (Batinić, 2015).

Tourism businesses allocate large budgets and conduct R&D studies in order to show a successful performance in the digital platforms they turn to for the main purposes of protecting their business identity and brand value, combating economic and social crises, expanding their market share etc. (Yazdanifard & Yee, 2014). Hotel reservations, flight and bus tickets, visuals, or videos for the region to visit, virtual guides, navigations and virtual reality applications in the digital environment will present the Metaverse virtual tourism market as an alternative tourism market in the near future. For example, a vacationer who wants to go on vacation will be able to contact the avatar sales representative of the vacation destination he wants to buy and make a hotel reservation and have the opportunity to wander around the hotel where he will spend his vacation. People who want to have a holiday will have the opportunity to reach any part of the country or hotel where they will spend their holidays, with their avatars, thanks to the hologram technology. Even tourism fairs, hotels, cruise ships, travel agencies and other tourism businesses will use the Metaverse virtual world to showcase their products. Those who want to travel will have the opportunity to visit two different cities by comparing and examining hotels, museums and historical sites using 3D images and hologram technology, without leaving their homes (Coban, 2021). In the digital era, where individualization and consumer orientation are dominant, potential tourists tend to alternative tourism types rather than mass tourism and give importance to the concept of sustainability. In addition, the desire of individuals to purchase tourism products, which are indicators of their own status and reflect their identities, has increased. Within the scope of all these, one of the biggest innovations in the tourism sector is product diversity. In this context, the metaverse network can be considered as a promising environment for tourism businesses to provide interactive communication to tourists, expand trust and enjoyable consumer experiences (Kozak, Evren, & Çakır, 2013)

In this context, it is possible that the lack of knowledge of tourism businesses on managing digital marketing channels and metaverse marketing may negatively affect the sustainable development of businesses. For this reason, it is foreseen that the department will show the way the sector and future academic studies on the use of digital and metaverse marketing.

CONCEPT OF DIGITAL MARKETING IN SUSTAINABLE TOURISM

Digital marketing is a system for increasing the benefits obtained by supporting traditional marketing tools with new methods that bring speed and flexibility developed as a result of technological developments. The Marketing Association (AMA) defines digital marketing as a marketing method conducted through electronic devices. Its most distinctive distinguishing features are its low cost, measurability, timeliness, and its ability to appeal to the entire market on a global scale (Murray & Waller, 2007). In addition, digital marketing is considered as commercial activities using internet-based platforms, unlike traditional media tools and methods, in order to promote and support the brand or company indirectly or directly (Altındal, 2013). Tourism businesses show a successful performance in digital platforms that they turn to for the main purposes such as protecting their business identity and brand value, combating economic and social crises, expanding their market share, developing new products and services, and increasing their profitability. According to Kotler and Keller (2006) the promotion aspect of the marketing mix provides a general framework for analyzing the social media content of tourism businesses and the impact of social media on their marketing efforts.

Since the main purpose of marketing communication is to reach the target audience and direct them to the product or service, one of the most important issues will be the cost. With the rapid development of the internet and the fact that it has become almost a necessity as a result of the COVID-19 pandemic, it has now become an indispensable element with the acceptance and use of very large masses. This widespread use has necessitated companies to use the innovations brought by the internet in order to reach and influence their target audiences. In addition, it is one of the important advantages of digital marketing that it can measure instantaneous consumer reactions, which cannot be analyzed in traditional media, and that communication can be mutual and that it can reach very large masses at a lower cost compared to traditional media. When we look at it from the point of view of touristic product consumers, we have been finding that specifically social platforms and therefore the Internet are used extensively for information resource and information sharing (Litvin, Goldsmith, & Pan, 2008).

In supply and demand markets, with the use of the internet in commercial activities, it has moved away from its traditional concrete structure and evolved into an abstract dimension and has taken on a virtual structure. This virtual structure has enabled businesses to reach any point with an internet connection in the desired time period, eliminating the need for concrete places such as stores and workplaces. When promotion is carried out with digital marketing tools, the creation of content by users who are considered potential customers, especially in social media, causes a transformation in marketing activities. For this reason, adapting the marketing mix to online environments within tourism businesses is a necessity (Chaffey & Smith, 2008).

In this context, in digital marketing, product information has replaced the product, face-to-face sales have been replaced by an electronic screen, and the market or store has been replaced by computers and other internet-connected devices. In this way, virtual markets have provided conveniences such as lower costs and instant access compared to traditional markets. Promotion mix elements created in line with the content created by the users in the predominantly preferred social media is a personalized marketing tool added to classical advertising, sales development, public relations, sponsorship, direct marketing, word-of-mouth marketing (Chaffey & Smith, 2008).

Social activity is to carry out daily conversations with potential customers, unlike professional information about the company's products and services. This includes all kinds of friendly conversations, from happy birthday wishes to good weekend wishes. In the social activity, which is carried out in a sincere and gentle language, it is envisaged to establish an emotional and social bond between users and businesses (Kozinets, 2009).

Developments in Internet technology have led to the emergence of new communication tools such as websites, blogs, and social networks, which can be classified as digital communication tools in parallel. The emergence of these new digital communication tools has caused existing communication tools such as radio, television, newspaper, and magazine to be named as traditional communication tools. It is the Internet that Internet users most prefer for actions such as choosing a destination for touristic travels, searching for hotels and reservations (Xie, Miao, Kuo, & Lee, 2011).

Benefiting from the advantages of digital marketing can only be possible by understanding and recognizing the target audience that is trying to influence. Messages about touristic pictures, videos and travel notes shared on social media play a decisive role in choosing destinations and places. The preferences of social media activists appear as key points that guide the marketing strategies of tourism businesses (Zhang, Fu, Cai, & Lu, 2014).

The development of the Internet, the widespread use of Internet-based devices and their accessibility have contributed to the development of digital media and have made undeniable contributions to people's access to and use of information. It has enabled people to communicate by removing physical distances from being an obstacle and creating a highly interactive environment. Digital media has created new opportunities and even reached dimensions that will affect people's lifestyles. With the emergence of social media, geographical and demographic barriers have become meaningless. Because social media allows people to connect and communicate and makes it easier for companies to innovate. Digital marketing relies more on consumer response than traditional marketing, the focus is on getting results. Traditional marketing, on the other hand, focuses on initiating customer interactions (Kotler, Kartajaya, & Setiawan, 2017).

User-based content means the collection of written text, audio and images created and uploaded by users consisting of individuals or organizations in the potential customer portfolio of tourism businesses on social media platforms and using them for promotion purposes. Online or virtual communities are technological platforms where interpersonal emotional bonds are developed and reinforced by user-based content creation. The groups created on social media can be considered as the best example of this. Social media is a convenient and simple method by which tourism businesses can communicate 24/7 not only with their customers, but also with their partners and suppliers, 24/7 (Boyd & Ellison, 2008).

Social Media Marketing (SMM) is a collective term that includes new media blogs, micro blogs, social media networks and marketing sites. It is the definition of marketing transformed by social media. SMM is a social structure that reflects a mix of different ICT tools, including podcasts, messaging applications or blogs, which the Internet has made globally popular for the realization of interactive social and economic structures with users (Markham, 2004).

The most popular microblog is Twitter. The social networking site is LinkedIn, Facebook, Instagram, and Plaxo. Popular sites for social marketing are YouTube, StumbleUpon, and Digg. In the tourism industry, all, or some of these can be used to maintain their existence and increase profits under the destructive competition conditions we are in. Social media describes an Internet increasingly influenced by intelligent web services that enable users to contribute to developing, rating, collaborating, and distributing Internet content and customizing Internet applications (Weinberg & Pehlivan, 2011).

Social media marketing (SMM) requires a clear understanding of the current strategies of tourism businesses. Because in order to effectively connect, communicate and build a community with social media, you need to communicate with your target audience and tell them about your business and products. For a properly oriented SMM, you must have a lively descriptive bio that describes your business and products in online chats. Platforms are environments where people can come together, share valuable information, establish relationships, and make business deals. Social media is a digital platform that conveys your message with words, videos, shares and replies to them (Kırçova, 2012).

In short, social media not only changes your business, but also complements it. The meaning of social media for a tourism business is exactly provide more benefits within the framework of a win-to-win strategy by establishing more frequent connections with more people and satisfying the wishes and needs of more customers.

USE OF OTHER DIGITAL MARKETING CHANNELS IN TOURISM

Search Engine Optimization (SEO) It is the position improvement process that enables a website or a page to rank higher in search engines' search results. A special algorithm is used, which searches for keywords entered in search engines in documents and scans the list of these documents, which is suitable for scanning only meaningful results for each query. By increasing the organic relevance of the website or web page with the search query made by the users with the search engine, it increases the success level of those websites and pages. Thus, both visibility increases, and visitor traffic accelerates. 80% of tourists use a laptop or smartphone when making a reservation and prefer to do all their research online. For example, Google, an SEO expert, has identified four "micro-moments" as key opportunities for the travel business to be discovered in search results relevant to the tourist journey. These are dream, planning, reservation, and experience (Nicole, 2021).

Search Engine Marketing (SEM) SEO is a type of internet marketing that aims to promote websites to gain high visibility in search engine results pages (SERP) through paid placements, contextual advertising, and paid engagements. SEM, which allows businesses to target consumers by placing ads on search engines, is a strategy to reach the target audience in a very effective way. It has constant traffic levels and tremendous return on investment. For example, Google-Ads, strategies for placing ads in search engines include the common use of websites, the sharing of banners and screens. When a specific keyword is entered in the search engine query, the most relevant paid ad will appear. The visitor clicks on that ad and book a flight. This is how Search Engine Marketing happens (Xiang, 2011).

Tourists worldwide are target customers of travel brands because 80% of tourists do online research before booking a flight for their tour. They choose a search engine for airlines to book flights. These travel companies rank higher in these search results that optimize their sites and generate many important inbound links (Khajawi, 2017).

Email Marketing is to make effective use of the internet ecosystem to promote the products of businesses via e-mail, to provide a wide range of users to be informed about the campaigns quickly, to increase the visibility of the e-mails and to enable access to the customers at minimum cost. The most important point to note here is that the e-mail marketing channel is used in accordance with the legal procedure in terms of advertising, promotion, and marketing. Travel email marketing is used to attract and retain new customers. An attempt is made to demonstrate the value of the business by providing tourists with details and updates about their travels. In addition, it can be tried to make a difference by sending promotional discounts and special offers that will help the travel and tourism business to stand out from the competitors (Jung, 2022).

Usually delivered in the form of email, **Remarketing** includes techniques such as sending a sales promotion to past customers to encourage future bookings. It also includes texting them on special occasions like birthdays, wedding anniversaries, and sharing memories on social media to re-shape with their friends and family. The purposes of using remarketing to strengthen tourism marketing programs can be listed; developing word-of-mouth strategies to attract more business, receiving repeat bookings by satisfying past customers, increasing brand visibility through popular online channels, and creating a loyal customer base (Calvino, 2021).

Mobile Marketing is the use of mobile media as a marketing communication tool. Today, the concept of mobile includes all portable and personalized wearable technologies, devices such as tablets, as well as mobile phones. In this context, mobile marketing differs by targeting highly contextual personification. Marketers with mobile marketing; They can present personalized notifications and advertisements that they design through mobile channels depending on the customer such as location, time, environment, movement (Inan, 2022). Consumers today are bombarded with messages, but because SMS offers travel and hospitality brands the opportunity to send personalized communications, it can provide a significant return on investment by improving the digital customer experience and increasing brand loyalty. Because there is perceived ease of use in mobile media. Users are ready to use this technology in their work or life/ease of service. In addition, users can feel happy and entertained while using the technology/service (Chang, 2017).

TECHNOLOGIES THAT ENABLE METAVERSE DEVELOPMENT

One-way communication in Web 1.0 has become two-way with Web 2.0, and customers and companies have had the chance to talk. Web 2.0, the current second generation of the World Wide Web, is a structure that moves users from a passive position that only receives information to an active position that can provide information, generate information, and share information. It is a much more interactive internet version compared to the previous version, Web 1.0.

The basis of Web 3.0, originally called the semantic web by the inventor of World Wide Web, Tim Berners-Lee, is a version of the Internet based on the open blockchain networks of the Internet, transparent, decentralized, and individual-oriented, with values such as privacy and ownership. In this system, also called the Internet of Things, users and machines will be able to interact with data. The decentralization of the system is an obstacle to its sale for data security and advertising purposes. In addition, it provides privacy assurance and creates resistance to possible censorship interventions.

Web 3.0 makes it possible for users to customize content and control the data they own. The risk of security vulnerabilities and system vulnerabilities in Web 2.0 has been minimized. Web 3.0, also known as the Decentralized Web, promises an area where users will control and manage their own areas of the internet instead of accessing their internet through services mediated by companies such as Google.

With the innovations offered by Web 3.0 technologies to consumers and companies, the concept of the intuitive web gains great meaning. The intuitive web makes it simple for consumers to find the word or product they are looking for by filtering the information they are looking for. Web 3.0 prioritizes personalization and behavioral experience. Digital marketing aims at a more immersive goal, not just to persuade consumers to buy a product (Calışır, 2018).

METAVERSE NETWORK AND TOURISM INDUSTRY

Although Metaverse is not fully functional as of now, it will be a new platform that covers a significant part of real life in the coming years, taking advantage of existing technologies such as augmented reality (AR), virtual reality (VR), mixed reality (MR), and games. In the context of digital marketing, Metaverse can be defined as a permanent 3D virtual space where users will be targeted with brand-related content and sales development tactics while spending time in the virtual environment. This three-dimensional virtual space: It connects the audience with tools such as virtual reality glasses and augmented reality glasses. It associates users with avatars (virtual identities) in the virtual universe. In order for the virtual universe to be perceived physically, human senses such as sight, hearing, touch and even smell are tried to be simulated through interface technologies. These virtual worlds accessed via Avatar allows her body to overcome its physical limitations as much as possible (Coban, 2021).

Social media platforms, where almost everyone uses at least one of them today, will be a good example for the Metaverse Concept to be embodied in the minds. Users can see tourism businesses through advertisements and sponsored content they encounter while interacting with other users by logging in to the relevant platforms. Social media, which was initially the areas where users only communicate, meet, share pictures, videos, or locations, has turned into a market full of stores where they can view touristic products, buy, and pay online (Büberkökü, 2022). Users will no longer just look at the products, they will also be able to enter 3D stores and try the touristic products themselves. While clicking the link on social media and going to the touristic product page, it will be possible to enter the store in the metaverse. Since tourism is a type of service that includes meeting both physiological and psychological needs of people, it is compatible with Metaverse at this point. Although there is no data on whether Metaverse can be used as a new tourism method, it is thought that the service will provide an advantage in terms of providing comparative benefits to consumers due to its features such as intangibility, simultaneous production, and consumption, and its features such as testing after purchase. In addition, with new industries, markets, and resources in the future new types of skills, occupations and certifications will emerge (Celikkol, 2022).

The main feature of web 3.0, which represents the next stage after Web 2.0, is that it does not have a center and therefore, unlike today, it will not need any intermediaries. Thanks to Web 3.0, also called the decentralized web, users will be able to use and manage their own internet areas. This concept of decentralization also covers the issue of finance. The fact that the Internet can be accessed without intermediaries and financial transactions can be carried out independently of banks or governments will provide users with a wide range of freedom. It enables users to communicate concretely with each other and with virtual objects through real-time interactions (Demir, 2022).

It will be useful to reveal the characteristics of the metaverse in order to answer the questions of how touristic products will be adapted to the Metaverse and what will be the advantages of promoting in the metaverse (Babur, 2022):

- **Multiple Technology:** Metaverse offers and connects various new technologies together. With augmented reality, it gives the feeling of really living in the virtual world. The metaverse will continue to exist and function, just like the real world, whether we exist or not. Therefore, even if the user logs out, the activities in the metaverse do not pause or end. It connects the real world and the virtual world with the economic system established with blockchain technology. Physical elements do not pose any problems in this universe. While a person who wants to go to the stadium for a football match within the scope of sports tourism cannot go because tickets are sold according to the capacity of the stadium, this is not a problem in the metaverse. The constraints of the real world are removed. An organization held in the Metaverse can host audiences of all sizes from all over the physical world, allowing participants to experience this event at the same time.

- **Sociability:** Metaverse features a new social form. It has a time zone synchronized with the real world. Something that exists in the real world exists simultaneously in the metaverse. The metaverse includes cultural, economic, and legal systems. It is related to reality but has its own characteristics. It is not a sandbox where time is spent with existing content, but a sandbox where users can create content and other users can benefit from it. It also provides the opportunity to pass from different virtual worlds to each other. They can even send the products they buy in the virtual world to the address they want in the real world. From this point of view, the metaverse both provides a transitivity between virtual worlds and connects with the physical world. Users can customize avatars with digital identities to reflect their personality or mood at any time. For example, they can customize their avatar by changing the color of their hair and increasing their height.
- **Space-Temporality:** This feature refers to there is a world of Metaverse parallel to the real world. An extraordinary experience by transcending users' space and time limits allows them to live. It is a functional universe that allows users to create, invest, own, sell something, eliminating the need for another intermediary. Metaverse is a mix of different platforms working together. For example, when it comes to video games, you can bring items from one game to another. Transition between universes is possible. Metaverse guarantees complete freedom without any restrictions. Users can freely enter and explore other sandboxes in the metaverse. This provides a wide range of action for brands.

Metaverse allows people to experience real-world activities in the virtual world and takes the current situation one step further. Social interaction changes dimensions, making communication even easier. This interaction is not only for individual purposes but can also be used by businesses. Thanks to Metaverse, virtual interactions become a more natural and real experience. People will have stepped into a large area with the metaverse, where they can express themselves differently, through avatars, a new self that is different from themselves. These great features of the metaverse cause various fields such as tourism, marketing, and education to recognize the metaverse as a new business model. For this reason, in the past the business world has expanded information technology from offline use to website-based online. Today, this usage is once again expanding from the website to the metadata store (Unal, 2022). See a metaverse tourist in Figure 1.

The applications that form the basis of the Metaverse universe are virtual reality and augmented reality. Augmented reality, for the purpose of adding digital information to the image of something viewed through a smartphone or a computer, it is defined as a reality that has emerged with the use of technology (Inan, 2022). Metaverse, a world beyond the physical universe created on the virtual platform, will bring about a great change in human relations. Aiming to rebuild the world, Metaverse will also go much further in transforming the existing relationship between people and the city.

While in the past the city was the symbol reflecting real-world experiences, with the advent of Metaverse, the real city is expected to remind of experiences from the virtual world. Today's social media presents the world in three dimensions in the minds of users and introduces Metaverse as a tool that can create more believable images than what actually exists. Recent developments in virtualization technol-

Figure 1. Metaverse tourist
(Unal, 2022)

ogy make it possible for tourists to experience the smart tourism city in the virtual world, and physical accessibility is not seen as a problem. The number of hotels that accept payments with cryptocurrencies is increasing. You can make choices by visiting the hotel with virtual reality and augmented reality possibilities (Hemmati, 2022).

Figure 2. 999 Ethereum is required to purchase Egypt's famous structure, the Pyramids and Necropolis of Giza
(Kaytmaz, 2022)

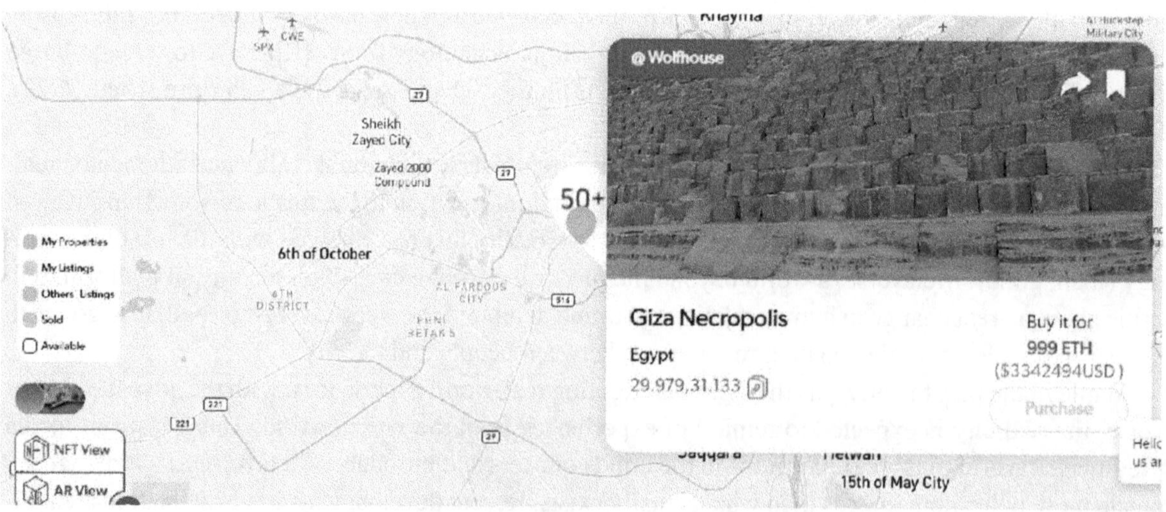

In augmented reality applications, a digital object is as if it were in the same environment with other physical world objects in the real-time physical world with the help of developing technology is transferred. For example, the virtual real estate company Metaverse Group in the USA invested in a virtual shopping store called SuperWorld. On the platform called SuperWorld, which invested 1 million dollars, many historical and important structures from Central Park to the Eiffel Tower, from the Great Wall of China to the Pyramids of Giza the domain can be purchased with cryptocurrency. As seen in Figure 2, when we look at the website of SuperWorld, the world map is seen digitally, and it also has an augmented reality feature (Kaytmaz, 2022).

THE EFFECTS OF METAVERSE TECHNOLOGIES ON THE TOURİSM SECTOR

Various technologies are used in Metaverse to provide a continuous and engaging experience delivery. These; Virtual Reality (VR), Augmented reality (AR) Mixed Reality (MR) Extended Reality (XR) technologies.

Virtual Reality (VR) Virtual reality is the imitation of real-world environment, image, hologram, sound, location through computers. The aim is to adapt to the virtual environment entered with the help of the necessary equipment and to move away from the real world. In virtual reality, the user experiences an artificial reality by isolating from the physical world. There are six main areas where VR benefits tourism: marketing; planning; sustainability and conservation; accessibility; education and entertainment. Many tourism researchers tend to focus marketing and sustainability from these areas. In terms of marketing, tourism companies and organizations perceive VR as a superior marketing tool. In fact, VR has been described as a technology that has revolutionized the way tourism products, services and experiences are promoted and sold. Tussyadiah et al. states that virtual reality offers potential tourists a "try before you buy" experience that allows them to virtually pre-experience a destination. This is beneficial as it can encourage potential tourists to physically travel to the actual destination (Verkerk, 2022).

In their study, Dodd and Johnson (2009) aimed to investigate whether virtual tourism provides a viable alternative to traditional tourism and to examine the relationship between the real universe and the virtual metaverse. As a result of the research, a significant number of those who have adopted virtual worlds, most of them under the age of 18. Thus, he concluded that the expansion potential of virtual tourism, which is a dynamic form of tourism, is possible. In addition, researchers have shown that most of the virtual world users will prefer virtual reality tourism as a complement to traditional tourism rather than a substitute.

In the studies of Marasco et al., the relationship between VR and its potential to increase destination competitiveness has been investigated. The study revealed that the visual appeal and emotional engagement of VR have a positive and significant impact on the behavior of tourists visiting a cultural heritage site. Marasco et al. reported that attractive visuals and emotional triggers in VR applications increase the likelihood of visiting cultural destinations (Babur, 2022).

In his study, Ashraf aimed to produce a virtual reality model of a Pharaoh tomb and to provide students in Architecture courses or tourists in general with the opportunity to try this virtual version. In the study, the walls, ceilings, floors, textures, drawings, and writings of the real tomb were adapted to the 3D digital model of the contents of the tomb. Thus, it became a place where everyone could walk around and touch any part of it where they could see its details, inscriptions, and writings. Via the head-mounted Oculus Quest 2 (a virtual reality headset developed by Facebook Reality Labs), viewers felt like they were in the original tomb, not a digital copy. It is envisaged that this application can be displayed in stereoscopic form on a large panoramic screen for group users and that these shows can also be used for tourism promotion by placing them in tourism offices abroad and visitor centers in touristic regions (Cilesiz & Aydın, 2022).

Augmented Reality (AR) The emergence of augmented reality as a research area as a result of technological developments has only been possible in the last two decades. Augmented reality (AR) is direct or indirect view of a new perception environment created by combining computer-generated elements such as audio, video, graphics, or GPS data, augmented and animated by sensory input, with the physical, real-world environment. In another definition, augmented reality is an exciting concept that allows us to feel and live the physical environment in the real world in a lively, dynamic, and real-time manner with computer-generated sensory inputs. Maybe it will take place in a large part of our life in the future. As the user of such a system moves in reality, virtual objects appear as if they actually exist (Inan, 2022).

With this technology, the user's reality and environmental experience can be improved by increasing the data obtained by people's senses of sight, hearing, smell, and taste, and even touch via a computer or mobile device. Augmented reality (AR) is a direct or indirect view of a new perception environment created by combining computer-generated elements such as audio, video, graphics, or GPS data, augmented and animated by sensory input, with the physical, real-world environment (Wei, Ren, & O'Neill, 2014).

Augmented reality is a kind of mixed reality in which the real environment is superimposed in a digital context. However, it is also defined as a digital extension of a person's vision and hearing senses. With augmented reality, the inputs that will appeal to the human senses and activate their feelings are modified and enriched by the computer, and the new reality that emerges is presented to the user's perception. Enrichment takes place in real time and interacts with surrounding elements. With Augmented Reality, the user can interact with the information and other elements that make up the reality environment. Artificial information and elements about the environment can be compatible with the real world. Augmented reality is the result of the interaction of the created sensory elements with the physical world by enriching them in real time. AR can also increase cultural tourism accessibility. Cultural sites welcome a more diverse target group with diverse interests. To remain attractive to visitors, new technologies are often the key to increasing visitor engagement (Yung & Khoo-Lattimore, 2019).

The first of the examples in the context of AR is the projects of Vlahakis et al. (2001) that enable users to reconstruct the ancient city of Olympia with augmented reality based on real-time image processing according to their location and direction. The project named ARKEOGUIDE (Augmented Reality Based Cultural Heritage Site Guide) is funded by the EU and aims to provide visitors of cultural sites with a personalized electronic guide and tour assistant. It includes a multimedia cultural material database for online access to cultural data, virtual visits, and restoration information. Multimodal user interfaces are used. It caters to professional and recreational users and applications ranging from archaeological research to education, multimedia broadcasting and cultural tourism.

Secondly, Papagiannakis et al. (2002) research is an augmented reality project called LIFEPLUS, which provides three-dimensional reconstruction of ancient fresco paintings through real-time animation of the fauna and flora of the ancient city of Pompeii and virtual animated characters. The aim of LIFEPLUS is to interactively explore the narrative design processes of visitors' imaginary spaces, pushing the limits of current Augmented Reality (AR) technologies. Although originally targeted at Heritage Centers, the paradigm is by no means limited to such topics, but encompasses all future Location Based Entertainment, E-visitor attractions and on set visualizations for the movie industry.

Figure 3. Reality-virtuality
Source: (Flavián, Ibáñez-Sánchez, & Orús, 2019)

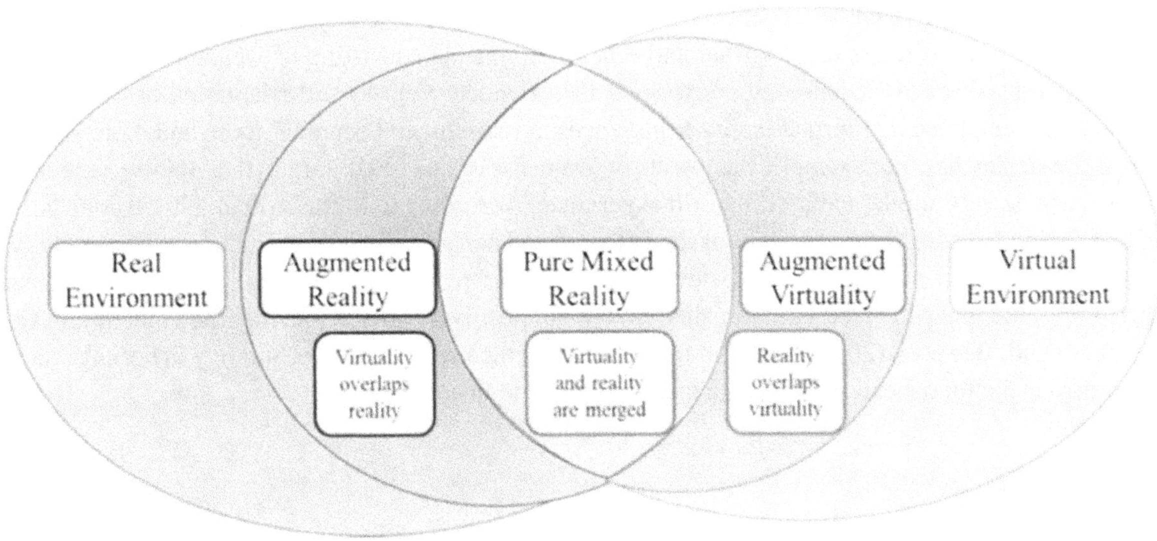

Mixed Reality (MR) Figure 3 shows the differences between virtual, augmented, and mixed realities. The term mixed reality was first defined by Paul Milgram and Fumio Kishino in 1994. Mixed reality is the merging of the physical and digital world under one roof by using virtual and augmented reality technologies together. In short it refers to the next stage of augmented reality (Carre, et al., 2022). So, it is a technology that allows virtual and real elements to interact in a single environment as in the physical world. While virtual reality is a computer-generated simulation of an environment, mixed reality; allows real and virtual worlds to coexist (Gaberli, 2019).

In augmented reality, user-directed computer-generated images are used in real time and the user is completely in the virtual environment. On the other hand, in mixed reality, evoked objects exist in a form that can interact with the real environment, such as a ball bouncing off the wall or new furniture rearranged in the room (Flavián, Ibáñez-Sánchez, & Orús, 2019).

Cultural tourism has been defined as an important economic and social contribution worldwide. The main drivers are linked to increased cultural awareness, the desire to create meaning and learn. More research in the field of VR, AR and MR development through purposeful design is highlighting types of cultural tourism. Hence, in recent years, many studies in the literature generally examine museum visits in terms of technology and tourism interaction. For example, the study, which proposes a conceptual model for user experience at the Ara Pacis Museum in Italy, concludes that the integration between museum heritage, physical components and mixed reality creates unforgettable and unique experiences. In the results, it was emphasized that the interaction of human and technology in the museum is important for the preservation of cultural heritage. In addition, human and technology interaction in the museum; It is among the findings that it is important for the protection of cultural heritage, virtual accessibility, and local development (Trunflo & Campana, 2020). Jung et al. (2016) examines the effects of technological innovations on the overall museum visitor experience in terms of the structural equation model and social presence and experience economy theory. According to the results of the research, it was concluded that social presence positively affects entertainment, education, aesthetics, and escape experience. It was also emphasized that visitor experience has a positive effect on the intention to revisit the tourist attraction.

Extended Reality (ER) It is a concept that includes all kinds of reality called Augmented, Virtual and Mixed Reality. It refers to all virtual and real environments consisting of wearable technologies. Technical support of 5G technology is expected for extended reality's full adaptation to life. In the literature, it is claimed that virtual reality experiences in tourism are becoming more and more similar to real life day by day. For example, the results of Wagler &Hanus (2018) argue that 360-degree video tourism can be very similar to the real-world experience. According to Marasco et al. (2018) concluded in their research that behavioral intentions to visit a cultural heritage site in a destination in Naples, Italy, virtual reality experience with wearable devices positively affects perceived visual appeal. Li and Chen (2019) revealed that perceived virtual reality enjoyment positively affects tourists' travel intention. On the other hand, Bec et al. (2019) argued in their study that the use of digital technology in heritage tourism provides the tourist experience and a greater interest in history.

CONCLUSION

Metaverse came to the fore in 1992 with Neal Stephenson's science fiction book Parasite and gained popularity in 2021 after Facebook founder Mark Zuckerberg announced that he would merge his companies under a company name called Meta. In fact, the concept of Metaverse has gained importance for the tourism sector based on simultaneous production and consumption, especially during the COVID-19 pandemic, when face-to-face communication became difficult. In this period, Metaverse has become a great marketing opportunity by combining with AR and VR, bringing tourists and touristic product manufacturers together in a 3D digital environment. Thus, a virtual universe is created over the internet. When the metaverse universe, which was originally developed for game/player communities, started to be supported by various investments, its importance increased in economic, commercial, and environmental areas. Therefore, the idea of visiting interesting places without traveling has become more attractive and increasingly possible. Virtual tours with VR, AR, MR and ER technologies can be expected to positively affect touristic experiences at the booking stage. In the future, thanks to virtual tours, it is possible that there will be no tourists who do not see the destination beforehand, and thus a certain awareness will be created in the tourists. For example, increasing interest in virtual tourism can raise awareness about problems that concern the whole world, such as recession in the global economy and global warming.

In short, future tourist experiences will be more conscious thanks to virtual realities. It can also be considered that virtual tourism may become a more attractive offer option for the disabled and families with children. Because the specific needs of disabilities and tourists with children in the visited destinations cannot be met, which turns the touristic trip into a disappointment.

In addition, the increasing variety of products and brands today necessitates using different and innovative applications to reach the target audience in order to gain competitive advantage. This emphasizes the importance not only of the quality of the product or message, but also of reaching the target audience. Beyond traditional tourism marketing, Metaverse strives to create brand experiences that are more engaging and engaging, as well as non-irritating, than those we currently experience in digital tourism. Since you naturally include yourself in the platform in the metaverse universe, it will be possible for you to immerse yourself in the experience in that universe in a natural process, perhaps willingly, and experience the brand's offer.

Although Metaverse Tourism opportunities are still in their infancy, global brands have already started to take their place in this brand-new marketplace with great potential. Metaverse's use of technologies that combine Virtual Reality (VR), Augmented Reality (AR) and Mixed Reality (MR) can be seen as a way to reach private tourists who value innovation.

REFERENCES

Altındal, M. (2013). Dijital Pazarlamada Marka Yönetimi ve Sosyal Medyanın Etkileri. Akademik Bilişim 2013- XV. Akademik Bilişim konferans Bildirileri 23-25 Ocak. Antalya.

Babur, Y. (2022). Metaverse Ağında Turizm Endüstrisinin Rolü. *Turizm Ekonomi ve İşletme Araştırmaları Dergisi*, *4*(1), 91–101.

Batinić, I. (2015). The Role and Importance of Internet Marketıng in Modern Hotel Industry. *Journal of Process Management- New Tecnologies International,* *3*(3), 34-38.

Bec, A., Moyle, B., Timms, K., Schaffer, V., Skavronskaya, L., & Little, C. (2019). Management of Immersive Heritage Tourism Experiences: A Conceptual Model. *Tourism Management*, *72*, 117–120. doi:10.1016/j.tourman.2018.10.033

Boyd, D., & Ellison, N. (2008). Social network sites: Definition, history, and scholarship. *Journal of Computer-Mediated Communication*, *13*(1), 210–230. doi:10.1111/j.1083-6101.2007.00393.x

Büberkökü, Ö. (2022). Financial Analysis Based on Metaverse Coins. *International Izmir Economics Congress*.

Calışır, E. Ç. (2018, January 9). *WEB 1.0-WEB 2.0-WEB 3.0-WEB 4.0 "NEDİR BU WEB?* Retrieved September 2022, from Esma Ç. Çalışır: https://esmacalisir.com/2018/01/09/web-1-0-web-2-0-web-3-0-web-4-0-nedir-bu-web/

Calvino, J. (2021, May). *The Guide to Remarketing for Tour & Activities -Part 1*. October 2022 tarihinde Blend: https://blend.travel/blog/the-guide-to-remarketing-for-tour-activities-part-1/adresinden alındı

Carre, A., Dubois, A., Partarakis, N., Zabulis, X., Patsiouras, N., Mantinaki, E., Zidianakis, E., Cadi, N., Baka, E., Thalmann, N. M., Makrygiannis, D., Glushkova, A., & Manitsaris, S. (2022). Mixed-Reality Demonstration and Training of Glassblowing. *Heritage*, *5*(1), 103–128. doi:10.3390/heritage5010006

Celikkol, Ş. (2022). Metaverse Dünyasının Tüketici Satın Alma Davranışları Açısından Değerlendirilmesi. *İstanbul Kent Üniversitesi İnsan ve Toplum Bilimleri Dergisi,* *3*(1), 64-75.

Chaffey, D., & Smith, P. (2008). eMarketing eXcellence: Planning and Optimizing Your Digital Marketing. Butterworth-Heinemann.

Chang, P. (2017). The Importance performance Analysis of Taiwan Tourism Mobile Marketing. *Journal of Tourism Management Research*, *4*(1), 12–16. doi:10.18488/journal.31.2017.41.12.16

Cilesiz, E., & Aydın, N. (2022). Metaverse and Tourism: A Conceptual Approach. *Journal of Academic Studies*, *3*(1), 32–44.

Coban, Ö. (2021). Turizmde Arttırılmış Gerçeklik. In Ş. Aydın & M. Boz (Eds.), *Turizmde Güncel Konu ve Eğilimler* (pp. 44–55). Detay.

Demir, Y. A. (2022, June 10). Retrieved August 2022, from Internetin Geleceği: Web 3.0 Nedir?: https://shiftdelete.net/internetin-gelecegi-web-3-0-nedir

Dodd, L., & Johnson, K. (2009). Virtual Tourism: Complementary or Contradictory to Traditional Tourism. *27th EuroCHRIE Annual Conference*, 672-679.

Flavián, C., Ibáñez-Sánchez, S., & Orús, C. (2019). The Impact of Virtual, Augmented and Mixed Reality Technologies on the Customer Experience. *Journal of Business Research*, *100*, 547–560. doi:10.1016/j.jbusres.2018.10.050

Gaberli, U. (2019). Tourism in Digital Age: An Explanation for the Impact of Virtual, Augmented and Mixed Reality Tecnologies on Tourist Experiences. *Journal of Tourism Intelligence and Smartness*, *2*(2), 61–69.

Hemmati, M. (2022). The Metaverse: An Urban Revolution Effect of the Metaverse on the Perceptions of Urban Audience. *Tourism of Culture*, *2*(7), 53–60.

Inan, Ü. S. E. (2022). Dijital Pazarlama Karması Bileşenleri ve Dijital Pazarlama Kanalları. In Ü. S. E. İnan (Ed.), *İşletme Bilimlerinde Dijital Dönüşüm* (pp. 1–18). Nobel Yayıncılık.

Inan, Ü. S. E. (2022). Use of Augmented Reality Technology in Marketing. In Y. Bayar (Ed.), *TechnologicalDevelopment and Impact onEconomic and Environmental Sustainability* (pp. 168–182). IGI Global. doi:10.4018/978-1-7998-9648-7.ch013

Jung,M.(2022,October21).*HowtoUseEmailMarketingtoPromoteYourTravel&TourismBusiness*.Retrieved October 2022, from Constant Contact: https://www.constantcontact.com/blog/email-marketing-travel-industry/#:~:text=Travel%20email%20marketing%20can%20be,busi ness%20stand%20out%20from%20others

Jung, T., Dieck, M. C., Lee, H., & Chung, N. (2016). Effects of Virtual Reality and Augmented Reality on Visitor Experiences in Museum. In A. Inversini & R. Schegg (Eds.), *Information and Communication Technologies in Tourism*. Springer International Publishing.

Kaytmaz, Y. C. (2022, January 13). *Metaverse'te Sultanahmet Camisini 0.1 Ethereum'a, Ayasofya Camisini 100 Ethereum'a Satıyorlar*. Retrieved August 2022, from Sözcü: https://www.sozcu.com.tr/2022/teknoloji/metaversete-sultanahmet-camisini-0-1-ethereuma-ayasofya-camisini-100-ethereuma-s atiyorlar-6885003/

Khajawi, A. (2017, December 24). *How Travel Brands Use SEM to Attract Potential Tourists?* Retrieved October 2022, from Brighton Gallery Blog: https://blogs.brighton.ac.uk/brightongallery/2017/12/24/how-travel-brands-use-search-engine-marketing-sem-to-attract-pot ential-tourists/

Khmiadashvili, L. (2019). Digital Marketing Strategy Based on Hotel Industry Study in Tbilisi. In *Proceedings of SOCIOINT 24-26 June 2019* (pp. 1032-1037). 6th International Conference on Education, Social Sciences and Humanities. 10.18769/ijasos.592628

Kingsnorth, S. (2017). Dijital Pazarlama Stratejisi. Nobel Yayınları.

Kırçova, İ. (2012). *İnternette Pazarlama*. Beta Yayınları.

Kotler, P., Kartajaya, H., & Setiawan, I. (2017). Pazarlama 4.0 Gelenekselden Dijitale Geçiş. Optimist Yayın Grubu.

Kotler, P., & Keller, K. L. (2006). *Marketing Management* (12th ed.). Prentice Hall.

Kozak, M. A., Evren, S., & Çakır, O. (2013). Tarihsel Süreç İçinde Turizm Paradigması. *Anatolia:Turizm Araştırmaları Dergisi, 24*(1), 7–22.

Kozinets, R. (2009). *Netnography: Doing Ethnographic Research Online*. Sage.

Li, T., & Chen, Y. (2019). Will Virtual Reality be a Double-edged Sword? Exploring the Moderation Effects of the Expected Enjoyment of a Destination on Travel Intention. *Journal of Destination Marketing & Management, 12*, 15–26. doi:10.1016/j.jdmm.2019.02.003

Litvin, S., Goldsmith, R., & Pan, B. (2008). Electronic word-of-mouth in hospitality and tourism management. *Tourism Management, 29*(3), 458–468. doi:10.1016/j.tourman.2007.05.011

Luck, D., & Lancester, G. (2003). E-CRM Customer Relationship Marketing in the Hotel Industry. *Managerial Auditing Journal, 18*(3), 213–231. doi:10.1108/02686900310469961

Marasco, A., Buonincontri, P., Van Niekerk, M., Orlowski, M., & Okumus, F. (2018). Exploring the Role of Next-Generation Virtual Technologies in Destination Marketing. *Journal of Destination Marketing & Management, 9*, 138–148. doi:10.1016/j.jdmm.2017.12.002

Markham, A. N. (2004). Internet communication as a tool for qualitative research. In D. Silverman (Ed.), *Qualitative Research: Theory, Method and Practice*. SAGE. doi:10.4135/9781848608191.d28

Murray, K., & Waller, R. (2007). Social networking goes abroad. *International Educator, 16*(3), 56–59.

Nicole. (2021, October 26). *The Growing Importance of SEO In Tourism Marketing*. Retrieved August 2022, from breaktheicemedia.com: https://breaktheicemedia.com/seo-in-tourism-marketing/

Papagiannakis, G., Ponder, M., Molet, T., Kshirsagar, S., Cordier, F., Thalmann, N., & Thalmann, D. (2002). LIFEPLUS: Revival of life in ancient Pompeii, Virtual Systems and Multimedia. *Proceedings of VSMM*, 1-11.

Trunflo, M., & Campana, S. (2020). A visitors' experience model for mixed reality in the museum. *Current Issues in Tourism, 23*(9), 1053–1058. doi:10.1080/13683500.2019.1586847

Unal, O. O. (2022, February 20). *Metaverse'te Tatil Deneyimi*. Retrieved September 2022, from YeniŞafak Teknoloji: https://www.yenisafak.com/teknoloji/metaversete-tatil-deneyimi-3755740

Verkerk, V.-A. (2022). Virtual Reality: A Simple Substitute or New Niche? In *Information and Communication Technologies in Tourism 2022*. Springer Nature Switzerland AG. doi:10.1007/978-3-030-94751-4_3

Vlahakis, V., Karigiannis, J., Tsotros, M., Gounaris, M., Almeida, L., Stricker, D., ... Ioannidis, N. (2001). Archeoguide: first results of an augmented reality, mobile computing system in cultural heritage sites. In *VAST November 28 - 30: Proceedings of the 2001 conference on Virtual reality, archeology, and cultural heritage* (pp. 131-140). Glyfada Greece: Association for Computing Machinery. 10.1145/584993.585015

Wagler, A., & Hanus, M. (2018). Comparing Virtual Reality Tourism to Real-Life Experience: Effects of Presence and Engagement on Attitude and Enjoyment. *Communication Research Reports*, *35*(5), 456–464. doi:10.1080/08824096.2018.1525350

Wei, S., Ren, G., & O'Neill, E. (2014). Haptic and audio displays for augmented reality tourism applications. *IEEE Haptics Symposium*, 485-488.

Weinberg, B. D., & Pehlivan, E. (2011). Social spending: Managing the social media mix. *Business Horizons*, *54*(3), 275–282. doi:10.1016/j.bushor.2011.01.008

Xiang, Z. A., & Pan, B. (2011). Travel queries on cities in the United States: Implications for search engine marketing for tourist destinations. *Tourism Management*, *32*(1), 88–97. doi:10.1016/j.tourman.2009.12.004

Xie, H. J., Miao, L., Kuo, P. J., & Lee, B. Y. (2011). Consumers' Responses to Ambivalent Online Hotel Reviews: The Role of Perceived Source Credibility and Predecisional Disposition. *International Journal of Hospitality Management*, *30*(1), 178–183. doi:10.1016/j.ijhm.2010.04.008

Yazdanifard, R., & Yee, L. T. (2014). Impact of Social Networking Sites on Hospitality and Tourism Industries. *Global Journal of Human-Social Science: E Economics, 14*(8), 1-5.

Yung, R., & Khoo-Lattimore, C. (2019). New realities: A systematic literature review on virtual reality and augmented reality in tourism research. *Current Issues in Tourism*, *22*(17), 2056–2081. doi:10.1080/13683500.2017.1417359

Zhang, H., Fu, X., Cai, L. A., & Lu, L. (2014). Destination Image and Tourist Loyalty: A Meta-Analysis. *Tourism Management*, *40*, 213–223.

Chapter 10
Examining Cryptocurrencies Within the Framework of Sustainability

Tolgahan Tuglu
Institute of Banking and Insurance, Turkey

Canan Dağıdır Çakan
Faculty of Financial Sciences, Turkey

Mehmet Hanifi Ateş
Institute of Banking and Insurance, Turkey

Aleyna Uca
Institute of Banking and Insurance, Turkey

ABSTRACT

Cryptocurrencies have been attracting a significant amount of attention in the world since they were first launched in 2009. Pretending to be a decentralized finance solution, it brought out a new era in technology called blockchain. Even though the benefits did not come into action in daily routines for many to be aware of, the market and its variety kept growing. On the other hand, there are also a lot of concerns and unpredictability about the future of this technology. Especially the high energy consumption while generating blocks for mining cryptocurrencies and completing transactions is commonly being criticised. In this study, blockchain technology and the basics of mining and validation procedures such as proof of work (PoW) and proof of stake (PoS) processes will be explained, and the environmental effects of bitcoin mining will be investigated. In the perspective of environmental sustainability of cryptocurrencies, the improvement in usage of renewable energy and its side benefits will be overviewed for a better prediction on the blockchain technology future.

DOI: 10.4018/978-1-6684-6620-9.ch010

INTRODUCTION

People used metals as a means of payment, coins and paper money, respectively, as a means of payment for the purchase of goods and services. In Prehistoric times, people used to mine precious metals for exchange purposes. In today's internet age, the concept of mining has gained a new meaning that Bitcoin's are started to be extracted from the internet, unlike the precious metals extracted from the earth. Bitcoin's are a digital currency that operates independently of central banks. Encryption algorithms are the formulas of the production of cryptocurrencies and exists as the key elements for verifying the transactions. To mine blockchain, a user needs to create new transactions and solve a complex mathematical puzzle. Thus, these transactions are verified by adding new blocks to the chain (Bitcoin Mining, Investopedia).

Bitcoin has been an important digital currency of the financial industry since it was conceived in 2008 by an author who names himself as Satoshi Nakamoto to "produce a medium of exchange that can be transferred electronically, securely, verifiably and immutably, independent of any central authority." Blockchain, which works on a decentralized public ledger, provides online transfers to be done directly between users without a need of a financial authority thanks to its pure peer to peer structure. (Nakamoto, 2008).

Generation of new bitcoins is limited with 21 million and this process is controlled by an algorithm. Until then, system is designed to a new bitcoin to be generated every hour. Since the peer - to - peer network limits the amount of Bitcoin in circulation thanks to the blockchain, the limited supply of Bitcoin instead of an unlimited amount of state-backed nominal currency makes it considered a more attractive asset (Cryptocurrency).

LITERATURE REVIEW

People have been using various currencies for thousands of years. Initially, currencies were produced directly from precious metals such as gold and silver, but later paper money was used. While cryptocurrency markets have grown in recent years, their sustainability continues to attract the attention of academic researchers because they require high energy consumption and create environmental concerns around the world.

Li et al. (2019), examined the electricity consumption of Monero mining and carbon emissions in China in the study in which they analyzed the power consumption and mining efficiency of 9 digital currencies. According to the results of the analysis, it was calculated that Monero had an electricity consumption of 30.34 GWh in China and 645.62 GWh at the global level in the April-December period of 2018, while the carbon dioxide emission in China was at the level of 19.12 thousand tons. According to the results, although blockchain technology and cryptocurrency mining are promising, it seems possible for it to reach alarming dimensions for energy consumption and sustainable development.

Stoll et al. (2019), calculated the carbon dioxide (CO_2) emissions of energy consumption by estimating the energy consumption of the Bitcoin blockchain. According to the findings presented, the electricity consumed was 45.8 TWh as of November 2018, leading to carbon dioxide (CO_2) emissions of around 22.0 to 22.9 MT. It is stated that the emission values obtained are equal to the amount of carbon dioxide (CO_2) emissions of Jordan and Sri Lanka.

Mora et al. (2018), reported that if its current technological use continues, bitcoin mining will cause a 2°C of global warming in next few ten years. While it is stated that miners will always chose the locations with cheaper energy, it is reported reducing Bitcoin's carbon footprint is only possible when the renewable energy is cheaper than traditional energy sources. In addition, they proposed modifying the system to reduce the difficulty and time needed to resolve the proof of work (PoW). They stated that if these measures are taken, Bitcoin's energy consumption will decrease.

Krause and Tolaymat (2018), reported that Bitcoin is one of the main cryptocurrencies responsible for carbon emissions of about 3 to 15 million tons from 2016 to 2018, while generating the same level of value compared to metal mining due to its energy-intensive nature. It is also stated that the energy needs of these currencies will keep increasing.

Examining Cryptocurrencies Within the Framework of Sustainability

Suazo (2021), claimed that it is more important to focus on the use of renewable energy to extract bitcoin from clean energy sources rather than calculating the energy consumption of bitcoin mining. Likewise, Corbet et al. (2021), have argued that traditional energy sources used in transaction processes, not digital currencies, should be blamed for carbon dioxide (CO_2) emissions. Therefore, they have argued that the environmental impact of crypto mining can be kept under control controlled through the shift to clean and renewable energy.

Polemis and Tsionas (2021) analyzed Bitcoin's impact on environmental quality using quantitative cointegrated vector autoregression (CQVAR) in their study. Using daily panel data from 50 economies over the period 2016-2018, they reported a causal relationship between Bitcoin and carbon dioxide (CO_2) emissions from increased energy use. They argued that there is a negative relationship between the incomes of Bitcoin miners and carbon emissions, and that emphasis should be placed on the need to use renewable energy and energy - efficient mining hardware to reduce the carbon footprint associated with Bitcoin.

Dogan et al. (2022), investigated the causal relationship between Bitcoin, clean energy and carbon emissions by applying Granger causality testing, which changes over time based on daily data from September 7, 2014 to October 12, 2021. The results from the analysis confirm that Bitcoin causes clean energy and carbon emissions. The study found that the use of renewable energy in Bitcoin mining is steadily increasing. As a result of the digitalization of the Bitcoin mining and energy sector through blockchain technology, it has been suggested that Bitcoin will reduce its carbon footprint as a result of increasing cost efficiency by bringing sellers and buyers together.

CRYPTOCURRENCY MARKETS

In this section, after giving information on methodology and data, authors present empirical results.

The first emergence of crypto coins basically coincides with the Global Financial Crisis in 2008, which resulted in a decrease in trust in many financial institutions, bankruptcy or loss. With this crisis, crypto coins have started to gain popularity with the article titled *''Bitcoin: A Peer-to-Peer Electronic Cash System''* published by unknown persons or persons under the name of Satoshi Nakamoto (2009). In this published article, reference is made to Bitcoin as a means of payment, which allows two parties to make electronic payments over a cryptocurrency without the need for a third party, and the classic element of trust with the ineffectiveness of a third party. The main factor that forms the infrastructure of this system is the blockchain.

Crypto assets are completely decentralized, unlike the centralized banking system. The decentralization of each cryptocurrency comes from a blockchain that functions as a distributed ledger. Thanks to their decentralized and limited supply compared to FIAT coins, crypto coins are more robust today in crisis and inflationary environments and protect their investors and users. One of the newest features that cryptocurrencies offer, unlike FIAT coins, is that they can be bought and spent by anyone in the world without the need for any bank or government.

The general characteristics of crypto coins can be briefly listed as follows:

- Allows interpersonal transfers without the need for third parties.
- Once the transactions are confirmed, they cannot be reversed and that transaction is deemed to have taken place.
- Compared to traditional FIAT money transfers, it happens much faster. Generally, the supply of cryptocurrencies is limited.
- Money transfers are secure thanks to strong encryption.

BLOCKCHAIN TECHNOLOGY AND CRYPTOCURRENCY MINING

In order to analyze the environmental impacts of cryptocurrency mining, we first need to understand the functioning of the blockchain system.

While FIAT currencies are issued by countries' central banks, bitcoins are created by individuals and this mining process keeps bringing new assets into the global digital market. (Martindale, 2018).

In order to send and receive bitcoin in the blockchain, using a bitcoin wallet is necessary. Traders can view their assets and make transactions by using this software application. Wallets have addresses and each of them has unique public and private keys. The name public comes from their visibility by everyone and people can send bitcoins to these accounts. Private keys can be thought as personal passwords and users will need these passwords to be able to spend Bitcoin's (unocoin.com, 2022). All Bitcoin transactions are secured by miners in groups of one block each.

Blocks in the blockchain are pieces of digitally stored data related to transactions, such as the date, time, and monetary amount. They also store the information of participants through registrations. Each block can be distinguished from other blocks by a unique code called a hash which is not possible to be copied. Miners have only 10 minutes to be able to legitimize the transaction by solving a challenging mathematical equation to seal a block to the blockchain, and at the end of this process they are rewarded with Bitcoin (bitcoinblockhalf.com, 2022; see Figure 1).

Many mechanisms are used to create new blocks on a blockchain. The most common mechanisms used to create blocks are PoW (Proof of Work) and PoS (Proof of Stake) (see Figure 2).

Examining Cryptocurrencies Within the Framework of Sustainability

Proof of Work (PoW) is a mechanism that prevents double spending, that is, the same funds from being spent on two different buyers at the same time. This mechanism is the original blockchain mechanism introduced by Satoshi in 2008 with his Bitcoin Whitepaper. At its most basic, PoW determines how to reach agreement between all distributed participants of a blockchain network without the need for third parties.

Figure 1. Shows the complete cycle of transaction
Source: https://www.investopedia.com/

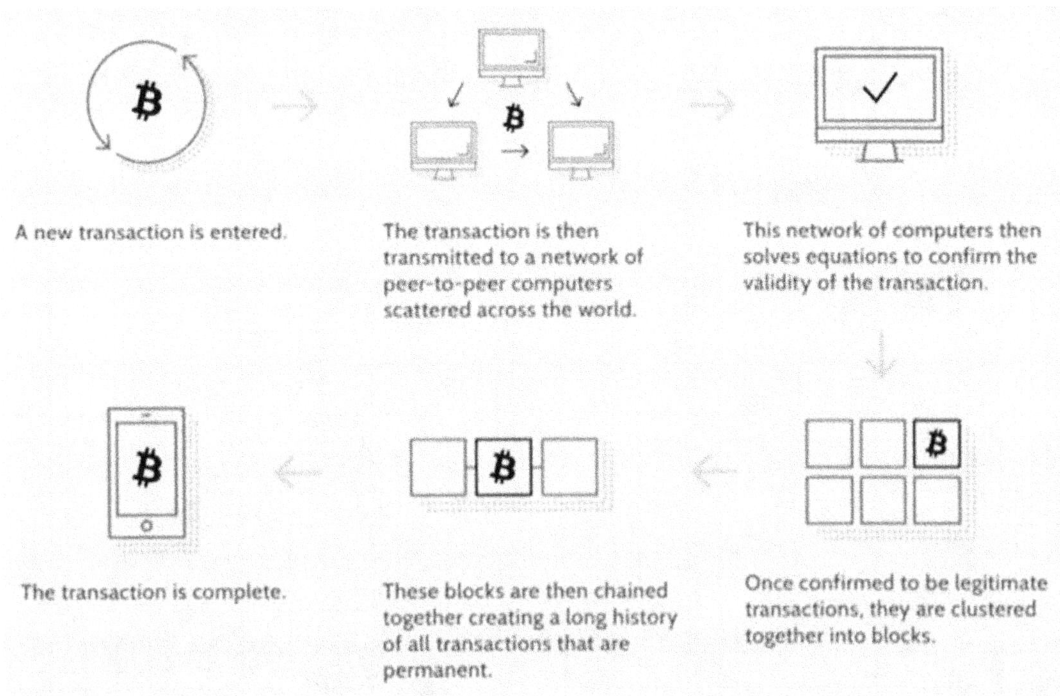

In PoS, that is, proof of shares, blocks are created instead of digging blocks. In this mechanism, pre-mined coins are usually sold or started with the PoW algorithm and then switched to the PoS algorithm. In PoW-based systems, new cryptocurrencies are created to reward miners, while in POS systems, transaction fees are distributed as a reward.

In the PoW network, transactions are verified by miners. They compete with miners' special mining equipment by solving complex puzzles to get the right to dig the next block. The first miner who finds the current solution broadcasts the transaction block to the blockchain and receives the block reward. But in PoS systems, there are validators instead of miners. There is no mining in this mechanism and there is no race to predict hashes. Instead, users are randomly selected and the selected user is asked to create a block. If the generated block is valid, the user receives a reward consisting of transaction fees paid for the transactions contained in the block.

On the other hand, the PoS mechanism has some advantages compared to the PoW mechanism. The most important advantage of the PoS system is that it has a smaller carbon footprint. The reason for this is that there is no need for high-power mining farms in the PoS mechanism and the amount of electricity consumed is much lower than the amount of electricity consumed in PoW.

Crypto mining, which refers to the production process of crypto coins, is important both for verifying the processes and for keeping records of those who do this work. Miners ensure that the blockchain system operates securely and that transactions are accurately recorded in distributed ledgers. Since blockchain technology is a public system, all users can control their transactions through this system.

Miners usually use computer equipment that allows them to mine Bitcoin efficiently. The hash rate is the primary feature in mining equipment. The number of calculations that hardware can perform every

Figure 2. Shows the difference between PoW and PoS
Source: https://leewayhertz.com/

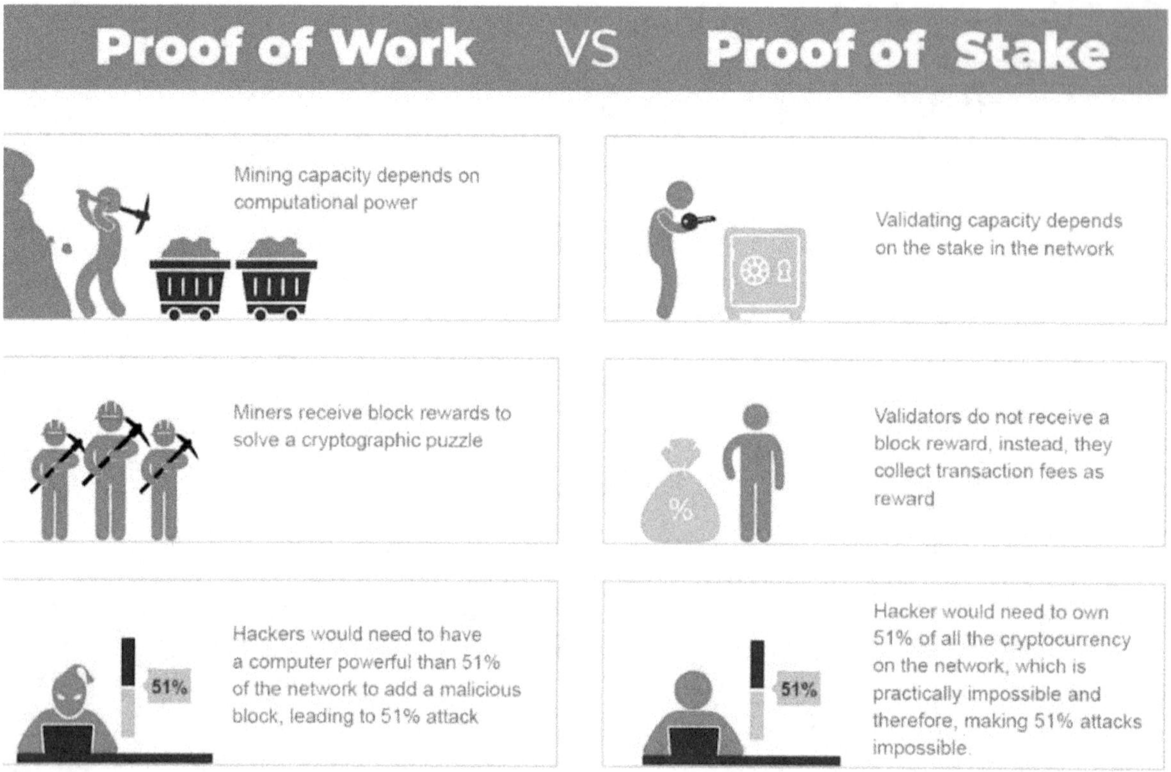

second is called the hash rate parameter. The higher the hash rate output, the morE-e likely a miner is to solve the calculation and therefore collect the reward. The higher the probability becomes. The higher the hash rate output, the more likely it is that a miner will solve the calculation and therefore collect the reward. Hash rates are measured in megahashes per second (MH/s). There are specially designed hardware for Bitcoin mining with hash speeds ranging from 336 MH/s to

Examining Cryptocurrencies Within the Framework of Sustainability

14,000,000 MH/s. The hardware used by Bitcoin miners has changed drastically over the years, requiring more strength to increases the likelihood of miners adding new blocks on the blockchain by solving the difficult calculation. Miners can only mine 1800 Bitcoins per day. So they have to have efficient technology to solve new blocks and compete with other miners.

Miners first used personal central processing units (CPU) to mine in 2009. After the CPU, all the things needed in high-end video games specially designed graphics processing units (GPU) are used to calculate complex polygons.

The field-programmable gate array (FPGA), that allows miners to purchase coins by bulk and customize them specifically for Bitcoin mining before adding them to their own equipment, lower the consumption of Bitcoin mining's technological devices from 400 watts to 80 watts in 2011, making power consumption more powerful. With the development of energy efficiency, Bitcoin mining farms began to

Figure 3. Distribution of bitcoin mining by country
Source: Cambridge Bitcoin Electricity Consumption Index (CBECI)

form, and FPGA was replaced in 2013 by a new technology called application-specific integrated circuit (ASIC), a microchip designed and manufactured solely to mine Bitcoins at top speed.

ASIC has achieved a 100-fold rise in hashpower while decreasing electricity consumption compared to all previous technologies. ASIC technology is the latest and highest level technology for Bitcoin mining software (cointelegraph.com, 2022; see Figure 3).

Blockchain Energy Consumption

Energy is generated from non-renewable sources or renewable sources. Non- sustainable resources include crude oil and crude oil yields, hydrocarbon gas liquids, natural gas, coal and nuclear energy. Examples of renewable sources include hydropower, biomass, biofuels, wind, geothermal, and solar power.

Electricity is a subsidiary source generated by many primary energy resources. Considering that the amount of energy consumed for Bitcoin mining is 31 TWh per year, the total annual energy use of the blockchain corresponds to more than the energy consumption of 150 countries (Gulli, 2020; see Figure 4).

Electricity consumption is measured in watts and the electricity consumed by a light bulb in an hour is measured as 15 watts. For larger consumptions, kilowatts are used, while megawatts are used to

Figure 4. Distribution of energy resources used in bitcoin mining

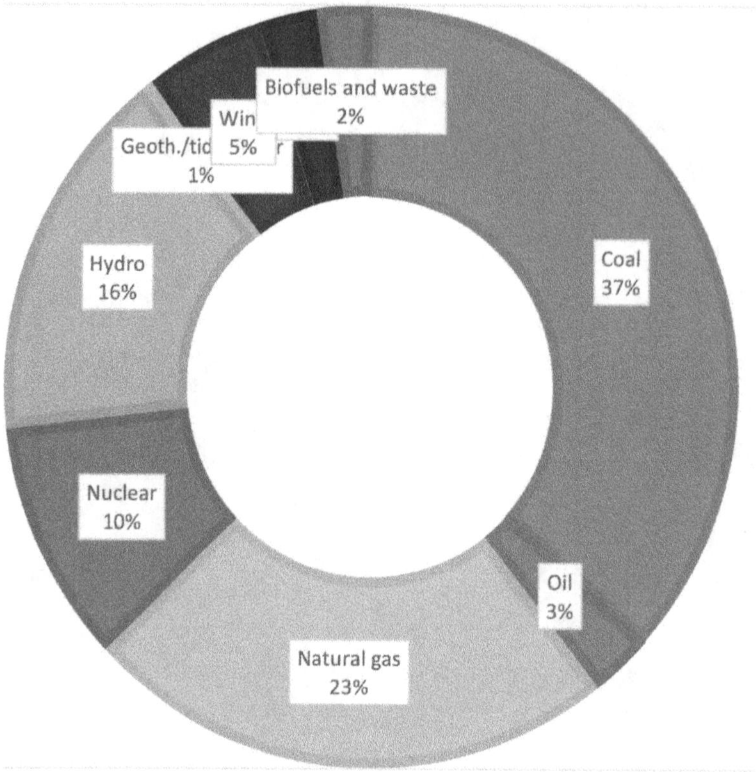

express the electricity consumption of power plants and cities, and gigawatts (GW), which is equal to 1,000 MW, are used to measure large power (IEA).

Terawatt-hours (TWh) are used to measure annual electricity production for countries and to describe large enough energy productions. One terawatt-hour is equal to one trillion watts for one hour, or 1000 GWh for one terawatt-hour. The United States consumed approximately 3930 TWh of electricity in 2021. (Statisca, Electricity end use in the U.S. 1975-2021). The United States generates electricity from the burning of coal or natural gas for 28% of greenhouse gas emissions. (Regulating Power Sector Carbon Emissions, C2ES). The power usage of Bitcoin mining is 30.14TWh per year, which is more than the electricity consumption of 19 European countries (Hern, 2017). Studies have suggested that Bitcoin mining could cross the 2ºC warming threshold set by the Paris Climate Agreement in 16-22 years twice as fast as the current global warming rate. (Berardelli, 2018).

The cost of cryptocurrency mining essentially consists of three factors. These are energy cost, maintenance cost (infrastructure cost and refrigeration facilities) and the cost of purchasing and replacing hardware (Clark & Greenley, 2019). The largest share of these costs is the cost of purchasing hardware and developing infrastructure with 45%, and the share of energy cost in total costs comes in second place with 42% (Blandin vd., 2020). Other cost items such as internet, maintenance, cabling remain at a lower level (De Vries, 2018).

Figure 5. Countries' annual electricity consumption and bitcoin (2022)
Source: Cambridge Bitcoin Electricity Consumption Index (CBECI)

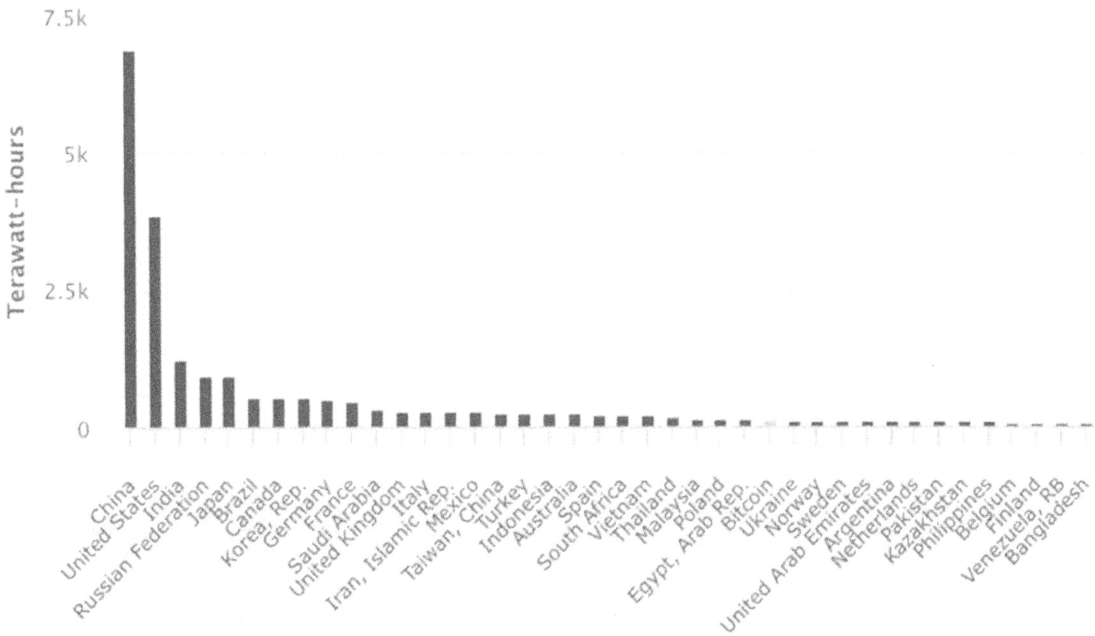

Examining Cryptocurrencies Within the Framework of Sustainability

Energy cost; the computational power of the hardware (with which device the production is made) is affected by various factors such as the hash rate, the degree of difficulty of proof of work, where the digging takes place and the amount of electricity consumed for the cooling systems (Küfeoğlu ve Özkuran, 2019).

In addition, the cost of electricity varies between 0.02-0.06 dollars / kWh depending on the countries. For example, Bitcoin is equivalent to an electricity expenditure of between \$2.000 and \$6.000 per country/region (Aydemir, 2020).

According to data published by the University of Cambridge, the amount of electrical energy used as a result of crypto mining reached 73.1 TWh at the beginning of 2019, 121.13 TWh at the beginning of 2021 and 191.11 TWh as of June 2022.

The amount of energy expended on Bitcoin mining is comparable to Thailand's power consumption (bitcoinenergyconsumption.com, 2022; see Figure 5).

According to data compiled by the University of Cambridge from the Cambridge Centre for Alternative Finance, the annual electricity consumption of Argentina, with a population of 45 million, is calculated as 121.36 TWh, which is more than the electricity consumption of the United Arab Emirates (113.2 TWh). Bitcoin, the cryptocurrency, consumes energy at the rate of %41 of Turkey's annual electricity consumption with 120 TWh of electricity consumed annually. In the ranking of electricity expenditure with countries,

Bitcoin is just ahead of Norway with 122.20 TWh and Ukraine with 133.20 TWh. According to the report, Bitcoin is among the top 30 countries that consume the most electricity in the world.

Table 1. Comparison of bitcoin electricity consumption with different countries and institutions

Name	Population	Annual Electricity Consumption (TWh)
China	1,443M	6,543
USA	330,2M	3,989
State of New York	19,3M	161
Bitcoin Network	-	**129**
Norway	5,4M	124
Google	-	12

Source: Cambridge Centre for Alternative Finance, Science Mag, New York ISO, Forbes, Facebook, Worldometer

The amount of energy used for the mining of the cryptocurrency in question is higher than the

Examining Cryptocurrencies Within the Framework of Sustainability

amount of energy consumed in Sweden as of today. Cryptocurrency miners have even begun to buy U.S. coal power plants.

One of the issues discussed in crypto mining is where and from which source is obtained from the energy used. According to a report published by the University of Cambridge in 2020, 76% of crypto miners use renewable energy sources. In the article published the most used energy source for mining is hydropower (see Figure 6 and Table 1).

Blockchain Carbon Emissions and Electronic Waste Amount

Global warming is based on carbon pollution, which is radically derived from burning of fossil fuels and lead to the change of the world's ecosystem negatively. Since the devices used for mining are unproductive and consuming great amounts of electricity, the dense nature of mining create an environmental distress. Because Bitcoin mining can potentially produce enough carbon dioxide emissions to push

Figure 6. Distribution of energy resources used for cryptocurrency mining
Source: University of Cambridge

Energy Source	Asia-Pacific	Europe	Latin America and the Caribbean	North America
Hydroelectric	65%	60%	67%	61%
Natural gas	38%	33%	17%	44%
Coal	65%	2%	0%	28%
Wind	23%	7%	0%	22%
Oil	12%	7%	33%	22%
Nuclear	12%	7%	0%	22%
Solar	12%	13%	17%	17%
Geothermal	8%	0%	0%	6%

global warming above 2 degrees Celsius in the next three decades, without innovation or government regulation, Bitcoin mining remains a dangerous carbon dioxide emitter that will negatively impact the world. (Mora, 2018).

It is estimated that by using the power used to process a Bitcoin, about 30 U.S. households were powered over the course of a day. This process also produces 434 kg of carbon dioxide, while 434 kg of carbon dioxide is equivalent to the amount of emissions that a Boeing 737 - 400 jet would produce in a five-hour flight. An analysis of 25,000 mining machines at the world's largest Bitcoin mining farm in Mongolia calculated that it emits between 8,000 and 13,000 kg of carbon dioxide per Bitcoin mined, which corresponds to 24,000 – 40,000 kg of carbon dioxide per hour (Bright, 2018).

In addition to this, as the energy used up in mining farms in China is generally coal, China dispose more carbon dioxide than the US and European Union combined. Bitcoin mining spends more energy than many smaller countries and consequently, it is a determiner in the improvement that is recorded in reducing global carbon emissions. (Rapier, 2018).

According to the study conducted by Izmir University of Economics in 2022, it was stated that the environmental effects were very large when Bitcoin was produced in Poland, where a large part of the electricity used for mining was produced from fossil jewelry. When we look at countries such as Denmark, where renewable resources are more intense, or France, where nuclear energy is used more predominantly, it is stated that Bitcoin's carbon footprint is 5 times that of gold, and in countries such as Poland it is 50 times higher.

As of June 2022, the amount of carbon emissions per single transaction in Bitcoin was 1158.77 kg of carbon dioxide. That's equivalent to 2,729,152 VISA transactions or 205,229 hours of YouTube video views. The electricity consumption per single transaction for Bitcoin generation is 2,077.54 kWh, which is equivalent to the 75.67 daily electricity consumption of an average U.S. household. The annual electricity consumed for Bitcoin mining is 191.11 TWh, which is comparable to Thailand's power consumption. On the other hand, Bitcoin's total annual carbon footprint is 106.58 Mt of carbon dioxide, which is equivalent to the carbon footprint of the Czech Republic.

When we look at the Ethereum Energy Consumption Indices, the amount of electrical energy per single transaction in June 2022 is 178.59 kWh and the amount of carbon emission is 99.61 kg of carbon dioxide. The electrical energy used for Ethereum mining corresponds to the average 6-day electricity consumption of a US household. The amount of carbon emissions is equivalent to 220,770 VISA transactions or 16,602 hours of video viewing on YouTube. The total Ethereum electrical energy used annually is 69.26 TWh. This amount is equivalent to the power consumption of the Czech Republic. The carbon footprint used on an annual basis is 38.63 Mt of carbon dioxide, which is comparable to Irlada's carbon footprint (digiconomist.net, 2022).

For Dogecoin, the amount of electrical energy per single process in June 2022 is 393.39 kWh and the amount of carbon emission is 219.42 kg of carbon dioxide. The electrical energy used for Dogecoin mining corresponds to the average 13-day electricity consumption of a US household. The amount of carbon emissions is equivalent to 486,311 VISA transactions or 16,602 hours of video viewing on YouTube. The total Dogecoin electricity used annually is 3.24 TWh. This amount is equivalent to Jamaica's power consumption. The carbon footprint used on an annual basis is 1.81 Mt of carbon dioxide, which is comparable to Guyana's carbon footprint.

The amount of electronic waste generated as a result of Bitcoin mining is 36.77 thousand tons per year, which is equivalent to the electronic waste production of the Netherlands. The amount of electronic

Figure 7. Electronic waste footprint
Source: http://digiconomist.net

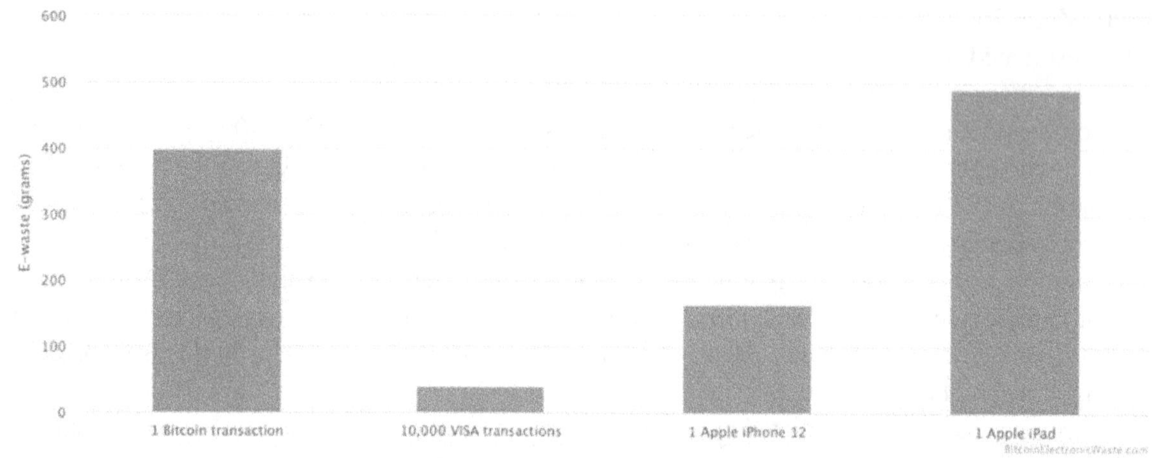

Figure 8. Bitcoin energy consumption (kt) (2017-2022)
Source: bitcoinelectronicwaste.com

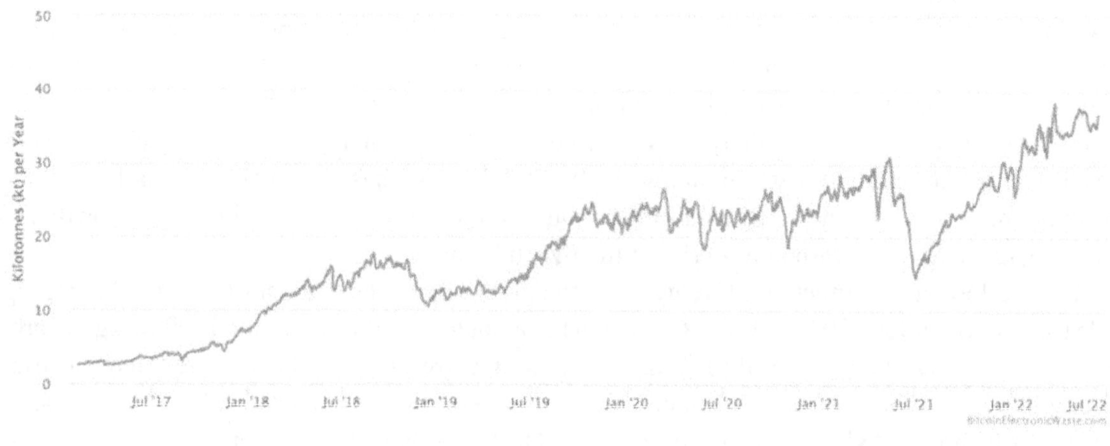

waste generated as a result of a single Bitcoin mining is 399.69 grams. This amount is equal to the weight of 2.44 Iphone 12 or 0.82 iPad (digiconomist.net, 2022; see Figure 7).

Examining Cryptocurrencies Within the Framework of Sustainability

As can be seen in Figure 8, the amount of electronic waste generated as a result of 1 Bitcoin mining process is approximately 400 grams, while the amount of electronic waste generated as a result of 10,000 VISA processes is approximately 45 grams.

As can be seen in Figure 7, while the amount of electronic waste in Bitcoin was approximately 0.65 kt at the beginning of 2017, there is an increase of about 15 times in the last 3 years in the amount of electronic waste, which triggers environmental pollution.

RENEWABLE AND CLEAN ENERGY

The energy consumption and carbon footprint caused by cryptocurrency mining is increasing day by day. If the energy required for cryptocurrency mining continues to be obtained from fossil fuels, the cryptocurrency system may not be sustainable. However, when the metaverse enters our lives, we will not be able to put crypto coins completely aside as our need for digital and global money will increase.

A study by Jiang (2021) estimates that the amount of energy consumption and carbon footprint used for cryptocurrency mining in 2024 is 296.59 TWh and 130.50 million metric tons for China, respectively. This amount of emissions is equal to the sum of the annual emissions of the Czech Republic and Qatar in 2016.

At this point, renewable energy sources rather than fossil fuels should be used to minimize the harmful environmental effects of crypto mining. In particular, crypto money mining is carried out in China mainly in regions where fossil fuels are concentrated. This increases the carbon footprint.

On the other hand, there are also cryptocurrencies that use different blockchains called "PoW (proof of business)", "PoS (proof of shares)" or "PoA (proof of ownership)" based on mining used by many cryptocurrencies such as Bitcoin and Ethereum. Proof of stake (PoS) is a system that takes digital asset ownership into account rather than a system based on computational power.

In this regard, the Ethereum Foundation, which is behind the Ethereum cryptocurrency, has stated that the energy cost of each transaction can be reduced by 99.95 percent by switching to Poof Stake, that is, the proof of share, method to verify transactions. If Etherium is transferred to the PoS system, people who hold a minimum of 32 Etherium in their wallet will be validators, that is, validators, will ensure that the transactions pass and will receive an annual reward of approximately 5%. There are basically 2 reasons why the Etherium cryptocurrency has switched to the PoS system. The first is that they want to give the appearance of a clean cryptocurrency by reducing energy consumption, and the other is that the supply of Etherium will decrease and its prices will increase because the majority of transaction fees will be burned.

The fact that Ethereum will switch from PoW to PoS protocol will have a very positive result in terms of environmental impact (Leising, 2021).

It is desired by many countries and sectors that any energy consumed as a result of cryptocurrency mining is completely carbon-free. In April 2021, the Energy Web Foundation, the Rocky Mountain Institute, and the Alliance for Innovative Regulations created the Crypto Climate Agreement, which is supported by organizations covering the climate, finance, NGO, and energy sectors. The goal of this deal is to decarbonize the industry in record time and achieve net-zero emissions in the global crypto industry by 2030.

ECONOMETRIC ANALYSIS

Purpose

In this study, it is investigated whether there is a casualty from the energy consumed in bitcoin mining in the world to the amount of carbon dioxide in the United States and it is also investigated if there is a causal connection among the energy used for Bitcoin mining activities and the amounts of electricity produced and consumed from renewable energy sources in the United States. By this investigation It is

Table 2. Dataset

VARIABLES	SYMBOL	SOURCE
BItcoIn hash RATE	LnHashRate	CBECI Cambridge*
RENEWABLE ELECTRIC ENERGY PRODUCTION (USA)	LnTotalProd	U.S. Energy Information Administration
TOTAL CONSUMPTION OF ELECTRICTY PRODUCED BY SOLAR POWER (USA)	LnSolarCon	U.S. Energy Information Administration
TOTAL CONSUMPTION OF HYDRO ELECTRICITY (USA)	LnHydroCon	U.S. Energy Information Administration
TOTAL ATMOSPHERIC CARBONDIOXIDE	LnAtmCo2	National oceanic and Atmospheric Administration

* Cambridge Bitcoin Electricity Consumption Index

aimed to see if there is a tendency to meet the electricity need for mining from renewable energy sources and if so, its affects to the production capacity of renewable energy sources.

In practice, Bitcoin hash rates, as the highest hash rate among crypto assets and the highest electricity consumption in mining are based for the analyze. (Hawson, Vries, 2022). The reasons for this preference are that it is mined through a proof-of-work (PoW) mechanism with the highest market share, and the highest total hash rates with Ethereum, which announced that it will switch to the PoS model in the near future.

Data and Methodology

In 2019, restrictions on crypto mining began to come in the People's Republic of China, where a large part of Bitcoin mining is carried out, and some mining pools have moved to continue their activities in the Americas (Yirka, 2022). So much so that the share of Bitcoin monthly hash rate,

Examining Cryptocurrencies Within the Framework of Sustainability

which was 4.06% in September 2019 in the US, increased to 31.57% by September 2021. In the People's Republic of China, the share of the hash rates on the same dates has decreased from 75.53% to 22.29% (CBECI, Cambridge).

Therefore, in order to examine the effects of the mining migration, the date range of February 2018 – January 2022, which covers this period, was selected and the data were used with monthly frequencies (see Table 2.

In the study, the energy needed for Bitcoin mining is represented by the bitcoin hash rate variable. Other variables were determined as the Total Renewable Energy Production of the USA, the total consumption of electricity from solar energy in the USA, the total consumption of electricity from hydroelectric power plants in the USA, and the amount of carbon dioxide in the atmosphere. All series were analyzed by taking their natural logarithms. The abbreviations and sources of the data used in the study are given in Table 3.

Analysis Method and Results

The Toda-Yamamoto (1995) Causality Test was applied to determine the presence and absence of causality relationships between the variables.

In this test, which is based on the VAR (Vector Autoregression) Model, unlike the granger causality test, it is not required for the series to be integrated at the same degree, but the maximum integration degrees (dmax) of the series should not exceed the optimal delay value (k) (Şadan, Mustafa, & Oytun, 2017).

The tests, predictions and models in this study were performed on the EViews 12 Package Program. The estimated VAR model is expressed by the following equation.

$$Y_t = \varpi + \sum_{i=1}^{k} \alpha_{1i} X_{t-i} + \sum_{i=1}^{k} \beta_{1i} Y_{t-i} + \sum_{j=m+1}^{d\,max} \delta_{1i} X_{t-i} + \sum_{j=m+1}^{d\,max} \theta_{1i} Y_{t-i} + \varepsilon_{1t}$$

$$X_t = \partial + \sum_{i=1}^{k} \alpha_{2i} X_{t-i} + \sum_{i=1}^{k} \beta_{2i} Y_{t-i} + \sum_{j=m+1}^{d\,max} \delta_{2i} X_{t-i} + \sum_{j=m+1}^{d\,max} \theta_{2i} Y_{t-i} + \varepsilon_{2t}$$

Examining Cryptocurrencies Within the Framework of Sustainability

In the Toda-Yamamoto Approach, the VAR model is estimated from the degree (k+dmax) for the Granger causality test and the Wald Test is applied to the first k of the coefficient matrix created (Yıldız, Çakan, Altınışık, 2022). In the applied Wald test, $\beta 1i = 0$ is the null hypothesis stating that there is no causality from the X variable to the Y variable, while the null hypothesis stating that there is no causality from the Y variable to the X variable is formed as $\beta 2i = 0$.

In the study, firstly, the stationarity levels of the time series were examined with Augmented Dickey-Fuller (ADF) and Philips Perron unit root tests, then it was seen that the series showed normal distribution from the descriptive statistics and it was also determined that they did not have autocorrelation and varying variance problems with Lagarge Multiplier (LM) and Heteroskedasticity tests. In Table 4 and Table 5, it is observed that all of the variables are stationary at the level or at the first difference.

The tests, predictions and models in this study were performed on the EViews 12 Package Program. The estimated VAR model is expressed by the following equation.

In the Toda Yamamoto approach, the maximum number of delays to be used in the estimation of the Vector Auto Regression model is calculated as k+dmax. As a result of unit root tests, Dmax=1 result was obtained and in the lag length criteria controls performed to determine the k value, the appropriate lag number was calculated as 2 for all information criteria (LR, AIC, SIC, FPE). Therefore, our estimation model is designed to be k+Dmax=3 delayed estimation of vector autoregression. Finally, the Wald test was applied to test whether there is a causal relationship between the variables and the chi-square values were obtained. For the interpretation of these chi-square values, the probability values were recalculated according to the 2-lagged equation and the hypotheses were interpreted according to this value as shown in Table 6.

Table 3. ADF unit root test results

Augmented Dickey-Fuller (ADF) Unit Root Test	T Statistic Value	Probability (%5 Meaning Level)	Result
LnHashRate (Stationary at the first difference)	-7.2881	0.0000	H_0 Rejected, There is no unit root
LnTotalProd (Stationary at the level)	-3.9562	0.0172	H_0 Rejected, There is no unit root
LnSolarCon (Stationary at the first difference)	-9.1138	0.0000	H_0 Rejected, There is no unit root
LnHydroCon (Stationary at the level)	-5.6470	0.0002	H_0 Rejected, There is no unit root
LnAtmCo2 (Stationary at the level)	-6.6794	0.0000	H_0 Rejected, There is no unit root

Table 4. PP unit root test results

Philips Perron Birim Unit Root Test	T Statistic Value	Probability (%5 Meaning Level)	Result
LnHashRate (Stationary at the first difference)	-9.4530	0.0000	H_0 Rejected, There is no unit root
LnTotalProd (Stationary at the level)	-4.0963	0.0120	H_0 Rejected, There is no unit root
LnSolarCon (Stationary at the first difference)	-3.5760	0.0101	H_0 Rejected, There is no unit root
LnHydroCon (Stationary at the first difference)	-6.0854	0.0000	H_0 Rejected, There is no unit root
LnAtmCo2 (Stationary at the first difference)	-3.2252	0.0248	H_0 Rejected, There is no unit root

Tablo 5. Causality test results

Causality Direction	x^2 Prob. Value	Level*	Result**
LnHashRate > LnAtmCo2	0.000218	0.05	H_0 Rejected
LnHashRate > LnTotalProd	0.02543	0.05	H_0 Rejected
LnHashrate > LnHydroCon	0.0327	0.05	H_0 Rejected
LnHashRate > LnSolarCon	0.0829	0.05	H_0 Accepted
LnTotalProd > LnHashRate	0.4702	0.05	H_0 Accepted

* %5 Meaning Level

** Ho is rejected if the Probability Value is less than 5% significance level..

Examining Cryptocurrencies Within the Framework of Sustainability

In Table 4 and Table 5, it is observed that all of the variables are stationary at the level or at the first difference.

According to the causality test results in Table 6; There is a causal relationship from Bitcoin Hash Rate to the amount of Carbon Dioxide in the Atmosphere.

There is a causal relationship from the Bitcoin Hash Rate to the total amount of electricity produced from US renewable power plants, but the total renewable electricity generation is not the granger cause of the bitcoin hash rate. Therefore, there is one-way causality.

There is a causal relationship from Bitcoin Hash Rate to total hydropower consumption in the USA. There is no causal relationship from the Bitcoin Hash Rate to the total consumption of solar electricity in the USA.

CONCLUSION

Cryptocurrencies and blockchain technology are often criticized for the environmental impact of bitcoin mining. The high amount of energy spent for Bitcoin mining, the huge greenhouse gas emissions resulting from the use of energy obtained from fossil fuel sources, and the electronic waste problem created by the equipment used for crypto mining losing its function, has been the subject of many academic studies and researches around the world. In addition, the transition to renewable energy sources and reducing the carbon footprint have a very important place in the policies created and implemented with the aim of sustainability in recent years. In this direction, the harmony of bitcoin mining, whose environmental sustainability is so discussed, is also of great importance for the future of crypto assets. The study, on the one hand, examines the environmental threat posed by bitcoin mining, similar to other studies in the literature, on the other hand, it examines the transition to the use of renewable energy in crypto mining and its potential contribution to the proliferation of renewable energy production in the world.

The analyzes applied in the econometric model of the research showed that there is a causal relationship from the bitcoin hash rate to the total amount of electricity produced from renewable energy sources in the United States and also from the bitcoin hash rate to the total hydroelectric consumption in the United States. While there is no causality from Bitcoin hash rate to the total consumption of electricity obtained from solar energy, it has been determined that there is a causality from the hash rate to the amount of carbon dioxide in the atmosphere. In the light of these results, the conclusion reached is that bitcoin miners' use or non-use of renewable energy sources in the United States affects the total renewable energy production. Hydroelectric energy comes to the forefront as the renewable energy source preferred by Bitcoin miners. Besides, bitcoin mining is the Granger cause of the amount of carbon dioxide in the atmosphere.

The causality test results also show us that crypto mining has a trend towards green energy use. Addressing the underlying causes of this finding in different studies and examining it from an economic and regulatory perspective may provide benefits for the faster transition of crypto mining to green energy use. The development of strategies on the dissemination and promotion of renewable energy production and usage to reduce environmental impacts will also directly affect the future of crypto assets and sustainability criteria. In this regard, business models to be planned and implemented between renewable energy sources and crypto mining pools should be closely followed. The spread of these applications should be considered not only for the reduction of greenhouse gas emissions of crypto mining, but also for its possible impact on the increase in green energy production on a global scale.

REFERENCES

Aguiar, M., & Hurst, E. (2013). Deconstructing Life Cycle Expenditure. *Journal of Political Economy*, *121*(3), 437–492. doi:10.1086/670740

Aydemir, A. (2020). *Bitcoin Yeşil Bir Dünyanın Neresinde?* 12 Haziran 2022. https://www.paribu.com/blog/ perspektif/bitcoin-yesil-bir-dunyanin-neresinde-akar-aydemir/

Berardelli, J. (2022). *Bitcoin Emissions Could Propel Planet To Critical Climate Threshold.* https://www.cbsnews.com/news/bitcoin-emissions-could-propel-planet-to-criticalclimate-threshold-study/

Bitcoin Block Reward Halving Countdown. (2022). https://www.bitcoinblockhalf.com/

Bitcoin Miners vs Bitcoin Nodes. (2022). https://blog.unocoin.com/bitcoin-miners-vs-bitcoin-nodes-6a4d35be9712

Blandin, A., Pieters, G., Wu, Y., Eisermann, T., Dek, A., Taylor, S., & Njoki, D. (2020). *3rd Global Cryptoasset Benchmarking Study*. University of Cambridge, Judge Business School. doi:10.2139srn.3700822

Bloomberght. (2022). https://www.bloomberght.com/bitcoin-cin-den-gelen-aciklamalarla-sert-dustu-2280652

Çalışkan, Ş., Karabacak, M., & Meçik, O. (2017). Türkiye Ekonomisinde Eğitim Harcamaları ve Ekonomik Büyüme İlişkisi: Bootstrap Toda-Yamamoto Nedensellik Testi Yaklaşımı. *Kocaeli Üniversitesi Sosyal Bilimler Dergisi*, (33), 45-56.

Cambridge Center For Alternative Finance. (2022). https://ccaf.io/cbeci/mining_map

Clark, C. E., & Greenley, H. L. (2019). *Bitcoin, blockchain, and the energy sector.* Congressional Research Service.

Coal Is Fueling Bitcoin's Meteoric Rise. (2022). https://www.bloomberg.com/news/articles/2017-12-15/turning-coal-into-bitcoindirty-secret-of-2017-s-hottest-market

CoinMarketCap. (2022). www.coinmarketcap.com

Corbet, S., Lucey, B., & Yarovaya, L. (2021). Bitcoin energy markets interrelationships-New evidence. *Resour. Pol., 70*(1–12), 101916.

De Vries, A. (2018). Bitcoin's growing energy problem. *Joule*, *2*(5), 801–880. doi:10.1016/j.joule.2018.04.016

Digiconomist. (2022). *Cambridge Bitcoin Energy Consumption Index.* 10 Haziran 2022, www.digiconomist.net

Dogan, E., Majeed, M. T., & Luni, T. (2022). Are clean energy and carbon emission allowances caused by bitcoin? A novel time-varying method. *Journal of Cleaner Production*, *347*, 347. doi:10.1016/j.jclepro.2022.131089

Foteinis, S. (2018). Bitcoin's alarming carbon footprint. *Nature*, *554*(7690), 169–170. doi:10.1038/d41586-018-01625-x

Frankenfield, J., Brown, J. R., & Kazel, M. (2022). *Bitcoin Mining*. 11 Mayıs 2022, https://www.investopedia.com/terms/b/bitcoin-mining.asp

Gulli, A. (2020). (un)sustainability of bitcoin mining. *Rutgers Computer & Technology Law Journal*, *46*(1), 95–123.

Hern, A. (2017). *Bitcoin Mining Consumes More Electricity A Year Than Ireland*. 12 Mayıs 2022, https://www.theguardian.com/technology/2017/nov/27/bitcoin-mining-consumeselectricity-ireland

Howson, P., & Vries, A. (2022). Preying on the poor? Opportunities and challenges for tackling the social and environmental threats of cryptocurrencies for vulnerable and low-income communities. *Energy Research & Social Science*, *84*, 102394. doi:10.1016/j.erss.2021.102394

Jiang, S., Li, Y., Lu, Q., Hong, Y., Guan, D., Yiong, Y., & Wang, S. (2021). Policy assessments for the carbon emission flows and sustainability of Bitcoin blockchain operation in China. *Nature Communications*, *12*(1), 1–10. doi:10.103841467-021-22256-3 PMID:33824331

Klebnikov, S. (2022). *Here's What The Crypto 'Bloodbath' Means For The Stock Market (It's Not Good)*. https://www.forbes.com/sites/sergeiklebnikov/2022/06/13/heres-what-the-crypto-bloodbath-means-for-the-stock-market-its-not-good/?sh=235b51af5fef

Krause, M. J., & Tolaymat, T. (2018). Quantification of energy and carbon costs for mining cryptocurrencies. *Nat. Sustain.*, *1*(11), 711–718. doi:10.103841893-018-0152-7

Küfeoğlu, S., & Özkuran, M. A. (2019). *Energy consumption of Bitcoin mining*. Cambridge Working Papers in Economics, University of Cambridge, Faculty of Economics.

Li, J., Li, N., Peng, J., Cui, H., & Wu, Z. (2019). Energy consumption of cryptocurrency mining: A study of electricity consumption in mining cryptocurrencies. *Energy*, *168*, 160–168. doi:10.1016/j.energy.2018.11.046

Martindale, J. (2020). *What is Bitcoin Mining?* 15 Mayıs 2022, https://www.digitaltrends.com/computing/what-is-bitcoin-mining/

Mora, C., Rollins, R. L., Taladay, K., Kantar, M. B., Chock, M. K., Shimada, M., & Franklin, E. C. (2018). Bitcoin Emissions Alone Could Push Global Warming Above 2'C. *Nature Climate Change*, *8*(11), 931–933. doi:10.103841558-018-0321-8

Polemis, M. L., & Tsionas, M. G. (2021). The environmental consequences of blockchain technology: A Bayesian quantile cointegration analysis for Bitcoin. *International Journal of Finance & Economics*, 1–20. doi:10.1002/ijfe.2496

Regulating Power Sector Carbon Emissions. (2022). https://www.c2es.org/content/regulating-power-sector-carbon-emissions/

Satoshi Nakamoto. (2008). *Bitcoin: A Peer-to Peer Electronic Cash System.* https://bitcoin.org/bitcoin.pdf

Stoll, C., Klaaben, L., & Gallersdorfer, U. (2019). The carbon footprint of bitcoin. *Joule, 3*(7), 1647–1661. doi:10.1016/j.joule.2019.05.012

Suazo, M. (2021). *The argument for mining Bitcoin from clean energy and waste energy streams.* https://www.emergingenergyinsights.com/2021/09/the-argument-for-mining-bitcoin-from-clean-energy-and-waste-energy-streams

Toda Y., & Yamamoto T., (1995). Statistical inference in vector autoregressions with possibly integrated processes. *Hournal of Econometrics,* (66), 225-250.

World gross electricity production by source. (2022). https://www.iea.org/data-and-statistics/charts/world-gross-electricity-production-by-source-2019

Yıldız, M., Çakan, C., & Altınışık, E. (2022). The Impact of Covid-19 on Financial Markets and Investment Preferences. *Çankırı Karatekin University Journal of Institute of Social Sciences, 13*(1), 121-140.

Yirka, B. (2022). *Bitcoin carbon emissions rise as mining moves to US and other countries.* 20 Mayıs 2022, https://phys.org/news/2022-02-bitcoin-carbon-emissions-countries.html

Chapter 11
The Impact of Digital Transformation on Personnel Number in the Banking Sector

Gokce Tekin Turhan
Nisantasi University, Turkey

ABSTRACT

Digital transformation affects the banking sector as it does every sector and enables significant innovations and changes in the banking and finance market. In this sense, it aims to ensure that banking transactions can be carried out quickly, securely, and at less cost, and to achieve the highest level of satisfaction in customer service. New solutions should continue to be developed to enable customers to benefit from digital banking products in the fastest way possible. There are also negative effects of digitalization. When evaluated within the framework of the banking sector, the increasing number of digital services and the number of customers using these services may lead to a decrease in the number of bank personnel. In the study, quantile regression was applied for the period of 2010.Q1-2020.Q4. This study examines the changes that may occur in the number of personnel as a result of the digitalization process in the banking sector and determines the decreasing effect of digitalization on the number of personnel.

INTRODUCTION

Thanks to technological progress, the banking sector is one of the sectors most affected by change and development in the modern digital marketing environment. In today's competitive economic conditions, businesses that adapt quickly to the radical changes in the marketing environment stand out in the race to achieve profitability, which is one of their main objectives (Shettar, 2019:3). The cost-cutting benefits of using digital channels, which provide consumers with access to financial services at the most cost-effective place and time, encourage banks to invest in this area (Hinings et al., 2018:53).

Banks aim to increase customer satisfaction in digital banking services through digitalization efforts. With the worldwide competition, banks are focusing on a system that enables their customers to receive banking services in a fast, practical and comprehensible way, and to intervene immediately in

DOI: 10.4018/978-1-6684-6620-9.ch011

the problems experienced by their customers (Cortet et al., 2016:14). Therefore, the number of employees, the consistency of the number of transactions and the number of branches have a very important place in digitalization in the banking sector. Digital transformation is defined as "a process which aims to improve a unit by combining information, information technology, communication and connectivity technology and bringing about significant changes in its structure". Digital transformation includes a broad concept in the banking sector, such as digitization of documents, electronic signatures for transactions, e-learning, teleconferencing, online trading platforms, digital stores, e-statements and m-payments (Ravi, 2017:173; Kitsios et al., 2021:4).

The fact that digitalization plays an important role in consumer preferences and reduces costs leads to an increase in technology investments made by banks and plays a decisive role in providing competitive advantage in the competition between banks (Omarini, 2017:3; Li, 2020:3). While the first stage of the digital transformation in the banking sector was the internet, the second stage was mobile applications. On the other hand, the increasing number of financial technology companies (FinTech) and large technology companies (BigTech), which have gained a significant share of banks' market share with the digital transformation, indicates that competition in the banking sector will be much more intense in the future. About fifteen years ago, more than half of banking transactions took place through the branch network, whereas today this proportion has fallen to less than 10%. By 2022, with the rapid growth of digital channels, it is predicted that at least one out of every two transactions will take place through a non-bank channel (Bakhareva, 2017:13; Gomber et al., 2017:539).

There are also negative effects of digitalization. When evaluated within the framework of the banking sector, the increasing number of digital services and the number of customers using these services may lead to a decrease in the number of bank personnel and branches. Therefore, the aim of this study is to reveal the impact of the digitalization process in the banking sector on the number of personnel with the help of quantile regression.

CONCEPTUAL FRAMEWORK

Digitalization and Employment in the Banking Sector

Today, there are still diverging views on the possible effects of digitalization on labor markets and employment. Because there is no doubt that the digital revolution, while increasing efficiencies and reducing costs, will destroy many existing professions and give rise to new ones (Hansen & Sia, 2015:53; Hagberg et al.,2016:698). While various surveys in this area reveal concerns that robots will displace blue and white collar jobs in the near future, there are also optimistic views that technology is always creating new employment opportunities and will not lead to further job losses (Lanzolla & Anderson, 2008:73; Buss et al., 2016:7). As a result, there is no consensus on the effects of technological developments and digitalization on the workforce. Some experts and academics argue that the developments will have partially or completely negative consequences, while others argue that they will have more positive effects (Bihari, 2011:35;White, 2012:207).

In this context, in order to better analyze the issue, when the results of technological advances to date are examined, it is observed that there are some common points. The first of these is that although the demand for labor initially decreased with the developments, it later increased (Umans et al., 2018:413). The other commonality is the increase in the productivity of the labor force. In this context, it can be said

Table 1. Number of personnel in the banking system (2016-2020)

	2016	2017	2018	2019	2020
Public Capital Banks	57586	58502	60195	60365	61592
Private Equity Banks	73742	73460	72025	69071	67523
Banks Transferred to SDIF (Saving Deposit Insurance Fund)	231	225	220	223	218
Foreign Capital Banks	59,804	56,079	54,478	54	52,155
Deposit Banks	191363	188266	186918	183659	181488
Development and Investment Banks	5336	5238	5395	5178	5124

Source: Turkish Banking Association

that technological developments increase the level of income and employment in the long run. However, some scholars argue that the Fourth Industrial Revolution will not lead to a similar outcome and that the power of labor will gradually weaken. According to supporters of this view, technologies that increase productivity do not necessarily increase the aggregate demand for labor (Berman, 2012:17; Matt et al., 2015:340). Some automation technologies can actually reduce the demand for labor. Wage increases commensurate with the increase in productivity brought about by automation are not always possible. Wages increase more slowly than productivity growth (Skinner, 2014:28).

In the past, too, there was a fear that new technological tools would displace labor to a large extent, a phenomenon known as "technological unemployment". Nevertheless, such fears did not prevail in the 19th and 20th centuries, as the productivity and income gains from technologically-induced new jobs exceeded the impact of labor savings (Bican & Brem, 2020:5240). But digitalization, artificial intelligence and automation are increasingly penetrating areas of reasoning, perception and decision-making that until recently were thought to be truly human tasks. For this reason, there are those who think that it may herald the end of employment (Hess et al., 2016:139; Osipovskaya & Mikhaylin, 2017:125).

In some sectors, the change in employment is evident. For example, with the implementation of new technologies in the banking sector, there has been a certain reduction in the number of employees in operational jobs. The digital transformation process is increasingly affecting the working life and employment in banking, and consequently the number of branches and employees (Padmaja et al., 2017:12; Pshenichnikov, 2018:49). According to the 2021 Report of the Banks Association of Türkiye, the total number of employees in the banking sector is on a downward trend when compared to the same periods of previous years.

The widespread use of digital banking has led to a decline in the number of branches and employees working in the sector, which is the physical distribution channel of banks. As of December 2020, there are 186612 bank employees in 9943 branches in the sector. In the last five years, the number of bank branches in Türkiye. has decreased by 1250 (11%) and the number of employees by 2853 (7%) (The Banks Association of Türkiye Report 2021).

Table 1 shows the number of employees of deposit banks and development and investment banks. In 2016, the number of employees was 57586 in state-owned banks, 73742 in privately owned banks, 231 in banks transferred to the SDIF, and 59804 in foreign-owned banks. In development and investment banks, this number was 5336. In 2020, the number of employees was 61592 in state-owned banks, 67523 in private-owned banks, 218 in banks transferred to the SDIF, 52155 in foreign-owned banks,

Table 2. Number of branches in the banking system (2016-2020)

	2016	2017	2018	2019	2020
Public Capital Banks	3702	3677	3718	3707	3701
Private Equity Banks	4132	4013	3963	3785	3644
Banks Transferred to SDIF (Saving Deposit Insurance Fund)	1	1	1	1	1
Foreign Capital Banks	2905	2809	2718	2644	2529
Deposit Banks	10740	10500	10400	10137	9875
Development and Investment Banks	41	50	54	62	64

Source: Turkish Banking Association

and 5124 in development and investment banks. As can be seen from Table 1, unlike publicly owned deposit banks, there has been a noticeable decrease in the number of employees in other types of banks over the years, especially since 2019, when the pandemic started.

Table 2 demonstrates the number of branches for the last 5 years. Accordingly, the number of branches of deposit banks, which was 10740 in 2016, was found to be 9875 in 2020. The number of branches of development and investment banks increased from 41 in 2016 to 64 in 2020. As a result of this situation, it is possible to say that the need for classical banking has started to decrease and both the number of branches and the number of employees will gradually decrease in the transition to digital banking.

Table 3. Number of customers using internet banking

	2016	2017	2018	2019	2020
Number of active retail customers	19077000	11856000	11258000	10714000	11074000
Number of active corporate customers	1321000	1270000	1286000	1271000	1369000
Total number of active customers	20399000	13125000	12544000	11985000	12444000

Source: Turkish Banking Association

Table 3 demonstrates the number of customers using internet banking. The number of active retail customers declined from 19077000 in 2016 to 11074000 in 2020. The number of active corporate customers increased from 1321000 in 2016 to 1369000 in 2020. The total number of active customers declined from 20399000 in 2016 to 12444000 in 2020. While the number of active retail customers using internet banking has decreased in the last 5 years, the number of active corporate customers has not changed. It is possible to say that the 42% decrease in the number of active retail customers using internet banking in the last 5 years is due to customers' preference for mobile banking.

Table 4 demonstrates the number of customers using mobile banking. While the number of active retail customers was 18649000 in 2016, it increased by 223% in the last 5 years to 60334000. The number of active corporate customers increased by 280% from 569000 in 2016 to 2166000 in 2020. In this case, the total number of active customers tripled from 19218000 in 2016 to 62500000 in 2020.

Table 5 demonstrates the number of transactions made through mobile and internet banking. In 2016, the number of mobile banking financial transactions was 158 million with a transaction volume

Table 4. Number of customers using mobile banking

	2016	2017	2018	2019	2020
Number of active retail customers	18649000	28712000	38385000	47799000	60334000
Number of active corporate customers	569	829	1167000	1479000	2166000
Total number of active customers	19218000	29541000	39552000	49279000	62500000

Source: Turkish Banking Association

of TL 330 billion; in 2020, the number of transactions was 859 million with a transaction volume of TL 3051 billion. In 2016, the number of financial transactions in internet banking was 149 million, with a transaction volume of TL 1057 billion. In 2020, the number of transactions was 123 million, while the transaction volume was TL 1987 billion. It is possible to interpret the increase in the number and volume of financial transactions made through both mobile banking and internet banking in the last five years as reflections of digitalization.

DATA AND ECONOMETRIC METHODOLOGY

The aim of the study is to reveal how the number of personnel has changed as a result of the digitalization process in the Turkish banking sector for the quarterly data of the 2010.Q1-2020.Q4 period. Based on this, the relationship between the number of customers using digital services and the number of bank personnel was analyzed with the help of quantile regression. All the series are quarter and obtained from Turkish Banking Association databases.

Linear quantile regression is related to linear least-squares regression in that both are interested in studying the linear relationship between a response variable and one or more independent or explanatory variables. However, whereas least-squares regression is concerned with modelling the conditional

Table 5. Number of transactions made via mobile and internet banking

		Mobile Banking Financial Transactions	Financial Transactions in Internet Banking
2016	Operation Quantitiy	158	149
	Operation Volume	330	1057
2017	Operation Quantitiy	280	137
	Operation Volume	649	1326
2018	Operation Quantitiy	393	133
	Operation Volume	933	1422
2019	Operation Quantitiy	559	120
	Operation Volume	1516	1492
2020	Operation Quantitiy	859	123
	Operation Volume	3051	1987

Source: Turkish Banking Association

Table 6. Dataset description

Variables	Variables Description	Data Source
DC	Number of customers using digital services	Turkish Banking Association
BP	Number of bank personnel	Turkish Banking Association

mean of the response variable, quantile regression models the conditional τth quantile of the response variable, for some value of τ (0,1). So, for example, τ= 0.5 would be the median. Given the vector y it is a well known result that the sample mean, y, solves the least squares problem

$$\underset{\mu\in\mathbb{R}}{\text{minimize}}\sum_{i=1}^{n}\left(y_i-\mu\right)^2 \qquad (1)$$

This result leads to least-squares regression where, given a design matrix X and defining the conditional mean of y as $\mu(X) = X\beta$, an estimate of β is obtained from the solution to

$$\text{minimize}_{\beta\in\mathbb{R}^p}\sum_{i=1}^{n}\left(y_i-x_i^{\mathrm{T}}\beta\right)^2 \qquad (2)$$

Quantile regression can be derived in a similar manner by specifying the τth conditional quantile as $Qy(\tau/X) = X\beta(\tau)$ and estimating $\beta(\tau)$ as the solution to

$$\text{minimize}_{\beta\in\mathbb{R}^p}\sum_{i=1}^{n}\rho_\tau\left(y_i-x_i^{\mathrm{T}}\beta\right) \qquad (3)$$

where ρ_τ (z) is a linear loss function defined as $z(\tau$ -1) if z< 0 and $z\tau$ otherwise (Koenker, 2005:97).

As quantile regression allows multiple quantiles to be modelled it can allow for a more comprehensive analysis of the data to be carried out compared to least-squares regression where only the mean is considered. This potentially enables more insight into the data and any underlying relationships, in addition, it will tend to be less sensitive to large outlying observations.

EMPIRICAL ANALYSIS

The analysis was carried out with the Bootstrap method with 500 iterations and 1000 replications. Random generator is obtained with Knuth (1969) algorithm. has been done. Bootstrap preference was due to the small number of data. GDP data is obtained as of 1998 with a new calculation method. bootstrap to produce robust (robust) predictions due to this constraint. method is preferred. Whether there is symmetry before the model's estimation measuring Wald Test was used, then Q1=0.25 and Q3=0.75 slices It was understood by accepting the Ho hypothesis. Each The model was estimated for three quantiles, the absolute deviation of the errors being the smallest The model is interpreted for Q2=0.5.

Table 7. Wald test results for symmetry

Symmetric Quantiles Test				
Specification: BP C DC				
Test Summary		**Chi-Sq. Statistic**	**Chi-Sq. d.f.**	**Prob.**
Wald Test		15.731	1	0.006
Restriction Detail: b(tau) + b(1-tau) - 2*b(.5) = 0				
Quantiles	**Variable**	**Restr. Value**	**Std. Error**	**Prob.**
0.25, 0.75	C	3.02E+11	1.99E+11	0.015

H_1 hypothesis accepted that there is no symmetry according to the Wald Test. In this case, according to the LS estimates of the quantile regression by many researchers in the literature that it will give more robust estimates determined.

Table 8. Quantile regression bootstrap estimation result

Dependent Variable: BP				
Method: Quantile Regression (Median)				
Sample: 2010Q1 2020Q4				
Bootstrap Standard Errors & Covariance				
Bootstrap method: XY-pair, reps=1000, rng=kn, seed=596242				
Sparsity method: Kernel (Epanechnikov) using residuals				
Bandwidth method: Hall-Sheather, bw=0.25694				
Estimation successfully identifies unique optimal solution				
Variable	**Coefficient**	**Std. Error**	**t-Statistic**	**Prob.**
C	9.368	1.509	6.208	0.000
DC	0.471	0.092	5.119	0.000
Pseudo R-squared	0.494	Mean dependent var		1285640
Adjusted R-squared	0.490	S.D. dependent var		17853.67
S.E. of regression	9.558	Objective		245.275
Sparsity	29.579	Quasi-LR statistic		42.644
Prob(Quasi-LR stat)	0.000			

The model's goodness-of-fit criterion is $p<0.05$ as a result of the Quasi LR test, and it is understood that the model is significant. Looking at the Pseudo R^2 value as a result of the model, the DC variable alone affects the BP value by 49.4%, that is; It is determined that there is a 49.4% relationship between them. Looking at the model coefficient, when the number of digital customers increases by 1 unit, the statistics that reduce the number of personnel by 0.471 have a significant effect.

CONCLUSION

The Turkish banking sector, like its global competitors, constantly updates its service offerings in line with market conditions through innovative digital products such as internet banking and mobile banking in line with the changing consumer profile. It is clear that by understanding the reasons for customers who have reservations about using digital banking tools and convincing these customers, the digital channel activation rate can be increased. In addition, improving the quality of the experience perceived by customers is vital for the success of digital banking applications.

The aim of the study is to reveal how the number of personnel has changed as a result of the digitalization process in the Turkish banking sector for the quarterly data of the 2010.Q1-2020.Q4 period. This study examines the changes that may occur in the number of personnel, as a result of the digitalization process in the banking sector and determines the decreasing effect of digitalization on the number of personnel.

In this study, it has been determined that the use of digital channels has increased at an extraordinary pace in recent years due to customer demand and industry supply, and that the Covid 19 pandemic reinforced the process of increasing digitalization. In addition, digitalization has changed the function of the traditional branch distribution channel in banking and the business model of the services provided through this channel, reducing the number of branches, employees and transactions. The sector has seen a decline in new hires, with additional hires shifting to audit and information technologies among candidates with new skill sets.

Digital banking solutions, also referred to as "unmanned" or "branchless banking", threaten the Branch distribution channel in a fundamental way, but it is understood that not all bank customers will use digital banking products in the near future. Therefore, in the financial sector, as in all service sectors, the Branch distribution channel, where a personal relationship with customers is established and maintained, will continue to exist and remain important for a long time. Because in the service sector, "human capital" is one of the most effective and irreplaceable competitive advantages of enterprises.

Developing innovative financial products with high added value will provide banks with a differentiation advantage and contribute to increasing profitability. It is vital for banks to develop synergies with financial technology companies, called Fintech, to ensure security, create innovative solutions and improve the perception of experiential value. Cooperation with Fintechs has gained importance in the areas of cyber security, cloud, artificial intelligence, augmented reality, machine learning, big data, robotics and the Internet of Things.

Due to digital transformation, the banking sector will provide great opportunity for technology-related jobs such as cyber security specialist, credit analyst, robot programmer, blockchain architect and process modeler specialist. Therefore, employers need to upgrade their skills to appreciate the role of analytics in each banking function, as the banking sector will recruit the candidate with the required skills and improve the quality of human resources through various skill sets. Therefore, banks in the true sense will employ qualified employees to meet the day-to-day requirements of the banking sector.

As a result, to keep pace with digitalization, banks must adapt their business models to change the way they interact with customers, manage middle and back office operations, be competitive and future-proof.

REFERENCES

Bakhareva, A. A. (2017). Prospects for the Development of the Banking Sector in the Context of the Introduction of Modern Financial Technologies. *Symbol of Science*, *1*, 12–14.

Berman, S. J. (2012). Digital transformation: Opportunities to create new business models. *Strategy and Leadership*, *40*(2), 16–24. doi:10.1108/10878571211209314

Bican, P. M., & Brem, A. (2020). Digital Business Model, Digital Transformation, and Digital Entrepreneurship: Is There a Sustainable "Digital". *Sustainability*, *12*(1), 5239–5240. doi:10.3390u12135239

Bihari, S. C. (2011). Internet Technology and Customer Profitability in Banks. *Journal of Contemporary Research in Management, 6*, 33-48.

Cortet, M., Rijks, T., & Nijland, S. (2016). Psd2: The digital transformation accelerator for banks. *Journal of Payments Strategy & Systems*, *10*, 13–27.

Gomber, P., Koch, J. A., & Siering, M. (2017). Digital Finance and FinTech: Current research and future research directions. *Journal of Business Economics*, *87*(5), 537–580. doi:10.100711573-017-0852-x

Hagberg, J., Sundstrom, M., & Egels-Zandén, N. (2016). The digitalization of retailing: An exploratory framework. *International Journal of Retail & Distribution Management*, *44*(7), 694–712. doi:10.1108/IJRDM-09-2015-0140

Hansen, R., & Sia, S. K. (2015). Hummel's digital transformation toward omnichannel retailing: Key lessons learned. *MIS Quarterly Executive*, *14*(2), 51–66.

Hess, T., Matt, C., Benlian, A., & Wiesböck, F. (2016). Options for Formulating a Digital Transformation Strategy. *MIS Quarterly Executive*, *15*(1), 137–139.

Hinings, B., Gegenhuber, T., & Greenwood, R. (2018). Digital Innovation and Transformation: An institutional perspective. *Information and Organization*, *28*(1), 52–61. doi:10.1016/j.infoandorg.2018.02.004

Kitsios, F., Giatsidis, I., & Kamariotu, M. (2021). Digital transformation and strategy in the banking sector: Evaluating the acceptance rate of e-services. *Journal of Open Innovation*, *7*(1), 1–14. doi:10.3390/joitmc7030204

Koenker, R. (2005). *Quantile Regression, Econometric Society Monographs*. Cambridge University Press. doi:10.1017/CBO9780511754098

Lanzolla, G., & Anderson, J. (2008). Digital transformation. *Business Strategy Review*, *19*(2), 72–76. doi:10.1111/j.1467-8616.2008.00539.x

Li, F. (2020). The digital transformation of business models in the creative industries: A holistic framework and emerging trends. *Technovation*, *92*(1), 1–14. doi:10.1016/j.technovation.2017.12.004

Matt, C., Hess, T., & Benlian, A. (2015). Digital transformation strategies. *Business & Information Systems Engineering*, *57*(5), 339–343. doi:10.100712599-015-0401-5

Omarini, A. (2017). The Digital Transformation in Banking and The Role of FinTechs in the New Financial Intermediation Scenario. *Int J Financ Econ Trade.*, *1*(1), 1–6.

Osipovskaya, A. V., & Mikhaylin, A. V. (2017). The Development of Financial Technologies in the field of banking services: The main directions. *Young Scientist*, *26*, 124–127.

Padmaja, R., Rifana Meera, M., & Valli Mayil, M. (2017). Employee's Attitude towards Adoption of Electronic Banking Services: A Case of Public Sector Banks, International Journal of Banking. *Risk and Insurance*, *5*(2), 11–14.

Pshenichnikov, V. V. (2018). Influence of Financial Technologies on Changing the Model of Customer Banking Services. Theory and Practice of Service: Economics, Social sphere. *Technology*, *1*(35), 48–52.

Ravi, C. S. (2017). Digital payments system and rural India: A review of transaction to cashless economy. *International Journal of Commerce and Management Research*, *3*(5), 170–173.

Shettar, R. (2019). Digital Banking An Indian Perspective. *Journal of Economics and Finance*, *10*(3), 1–5.

Skinner, C. (2014). *Digital Bank: Strategies to Launch Digital or Become a Bank*. Marshall Cavendish Business Pbc.

Umans, T., Kockum, M., Nilsson, E., & Lindberg, S. (2018). Digitalisation in the banking industry and workers subjective well-being: Contingency perspective. *International Journal of Workplace Health Management*, *11*(6), 411–423. doi:10.1108/IJWHM-05-2018-0069

White, M. (2012). Digital workplaces: Vision and reality. *Business Information Review*, *29*(4), 205–214. doi:10.1177/0266382112470412

ADDITIONAL READING

Belous, A. P., & Lyalkov, S. Yu. (2017). The Vector of Development of Banks in the Digital Revolution. *Banking*, *10*, 16–19.

Gupta, R., Mejia, C., & Kajikawa, Y. (2019). Business, innovation and digital ecosystems landscape survey and knowledge cross sharing. *Technological Forecasting and Social Change*, *147*, 100–109. doi:10.1016/j.techfore.2019.07.004

Kiron, D., & Unruh, G. (2018). The convergence of digitalization and sustainability. *MIT Sloan Management Review*, *1*(1), 1–15.

Kupiainen, J. (2006). Translocalisation over the Net: Digitalisation, information technology and local cultures in Melanesia. *E-Learning and Digital Media*, *3*(3), 279–290. doi:10.2304/elea.2006.3.3.279

Liu, D. Y., Chen, S.-W., & Chou, T. C. (2011). Resource fit in digital transformation: Lessons learned from the CBC Bank global e-banking project. *Management Decision*, *49*(10), 1728–1742. doi:10.1108/00251741111183852

Westerman, G., Bonnet, D., & Mcafee, A. (2014). The nine elements of digital transformation. *MIT Sloan Management Review*, *1*(1), 1–6.

Chapter 12
The Impact of Knowledge Management on Organizational Learning in Universities

Gamze Sart
Istanbul University-Cerrahpasa, Turkey

ABSTRACT

In today's information and technology-driven markets, knowledge has an important role to play in gaining and maintaining competitive advantage and improving organizational performance. Since knowledge is the most important competitive advantage tool, effective management of knowledge will increase organizational performance. The knowledge management process is a process that includes activities related to the systematic acquisition, transformation, use, and protection of knowledge in order to better achieve organizational goals. This process develops depending on technology, culture, organizational structure, and human factors. The aim of this study is to determine the effects of knowledge management on organizational learning for 650 academicians working at universities in Istanbul. As a result of the structural equation model, a positive 68.1% increasing relationship was determined between the knowledge management and organizational learning.

INTRODUCTION

Today, knowledge is recognized as an important organizational and individual asset. For this reason, individual and organizational learning, which are ways of acquiring knowledge, are becoming increasingly important. In business and organizational life, especially with the effect of globalization and competition, information is used effectively in the production of products and services (Nouri et al., 2017:195). In this context, it has emerged as a necessity to know the methods of accessing information and integrating information with activities in order for the organization to survive and compete. The phrase "knowledge is power" refers to an understanding that brings the individual to the forefront today, as well as an understanding that realizes the acquisition, sharing, transfer and effective use of knowledge for organizations through appropriate channels. In an organization, knowledge obtained from the internal and external

DOI: 10.4018/978-1-6684-6620-9.ch012

environment is the common value of that organization and needs to be shared (Yang, 2011:17). Shared knowledge creates individual and organizational synergy that will lead the organization to success and creates the idea that knowledge is the common property and/or value of the organization.

Today's organizations' constant need for information and the necessity to use information in their activities have led to the emergence of knowledge and information-based practices such as knowledge management, information management, document management and organizational learning (Ainissyifa, 2012:81). Identifying, obtaining, processing, storing and sharing information related to activities and using it in organizational activities is possible through the conscious and programmed implementation of knowledge management (Cheng et al., 2017:179). For organizations, knowing where knowledge is located is of little use; knowledge becomes a valuable company asset only when it can be accessed. As the level of accessibility increases, the value of information undoubtedly increases at the same rate (Donate & Pablo, 2015:364).

Through the act of learning, which is one of the ways of accessing knowledge, both individuals and organizations make efforts to obtain the necessary knowledge for their activities. As the world changes rapidly, the level of uncertainty increases at the same rate. Past experience is therefore often not reliable enough for decision-making. Continuous learning is needed to anticipate and respond to changes and achieve efficiencies (Watanabe & Senoo, 2011:112). For this reason, it is a natural necessity for organizations to become learning organizations and to make organizational learning an organizational policy (Thambi & O'Toole, 2012:94). In this context, a learning organization is a knowledge-based organization that adapts all its elements and possibilities to the changing environmental conditions it is in, obtains the necessary information in various ways and disseminates it within the organization within a system, and uses the information effectively and efficiently in its activities (Stylianou & Savva, 2016:1517).

CONCEPTUAL FRAMEWORK

Knowledge Management

The fact that organizations that structuralize the process of knowledge creation, in other words, that give it an institutional quality, will have the most important competitive tool of the information age as organizations that can manage knowledge, makes the management of organizational knowledge mandatory. In this sense, knowledge management is a strategic concept that today's organizations need to focus on and gain skills in. The aim of knowledge management is primarily to identify valuable new information, to deliver it to the relevant units and to keep the accumulation of knowledge alive. In this sense, searching for, finding and accessing new information related to the work done, recording and accurate use of information, communicating and sharing it to other employees, updating and archiving outdated and unused information are the activities of knowledge management (Dahou et al., 2019:5). The aim of knowledge-age organizations, which focus on making better use of their knowledge capacities by improving the flow of knowledge among organizational members, is to gain competitive advantage through the effective dissemination of knowledge. In the last decade, driven by rapid advances in information technology and a dynamic global environment, knowledge has become a crucial resource that facilitates and enhances sustainable competitive advantage (Mohajan, 2017:8). Knowledge management is an important capability of the learning organization. Knowledge management is the acquisition, evaluation and sharing of knowledge in a way that achieves organizational value. Since the idea of producing and sharing knowledge lies at the basis of the learning organization the learning organization and knowledge management are mentioned or examined together in the studies. Garvin's definition of "A learning organization is an organization that has the ability to create, acquire and transfer knowledge and also has the ability to adjust its behavior to reflect new knowledge and views" (Garvin, 1993:80) leads to the conclusion that a learning organization is an organization that has the ability to manage knowledge.

From these definitions, it is observed that knowledge management is seen as a strategy and process that enables the creation and sharing of knowledge in order to create value. This process is a highly complex and challenging task, involving activities such as the creation of knowledge, the evaluation of knowledge from internal and external sources, the formatting, storing and sharing of the information obtained and the knowledge created. This is because knowledge is dynamic, constantly evolving and knowledge that is valid today may be invalid tomorrow (López, 2013:664).

Knowledge management is often referred to as the process of collecting, organizing and sharing data and information assets, but it is not only the process of acquiring, storing and organizing information. It also includes the creation of knowledge through learning and sharing across the organization (Kordab et al., 2020:5). In this case, Knowledge Management, when considered as a process, can be expressed as a process that helps to find, select, organize, disseminate, and transfer the knowledge required for organizational activities and to create new knowledge (Alsabbagh & Al Khalil, 2017:565). Similarly, another definition defines knowledge management as the explicit and systematic management of knowledge and its combination with the process of creating, organizing, distributing, using and processing knowledge (Al Ahmar et al., 2014:31). As can be seen, some researchers working on knowledge management in organizations have focused on the open and systematic management of knowledge as a process. Then we can say that knowledge management as a discipline consists of a certain process. The knowledge management process consists of the steps of acquiring/generating knowledge, storing/organizing knowledge, distributing/sharing knowledge and using/applying knowledge to organizational processes (Davenport & Völpel, 2001:219; Lee & Choi, 2003:180).

Organizational Learning

Today, organizational learning, which is one of the core capabilities of the learning organization, is seen as a very important process due to the rapid changes affecting all organizations. Because organizations have to learn new ways to survive (Hsu, 2014:52). For an organization to learn means to prepare an environment that enables the creation of new knowledge, to use the new knowledge developed in the production of new ideas and systems, and to encourage the creation of new knowledge by considering all the skills gained from this as a learning opportunity (Argote, 2013:55).

Organizational learning is the development of behavior, values and norms within the organization, rather than the acquisition of knowledge through education or experience; organizational learning is a process that requires the active participation of all employees. It is based on shared experiences, decisions, values. The collective values and norms formed in this way become shared by all employees. Organizational learning, which ensures the continuous change of an organization and its environment, is a never-ending process that continuously moves the organization to better situations (Alavi et al.,2006:195). Just as every result leads to a new problem, solutions to organizational problems will also lead to new problems after a while. Here, the basic feature of organizational learning comes into play, which is not only its ability to renew or change itself and its environment, but also its ability to ensure the continuity of this change (Ke & Wei, 2006:7).

Organizational learning is the process of improving organizational actions through better knowledge and understanding. Organizational learning is the conscious use of learning processes at the individual, group and system level with a management approach that envisages continuous change of the organization in a way to provide continuous satisfaction to the employees of the organization. Organizational learning is the change and expansion of knowledge and value systems within the organization; the development of problem-solving and action capacities; and the change of the common frame of reference of employees (Danish, 2012:5).

Learning in an organization often takes place due to the influence of various factors such as culture, technology, internal and external environment, strategy, policy, management approach. In a dynamic environment, the ability of organizations to adapt and respond to their internal and external environments requires continuous and conscious learning. For organizations that respond to their environment like a living organism, their ability to respond positively to the events occurring in their environment depends on their openness and flexibility to changes and accepting that continuous learning is important for the future of the organization. The organizations that will truly stand out in the future will be those who discover how individuals can assess the commitment and capacity to learn at all levels of an organization (Menezes et al., 2011:12).

Organizational learning is nowadays becoming increasingly popular with organizations as they seek to become more adaptable to change, as it relates to enhancing competitive advantage such as effectiveness and innovation. Achieving the goal(s) set in today's organizational life requires continuous and effective learning along with other necessary practices (Argote, 2013:61). Because in today's world of rapid change, organizations can survive, maintain their activities and compete only if they keep themselves in a continuous movement of change, improvement and innovation. One of the most important ways to achieve this is through organizational learning. Indeed, continuous progress and development requires that the organization and employees commit themselves to learning (Giniuniene & Jurksiene, 2015:988).

Knowledge Management in Universities and the Importance of Organizational Learning

First of all, our universities must fulfil the requirements of being a learning organization, provided that they adapt to the concept of "lifelong learning". If universities encourage each of their employees to become more sophisticated and successful, they can succeed in making the students they prepare to work in the private sector more competent. Because these students will take their place in business life actively after a short time and they will want the system they work in to be learning organizations similar to the universities where they were educated, and they will be able to work in this direction (Laal, 2011:546). As a result, a multiplier effect will occur when an individual who understands the importance of the learning organization shares this with other colleagues and directs them towards this goal. Learning can even extend from organizational learning to inter-organizational learning. It is quite surprising that the people who give students the training that will benefit them in their working lives are satisfied with themselves and do not feel any need to make serious changes. Today, however, one of the criteria by which the quality of an educational institution can be judged is the presence of leaders and staff who are in the habit of asking themselves the question "What did you learn today?" (Jamalzadeh, 2012:1163).

The most fundamental task of a higher education institution structured as a learning organization is to create the necessary infrastructure for learning to become a part of the system and to provide the opportunities at the right time to unlock the continuous learning potential of its administrative and academic staff. A higher education institution should do its best to provide its staff with basic services such as easy access to resources, uninterrupted internet, support in attending scientific meetings, encouragement in postgraduate education, and the opportunity to get to know and compare their disciplines and other disciplines through domestic and international trips. Continuous improvement is not impossible (Salleh, 2012:5). However, a supportive attitude towards learning and openness to change in higher education institutions will be a mechanism that will accelerate the process of becoming a learning organization. Transforming universities into learning organizations will not solve all problems. But it could be a first step towards expanding their scope of action and enabling them to become modern renewable institutions. This first step will enable the university that succeeds in becoming a learning organization to stand out among the increasing number of universities and become an educational institution that educates more successful students (Sangeeta & Librarian, 2015:166).

Although the structure of universities is slightly different from that of primary and secondary schools, they are not very different from primary and secondary schools in terms of being a learning organization. Educational institutions, whose main purpose is to teach, especially universities, whose next stage requires putting theory into practice, need to be able to transform themselves into learning organizations so that the flexibility and continuous learning understanding required by our age can become a philosophy of life and a basic need for the people they educate (Wu et al., 2013:9). When universities strive to become learning organizations, the people in their staff, i.e. teaching staff, will also become learning teams and individuals. Only then will they be able to change their students' perspectives and turn them into learner individuals. Because it is observed that individuals in universities, especially lecturers, reflect the characteristics of being a learning organization in the classroom and research areas. However, the system as a whole does not reflect the definition of a learning organization, nor do most organizations, except for some rarely encountered ones, reflect the characteristics of a learning organization (Kidwell et al., 2000:29).

STATISTICAL ANALYSIS

The Purpose and the Importance of the Research

Knowledge management and organizational learning processes are also part of organizational culture. For the implementation of these interrelated processes, senior managements need to create a suitable environment and exercise commitment for their continuity. In order to change perception and thinking habits, an environment must be created that allows formal education, sharing and work to develop spontaneously. In addition, the environment should have an educator-centered process, such as learning, lectures and demonstrations. Shared learning focuses on parallel discussion such as topic work, study groups and seminars. As for self-directed learning, he believes that workers should seek out non-interactive learning materials and ways of learning themselves, such as using the internet, reading newspapers and books. In other words, while training, sharing and self-study are attractive forces in a learning organization, job expansion, job enrichment and job rotation are driving forces in the knowledge management process. The learning organization is an understanding of management thinking within organizational behavior, while knowledge management is an activity practiced by organizations concerned with structures and processes for creating, storing, sharing and disseminating knowledge. Organizational learning and knowledge management are interdisciplinary in nature. The real organizational problems of organizational

Table 1. Research hypotheses

	Hypotheses
H_1	Knowledge management and organizational learning are statistically significantly correlation
H_2	Knowledge Acquisition and organizational learning are statistically significantly correlation
H_3	Knowledge Conversation and organizational learning are statistically significantly correlation
H_4	Knowledge Application and organizational learning are statistically significantly correlation

learning and knowledge management are not strictly disciplinary, but involve many disciplines and many perspectives, such as psychology, sociology, operational management, organizational behavior, strategic management, economics and information systems. Therefore, understanding organizational learning and knowledge management requires a multidisciplinary approach. Universities are institutions where knowledge management and organizational learning are the most important. The aim of the study is to determine the effects of knowledge management on organizational learning for 650 academicians at universities in Istanbul.

Research Sample

The universe of the study consists of academic staff working in research universities operating in Istanbul. According to the statistics of the number of academicians of the Higher Education Council (HEC) in 2021, the total number of academicians (research assistants, lecturers, doctor lecturers, associate professors, and professors) in state universities in Istanbul is 36,896. "Simple random sampling" was used as the sampling method. Surveys were conducted via e-mail between 15.04.2022 and 15.06.2022. Since some participants did not answer most of the questions in the questionnaire during the data entry process, these questionnaires were not included in the analysis process, and a total of 650 questionnaires were used in practice. In the sample size table developed by Yazıcıoğlu and Erdoğan (2004), it was determined that 384 people should be studied for p=0.50 and q=0.50 with a sampling error of 0.05. In this application, analyzes were carried out with 650 samples.

Research Hypotheses

Given the aforementioned conceptual arguments and empirical evidence, we hypothesize that knowledge management has an impact on organizational learning at public and private universities in Istanbul. The developed hypotheses are presented in Table 1.

Data Collection Tool

The study is built on a non-experimental quantitative research design and is a relational screening model according to the method of practice. The questionnaire used in the study was prepared using the scales whose validity and reliability were approved in previous studies as a result of a wide-scope literature review. We can explain these scales as follows:

Organizational Learning: The organizational learning scale was adapted from the study of Marsik and Watkins (2003). The scale is unidimensional and consists of 9 articles. It contains no reversing item. Questions prepared in a 5-point Likert style have scores ranging from 1 to 5, with options listed as "(1)-strongly disagree to (5)-strongly agree".

Knowledge management: The knowledge management scale was developed by Gold et al., (2001). As a result of this study, 3 sub-dimensions were obtained as Knowledge Acquisition (8 items), Knowledge Conversation (7 items) and Knowledge Application (7 items). It contains no reversing item. The prepared scale questions were asked to the participants using the 5-point Likert method, which scales as 1 = Strongly Disagree to 5 = Strongly Agree.

Statistical Methods

The data acquired from the scale utilized in the study were analyzed via IBM SPSS 27.0 and AMOS 25.0 programs. In the first stage, percentage distributions regarding demographic and general information were presented. In the second stage, exploratory factor analysis (EFA) was applied for the scales, and besides, confirmatory factor analysis (CFA) was performed to test construct validity. In the last stage, correlation analysis and structural equation model estimation results were included in order to identify the relationships.

Assumptions and Limitations of the Study

It was assumed that the people participating in the research acted with their true feelings and thoughts while answering the propositions in the scales. It was accepted that each participant answered the survey study voluntarily and answered each statement in the survey correctly and completely. It was also assumed that each participant understood the real meanings of all the words while answering the propositions. There were some difficulties in increasing the sample size in the survey; people were reluctant to participate due to different reasons. The fact that they did not have a positive attitude towards participation in the survey can be considered as an important limitation. In addition, employees who were not aware of the survey and did not check their emails on the specified dates were excluded from the sample. The research is limited to the province of Istanbul.

FINDINGS AND RESULTS

Survey Reliability

For the data obtained from the survey study, Cronbach Alpha, Split, Parallel, Absolute Precise Parallel (strict) tests were carried out as a reliability test. A Cronbach Alpha value above 70% indicates that the survey was successful. Some researchers take this value as 75% (Sezgin, 2016). The fact that other criteria are above 70% indicates that the questionnaire has internal consistency and that the results can be trusted (Sart et al., 2018). In this study, the results of the reliability analysis of the questionnaire are Cronbach-Alpha = 0.918, Parallel = 0.918, Split = 0.916-0.919, and Strict = 0.917.

Table 2. Exploratory factor analysis results

Knowledge Management	PVE	CA	AVE	CR
Knowledge Acquisition	28.45%	0.910		
Knowledge Conversation	23.59%	0.909	0.753	0.909
Knowledge Application	20.07%	0.908		
KMO= 0.931; Bartlett χ2= 8452.12 and p= 0.000; Percent of Variance Explained: 72.11%				
Organizational Learning	**PVE**	**CA**	**AVE**	**CR**
Organizational Learning	68.47%	0.912	0.784	0.911
KMO= 0.925; Bartlett χ2= 6832.47 and p= 0.000; Percent of Variance Explained: 68.47%				

PVE: Percent of Variance Explained, CA: Cronbach's Alpha Coefficient; AVE: Average Variance Extracted; CR: Construct Reliability

Descriptive Statistics

53.8% of the participants were female, and 46.2% were male. 40.9% of the academician were aged 20-30, 25.7% were aged 31-40, 11.5% were aged 41-50, and 21.9% were aged 51 years and over. In terms of marital status, 52.3% of the individuals were married, while 47.7% were single. Looking at the participants' employment terms, 4.2% worked less than 1 year, 20.1% between 1-5 years, 26.3% between 6-10 years, 22.6% between 11-20 years, and 31,0% worked for 21 years or more. According to their job positions, 17.2% were research assistant, 26.8% assistant professor, 25.5% associate professor, 24.3% professor and 6.2% lecturer.

Exploratory Factor Analysis

Factor analysis is a multivariate statistics method intended for finding or discovering a small number of unrelated and conceptually significant new variables (i.e. factors, dimensions) by bringing together p number of interrelated variables. After the suitability of the data set was confirmed by the tests, the "Principal Components Analysis" approach was applied via the "Varimax" rotation technique as the factor retention method in order to reveal the factor structure.

As a result of the exploratory factor analysis (EFA): Regarding the factor structure, for the knowledge management scale, 4 factor structure was obtained, which explained 72.11% of the total variance; for the organizational learning scale, 1 factor structure was obtained which explained 68.47% of the total variance. Following the EFA, there was no need to eliminate any questions since there was no item below 0.20 in the inference column and no item with anti-image matrix diagonal values below 0.50.

Confirmatory Factor Analysis

Confirmatory factor analysis (CFA) is utilized to express multivariate statistical analyses containing latent structures denoted by a large number of observable or measurable variables. CFA is a factor analysis technique utilized to question whether the factors revealed as a result of Exploratory Factor Analysis (EFA) are appropriate for the structure of the factors revealed as a result of the hypotheses. While EFA is utilized to examine which variable groups are highly correlated with which factors, CFA is employed to

Table 3. CFA model fit indices

Measurement (Fit Statistic)	Good Fit	Acceptable Fit	Research Model Value	Fit Status
General Model Fit				
X^2 /sd	≤3	≤4-5	2.18	Good fit
Comparative Fit Statistics				
NFI	≥0.95	0.94-0.90	0.936	Acceptable fit
TLI (NNFI)	≥0.95	0.94-0.90	0.982	Good fit
IFI	≥0.95	0.94-0.90	0.976	Good fit
CFI	≥0.97	≥0.95	0.961	Acceptable fit
RMSEA	≤0.05	0.06-0.08	0.014	Good fit
Absolute Fit Indices				
GFI	≥0.90	0.89-0.85	0.944	Good fit
AGFI	≥0.90	0.89-0.85	0.957	Good fit
Residual-Based Fit Index				
RMR	≤0.05	0.06-0.08	0.025	Good fit

determine whether the variable groups contributing to the k number of identified factors are sufficiently represented by the said factors.

In Table 3, X^2 /sd= 2.18, a "good fit" decision was made since it met the condition ≤3. NFI=0.936 resulted in the range of 0.94-0.90, thus an "acceptable fit"; other results were obtained as TLI (NNFI)= 0.982 was ≥0.95, a "good fit"; IFI= 0.976 was ≥0.95, a "good fit"; CFI= 0.961 was ≥0.97, an "acceptable fit"; RMSEA= 0.014 was ≤0.05, a "good fit"; GFI= 0.944 was ≥0.90, a "good fit"; AGFI= 0.957 was ≥0.90, a "good fit"; and RMR= 0.025 was ≤0.05, thus it was a "good fit".

Structural Equation Modeling

Structural equation modeling (SEM) is a statistical technique used to test models in which causal relationships and correlation relationships between observed variables and latent variables coexist; it is a multivariate method that is formed by combining analyses such as variance and covariance analyses, factor analysis, and multiple regression in order to predict interdependency relationships. Considering the goodness of fit criteria for SEM, it was found that X^2 /sd= 2.45, a "good fit" decision was made because it met the condition ≤3. NFI=0.942 resulted in 0.94-0.90, thus an "acceptable fit"; other results were obtained as TLI (NNFI)= 0.989 was ≥0.95, a "good fit"; IFI = 0.970 was ≥0.95, a "good fit"; CFI= 0.982 was ≥0.97, an "acceptable fit"; RMSEA= 0.018 was ≤0.05, a "good fit"; GFI= 0.952 was ≥0.90, a "good fit"; AGFI= 0.960 was ≥0.90, a "good fit"; and RMR= 0.022 was ≤0.05, thus it was a "good fit". SEM estimates fit criteria yielded the result "acceptable" for only one criterion; others revealed that a "good fit" was attained and the model was appropriate for interpretation.

According to the SEM results, significant relationships were obtained among the variables. KM has a statistically significant positive effect on OL (β=0.681; p<0.01). KAC has a statistically significant positive effect on OL (β= 0.664; p<0.01). KC has a statistically significant positive effect on OL (β=

Table 4. SEM model estimation results

Structural Relation	Direction	Estimated Coefficient	Std. Error	t-Statistic	p	Result
KM → OL	+	0.681	0.126	5.404	0.000*	Significant relationship
KAC → OL	+	0.664	0.106	6.264	0.000*	Significant relationship
KC → OL	+	0.613	0.094	6.521	0.000*	Significant relationship
KAP → OL	+	0.645	0.122	5.286	0.000*	Significant relationship

*p<0.01, *KM: Knowledge Management, KAC: Knowledge Acquisition KC: Knowledge Conversation*
KAP: Knowledge Application OL: Organizational Learning

0.613; p<0.01). KAP has a statistically significant positive effect on OL (β= 0.645; p<0.01). According to the coefficient sizes for sub-dimension, the KAC variable is most effective on the OL variable.

CONCLUSION

As knowledge management practice and/or process is a requirement of organizational learning practice, organizational learning activity and the effort to become a learning organization that it brings with it can be considered as an approach specific to organizations that implement knowledge management. In organizations, obtaining and producing the necessary information about the activities, and using the obtained and produced information effectively and efficiently requires organizational learning effectiveness; because there is a direct and reciprocal relationship between knowledge management and organizational learning practices as well as between knowledge and learning concepts (Chou & Ramser, 2019:136). As new, different and necessary knowledge can be acquired through the act of learning, the sum of the knowledge acquired through learning constitutes organizational and individual knowledge (Singh, 2011:355). However, the regulation of uninterrupted and unhindered information sharing within the organization and the provision of easy access to information by employees, that is, the management of knowledge, often necessitates the application of organizational learning (Yang et al., 2014:43). Knowledge management, in the context of its implementation, is a requirement of being a learning organization. In other words, a learning organization is an organization that can take knowledge management as a strategic organizational practice and apply it at a level that will contribute to organizational learning.

The aim of this study is to determine the effects of knowledge management on organizational learning for 650 academicians working at universities in Istanbul. As a result of the structural equation model, a positive 68.1% increasing relationship was determined between the knowledge management and organizational learning. Sub-dimension has a statistically significant positive effect on OL. According to the coefficient sizes for sub-dimension, the KAC variable is most effective on the OL variable.

The findings of the study are similar to JerezGomez et al., (2005), Rose et al., (2009), Aktharsha & Anisa (2011), Jamalzadeh, (2012), Nodehi, et al., (2013), Al Ahmar et al., (2014), Minghui, (2016), Alrefaai & Khalil (2019) and Karasneh (2019) studies.

Based on the results of the study, some recommendations can be made. In order to be more resilient in the face of a possible bad situation or crisis, organizations should build their structures in a way that is flexible and can adapt to change immediately. This requires a balanced distribution of authority and responsibility, with the authority to make and implement decisions resting with the person doing the work, in other words, a people-centered structure. Within this structure, multi-directional communication should be encouraged and interaction between units should be established to ensure that the whole, not the part, is seen. It is very important that knowledge is dominant in all units of the organization, from the simplest to the most complex tasks, and that producing and sharing knowledge becomes a culture. In this process, the human resources function has an important role to play in recruiting and motivating staff who are open to learning and sharing. Some of these roles include employing employees with different experiences, different perspectives and different abilities, keeping rules and procedures to a minimum, creating an organizational climate of open communication and cooperation, promoting problem solving and decision making through organizational learning, and supporting social interactions inside and outside the organization. Another issue that is essential for the preservation and application of the knowledge acquired or generated at universities is the provision of financial resources. In this context, universities should purchase the necessary software and hardware to ensure the integration of up-to-date information technologies into systems, and provide the necessary support to R&D units that will ensure the transformation of knowledge into production. Some suggestions that can be used to transform the university into a learning organization are as follows:

- A message system should be established so that researchers in different fields can communicate with each other when needed.
- There should be an online archive where researchers can submit their work to be critiqued by their colleagues.
- There should be a website for each course, especially for multi-section courses taught by some different faculties with separate instructors.
- Knowledge sharing should not be limited to the university, but should also be shared with other segments of society.
- University administrators need to realize that pyramidal organizational structures do not make sense for academic institutions.
- For universities to become true learning organizations, students must be part of knowledge sharing.
- Many future researchers believe that interdisciplinary majors will become vital for the future of universities. New departments should be interdisciplinary.
- A learning organization cannot last very long if its members are not interested in learning. Lifelong learning should be the core philosophy of university employees
- Educational institutions should appoint a responsible person (leader) to formally guide and support organizational learning, and this person should develop appropriate learning measurement methods to measure the learning levels of employees.

REFERENCES

Ainissyifa, H. (2012). The influence of knowledge management implementation toward the quality of high schools. *International Journal of Development and Sustainability*, *1*(2), 79–84.

Aktharsha, S. U., & Anisa, H. (2011). Knowledge Management System and Learning Organization: An Empirical Study in an Engineering Organization. *The IUP Journal of Knowledge Management, 6*(2), 33–51.

Al Ahmar, G. O., Rofiqb, A., & Hadiwidjojoc, D. (2014). The Impact of Knowledge Management, Learning Organization, and Educations Organization on Organization Performance: A Case in Brawijaya University. *Asia-Pacific Management and Business Application, 3*(1), 28–47. doi:10.21776/ub.apmba.2014.003.01.3

Alavi, M., Kayworth, T. R., & Leidner, E. L. (2006). An empirical examination of the influence of organizational culture on knowledge management practices. *Journal of Management Information Systems, 22*(3), 191–224. doi:10.2753/MIS0742-1222220307

Alrefaai, M. R., & Khalil, A. H. A. L. (2019). The Impact of Knowledge Management on the Learning Organization: An Empirical Study on the Telecommunications Sector in Damascus City. *International Journal of Academic Research in Business and Social Sciences, 9*(12), 774–783.

Alsabbagh, M., & Al Khalil, A. H. (2017). The Impact of Knowledge Management on Organizational Learning (An Empirical Study on the Education Sector in Damascus City). *International Journal of Academic Research in Business and Social Sciences, 7*(4), 560–578. doi:10.6007/IJARBSS/v7-i4/2833

Argote, L. (2013). *Organizational learning: Creating, retaining and transferring knowledge*. Springer. doi:10.1007/978-1-4614-5251-5

Cheng, E. C. K., Wu, S. W., & Hu, J. (2017). Knowledge management implementation in the school context: Case studies on knowledge leadership, storytelling, and taxonomy. *Educational Research for Policy and Practice, 16*(2), 177–188. doi:10.100710671-016-9200-0

Chou, S. Y., & Ramser, C. (2019). A multilevel model of organizational learning Incorporating employee spontaneous workplace behaviors, leadership capital and knowledge management. *The Learning Organization, 26*(2), 132–145. doi:10.1108/TLO-10-2018-0168

Dahou, K., Hacini, I., & Burgoyne, J. (2019). Knowledge management as a critical success factor in developing international companies' organizational learning capability. *Journal of Workplace Learning, 31*(1), 2–16. doi:10.1108/JWL-12-2017-0118

Danish, K. (2012). Impact of Knowledge Management Practices on Organizational Performance, evidence from Pakistan. *International Journal of Scientific and Engineering Research, 3*(8), 1–7.

Davenport, T. H., & Völpel, S. C. (2001). The rise of knowledge towards attention management. *Journal of Knowledge Management, 5*(3), 212–221. doi:10.1108/13673270110400816

Donate, M. J., & Pablo, J. D. S. (2015). The role of knowledge-oriented leadership in knowledge management practices and innovation. *Journal of Business Research, 68*(2), 360–370. doi:10.1016/j.jbusres.2014.06.022

Garvin, D. A. (1993). Building a Learning Organization. *Harvard Business Review, 71*, 78–91. PMID:10127041

Giniuniene, J., & Jurksiene, L. (2015). Dynamic capabilities, innovation and organizational learning: Interrelations and impact on firm performance. *Procedia: Social and Behavioral Sciences*, *213*, 985–991. doi:10.1016/j.sbspro.2015.11.515

Gold, A. H., Malhotra, A., & Segars, A. H. (2001). Knowledge management: An organizational capabilities perspective. *Journal of Management Information Systems*, *18*(1), 185–214. doi:10.1080/07421222.2001.11045669

Higher Education Council (HEC). (2021). Retrieved from: https://istatistik.yok.gov.tr/

Hsu, T. (2014). Effects of organization culture, organizational Learning and IT Strategy on Knowledge Management and Performance. *Journal of International Management Studies*, *9*(1), 50–57.

Jamalzadeh, M. (2012). The relationship between knowledge management and learning organization of faculty members at Islamic Azad University, Shiraz Branch in Academic year (2010-2011). *Social and Behavioral Sciences*, *62*, 1164–1168. doi:10.1016/j.sbspro.2012.09.199

Jerez-Gómez, P., Cespedes-Lorente, J., & Valle, R. (2005). Organizational Learning Capability: A Proposal of Measurement. *Journal of Business Research*, *58*(6), 715–725. doi:10.1016/j.jbusres.2003.11.002

Karasneh, A. A.-F. (2019). Reinforcing Innovation through Knowledge Management: Mediating Role of Organizational Learning. *Interdisciplinary Journal of Information, Knowledge, and Management*, *14*(1), 235–252. doi:10.28945/4427

Ke, W. L., & Wei, K. K. (2006). Organizational Learning Process: Its Antecedents and Consequences in Enterprise System Implementation. *Journal of Global Information Management*, *14*(1), 1–22. doi:10.4018/jgim.2006010101

Kidwell, J. J., Linde, K. M. V., & Johnson, S. L. (2000). Applying corporate knowledge management practices in higher education. *EDUCAUSE Quarterly*, *4*, 28–33.

Kordab, M., Raudeliuniene, J., & Meidute-Kavaliauskien, I. (2020). Mediating Role of Knowledge Management in the Relationship between Organizational Learning and Sustainable Organizational Performance. *Sustainability*, *12*(23), 1–20. doi:10.3390u122310061

Laal, M. (2011). Knowledge management in higher education. *Procedia Computer Science*, *3*(1), 544–549. doi:10.1016/j.procs.2010.12.090

Lee, H., & Choi, B. (2003). Knowledge management enablers, processes, & organizational performance: An integrative view & empirical examination. *Journal of Management Information Systems*, *20*(1), 179–228. doi:10.1080/07421222.2003.11045756

López, V. M. (2013). Leadership in organization knowledge to Mexico. *Procedia: Social and Behavioral Sciences*, *73*, 661–668. doi:10.1016/j.sbspro.2013.02.103

Marsick, V. J., & Watkins, K. E. (2003). Demonstrating the Value of an Organization's Learning Culture: The Dimensions of the Learning Organization Questionnaire. *Advances in Developing Human Resources*, *5*(2), 132–151. doi:10.1177/1523422303005002002

Menezes, E. A. C., Guimarães, T. A., & Bido, D. S. (2011). Dimensions of learning in organizations: Validation of the Dimensions of the Learning Organization Questionnaire (DLOQ) in the Brazilian context. *RAM. Revista de Administração Mackenzie*, *12*(2), 4–29. doi:10.1590/S1678-69712011000200002

Minghui, L. (2016). The Construction of Learning Oriented Enterprise Organization Based on Knowledge Management. *International Journal of Elementary Education.*, *5*(4), 40–46. doi:10.11648/j.ijeedu.20160504.11

Mohajan, H. (2017). The roles of knowledge management for the development of organizations. *Journal of Scientific Achievements*, *2*(2), 1–27.

Nodehi, H., Nehardani, H., & Keavanlou, F. (2013). The Relationship between Learning Organization and Knowledge Management. *International Research Journal of Applied and Basic Sciences*, *5*(10), 1303–1306.

Nouri, B. A., Ghorbani, R., & Soltani, M. (2017). The Effect of Knowledge Management on Organizational Innovation with the Mediating Role of Organizational Learning (Case Study: Agricultural Bank in Iran). *Journal of Applied Economics and Business Research*, *7*, 194–211.

Rose, S., Feldman, J., & Jankowski, J. (2009). A Cognitive Approach to the Development of Early Language. *Child Development*, *80*(1), 134–150. doi:10.1111/j.1467-8624.2008.01250.x PMID:19236397

Salleh, K. (2012). Knowledge Management in Public University: Empirical Relationships Between Learning Organization and Knowledge Management Process of Tacit Knowledge. *Knowledge Management International Conference (KMICe) 2012*, (July), 4-6.

Sangeeta, M., & Librarian, N. D. (2015). The importance of Knowledge Management in the Higher Educational Institutes. *Turkish Online Journal of Distance Education*, (January), 162-183.

Sart, G., Sezgin, F. H., & Demir, N. (2018). Mobbingin Mesleki Tükenmişlik Algısı Üzerine Etkileri: Kadın Akademisyenler Örneği. *Beykoz Akademi Dergisi*, *6*(1), 118–135. doi:10.14514/BYK.m.21478082.2018.6/1.117-135

Sezgin, F. H. (2016). Bayesci Faktör Analizi ve Maslach Tükenmişlik Envanteri Uygulaması, *International Conference on Scientific Cooperation for the Future in the Social Sciences (USAK) Bildiriler Kitabı*, 1283-1296.

Singh, S. K. (2011). Leadership and organizational learning in knowledge management practices in global organizations. *Indian Journal of Industrial Relations*, *47*(2), 353–365.

Stylianou, V., & Savva, A. (2016). Investigating the knowledge management culture. *Universal Journal of Educational Research*, *4*(7), 1515–1521. doi:10.13189/ujer.2016.040703

Thambi, M., & O'Toole, P. (2012). Applying a knowledge management taxonomy to secondary schools. *School Leadership & Management*, *32*(1), 91–102. doi:10.1080/13632434.2011.642350

Watanabe, R. M., & Senoo, D. (2011). The role of a strong corporate mission for knowledge management. *Journal of Information & Knowledge Management*, *10*(2), 109–122. doi:10.1142/S0219649211002869

Wu, W.-L., Lee, Y.-C., & Shu, H.-S. (2013). Knowledge management in educational organizations: A perspective of knowledge spiral. *The International Journal of Organizational Innovation*, *5*(4), 7–13.

Yang, D. (2011). The Effect of Knowledge Management on Product Innovation-Evidence from the Chinese Software Outsourcing Vendors. *iBusiness, 3*, 16-22.

Yang, L., Huang, C., & Hsu, T. (2014). Knowledge leadership to improve project and organizational performance. *International Journal of Project Management*, *32*(1), 40–53. doi:10.1016/j.ijproman.2013.01.011

Yazıcıoğlu, Y., & Erdoğan, S. (2004). *SPSS Uygulamalı Bilimsel Araştırma Yöntemleri*. Detay Yayıncılık.

ADDITIONAL READING

Flores, L., Zheng, W., Rau, D., & Thomas, C. (2012). Organisational learning: Subprocess identification, construct validation, and an empirical test of cultural antecedents. *Journal of Management*, *38*(2), 640–667. doi:10.1177/0149206310384631

Fong, P. S. W., & Choi, S. K. Y. (2009). The processes of knowledge management in professional services firms in the construction industry: A critical assessment of both theory and practice. *Journal of Knowledge Management*, *13*(2), 110–126. doi:10.1108/13673270910942736

Newman, N., & Newman, D. (2015). Learning and knowledge: A dream or nightmare for employees. *The Learning Organization*, *21*(1), 58–71. doi:10.1108/TLO-02-2013-0002

Patnaik, B., Beriha, G. S., Mahapatra, S. S., & And Singh, N. (2003). Organizational Learning in Educational Settings (Technical): An Indian Perspective. *The Learning Organization*, *20*(2), 153–172. doi:10.1108/09696471311303782

Puncreobutr, V. (2016). Education 4. 0: New challenge of learning. St. Theresa. *Journal of the Humanities and Social Sciences*, 2(2), 92–97.

Wang, Y., & Ellinger, A. D. (2011). Organizational learning: Perception of external environment and innovation performance. *International Journal of Manpower*, *32*(5/6), 512–536. doi:10.1108/01437721111158189

Chapter 13
The Effect of Knowledge Sharing on Innovative Behavior in Organizations:
A Case of Academic Members

Pınar Tokal
Netkent Mediterranean Research and Science University, Cyprus

ABSTRACT

Knowledge sharing is the transfer of knowledge owned by individuals to other individuals within the organization within the scope of knowledge management. Knowledge sharing is vital for all organizations that aim for success. The concept of knowledge includes concepts such as innovation, development, research, learning, effectiveness, progress, and replacing what is wrong with what is right. Today, one of the common characteristics of developed countries is the updating of information and the effective use of the information network. Knowledge and the transfer and sharing of this knowledge between individuals affect the development of societies and countries, as well as businesses and organizational structures within businesses. The aim of this study is to determine the effect of knowledge sharing on innovative behavior for 890 academics. As a result of the structural equation model analysis, it is seen that knowledge sharing is positively and significantly related to innovative behavior (β=0.593; p<0.01).

INTRODUCTION

Knowledge sharing is the phenomenon that makes the greatest contribution to the modern age's greatest passion - learning. As information is shared, it enriches and develops itself and gains a different dimension in reaching the right information. The concept of knowledge includes innovation, development, research, learning, effectiveness, and progress (Cummings, 2004:355). When the sociocultural structures of many developed countries are examined today, it is possible to see that information is updated and the information network is used effectively. Knowledge and the transfer and sharing of this knowledge

DOI: 10.4018/978-1-6684-6620-9.ch013

between individuals affect the development of societies and countries, as well as businesses and organizational structures within businesses (Hanif and Bukhari, 2015:24).

Today, businesses have understood that in order not to fall behind the global market, they need to create the necessary structures to obtain information quickly and effectively and to channel that information into the organization. Knowledge is not only a part of the management system in organizations, but it is also accepted as the main part of the system. In the past, organizations have traditionally managed three types of resources effectively: financial, human, and material (Pittino et al., 2018:314). Today, the asset of information has become the most important organizational resource.

The phenomenon of innovation is accepted as one of the most important phenomena that create competitive advantage today. One of the main reasons for this is the very short production and consumption time; the other is the increasing share of technological products and services in total consumption (Zhang, 2019:65). In order for their businesses to survive and be successful in the future, managers find solutions to business problems and challenges through innovation (Huang & Liu, 2015:208). In Kim & Park (2017) study, behaviors such as researching new technologies, proposing new ways to achieve goals, researching and securing resources to implement new business methods and new ideas by employees are expressed as innovative work behavior.

As knowledge is shared among employees in business life, it paves the way for learning something new. Thus, it facilitates employees to develop innovative work behaviors for their job roles. It is possible for employees to learn or apply new methods and approaches in business processes through knowledge sharing (Akram, 2018:188). Knowledge sharing and the impact of knowledge on innovative behaviors constitute a very important issue for academic members in universities, which are knowledge producing institutions. Considering the need for well-equipped and trained human resources in our age, the importance of universities that train these human resources is indisputable. In today's competitive environment, it is known that innovative behaviors in universities will create input for innovation and entrepreneurial activities. From this point of view, the aim of the study is to determine the effect of knowledge sharing on innovative behaviors for 890 academics working in universities in Istanbul with the help of structural equation modeling.

CONCEPTUAL FRAMEWORK

Knowledge Sharing

Knowledge sharing is the process by which knowledge between individuals and organizations is transferred to the organizational level and transformed into an economic and competitive value for the organization (Casimir et al., 2012:742). The focus of knowledge sharing is on individuals who can explain, encode and communicate knowledge to other individuals, groups and organizations (Hau et al., 2013:258). In their study, Liao et al. (2007) stated that knowledge sharing is a synergistic process and that engaging in a dialog with others enables an individual to benefit from new insights and ideas and allows the individual to develop their ideas. Benefiting from knowledge is only possible when people share their knowledge and build on the knowledge they have learned from others. In this context, an organization's ability to make effective use of knowledge depends on the employees who produce, share and use knowledge (Mirzaee & Ghaffari, 2018:503). The ultimate goal of sharing the knowledge that employees also possess is to transfer this knowledge into organizational assets and resources. The process of knowledge sharing can be defined as the generation of new knowledge as a result of the mutual exchange of knowledge among employees (Ouakouak & Ouedraogo, 2019:758). Sharing behavior in the knowledge sharing process is the sharing of explicit knowledge (that which is written or codified in some way) and the sharing of tacit knowledge (that which exists in an individual's mind) or a combination of two distinguishable types of knowledge sharing (Akhavan et al., 2015:564). Davenport (1997) defined sharing as a voluntary act and separated it from reporting. Reporting involves the exchange of information based on some routine or structured format. Sharing, on the other hand, refers to the conscious act of a person who exchanges information, but who participates voluntarily in this process without any compulsion to do so. Sharing facilitates coordination of activities, but effective knowledge integration requires that each individual is aware of everyone else's knowledge repertoire.

The knowledge sharing process is defined as the generation of new knowledge as a result of the mutual exchange of knowledge among employees (Choi et al., 2010: 857). The knowledge sharing process is named as two dimensions: knowledge donating and knowledge collecting (Khorakian et al., 2019:166). In the process of knowledge sharing, organizations need not only to acquire knowledge but also to implement and manage the acquired knowledge efficiently (Kuo et al.,2014:701). The knowledge sharing process refers to the way in which employees share their work-related experience, expertise, know-how and contextual knowledge with other colleagues. Employees' willingness to actively communicate with colleagues (i.e. knowledge sharing) and actively consult (i.e. knowledge gathering) to learn from their colleagues encompass the knowledge sharing process (Lin, 2007:317).

The literature on knowledge sharing shows the existence of different influences on employees' knowledge sharing processes such as individual, organizational and technology factors. Individual factors include such things as employees helping each other, sharing their knowledge, encouraging each other in knowledge sharing, recognizing the effectiveness of knowledge and emphasizing this at every opportunity. Organizational factors include issues such as top management's perspective on knowledge and knowledge sharing, encouraging employees to share knowledge, enriching knowledge sharing channels, and knowledge sharing culture. Technological factors are the use of internal and external information technologies to access information (Sankowska, 2013:90; Liu & Phillips, 2011:46). As a result, by increasing the level of knowledge sharing within the organization, employees are made more open or ready for innovation, because knowledge sharing improves individual and organizational perspectives. Through knowledge sharing, employees can have the advantage of developing new thoughts and ideas, new methods and techniques, new products and services (Zhao et al., 2021:375).

Innovative Work Behavior

The concept of innovative behavior first appeared in the study conducted by Scott and Bruce (1994) to determine the antecedents and mediators of individual innovativeness based on the views of Kanter (1988) and Van de Ven (1986). In subsequent studies, innovative behavior was mainly found in organizational behavior studies under the name of innovative work behavior.

The phenomenon of innovation is accepted as one of the most important phenomena that create competitive advantage today. One of the main reasons for this is the very short production and consumption time; the other is the increasing share of technological products and services in total consumption (Hughes et al., 2018:754). In order for their businesses to survive and be successful in the future, managers find solutions to business problems and challenges through innovation (Spanuth & Wald, 2017:1305). In Yuan and Woodman's (2010) study, behaviors such as researching new technologies by employees, proposing new ways to achieve goals, researching and securing resources to implement new business methods and new ideas were expressed as innovative work behavior.

In general, innovative work behavior can be defined as discretionary and extra-role behavior that is not specified as part of employees' official job descriptions (Radaelli et al., 2014: 403). Therefore, it can be said that innovative work behavior is mostly the result of employees' discretionary efforts and intuition as a result of their cognitive evaluations of factors such as organizational environment and manager relationships (Akram et al.,2018:189). Rao et al., (2019) stated that innovative work behavior typically involves the exploration of opportunities and the generation of new ideas (creativity-driven behavior), but it can also include behaviors applied to the application of new knowledge or the improvement of processes to enhance personal and/or business performance (application-driven behavior).

Innovative behavior also refers to the change that transforms new ideas and thoughts from abstract concepts into concrete activities and affects development in individuals and organizations. It is very important for organizations, especially for employees to demonstrate innovative behaviors (Zhou and Shalley,2003:167). Because in order for organizations to ensure their competitiveness in the market they exist in, they must not lag behind developments and be open to innovations. The innovative behavior of employees also represents an important act that leads them to productivity and effectiveness (Majchrzak et al., 2004:178).

In order to systematically develop and improve their activities, organizations must have and manage innovative behavior capability, which is a strategic and dynamic competitive advantage (Michna, 2018:96). Innovative behavior can enable employees to develop themselves to adapt to job requirements by generating and applying ideas by changing themselves or the work environment (Anser et al., 2020:1834).

Examples of innovative behaviors of employees include inventing new technologies, suggesting new ways to achieve goals, implementing new working methods, securing resources to implement new ideas and searching for new resources (Yuan & Woodman, 2010:328). Innovative behavior includes both the generation of new ideas by employees and the internalization of new ideas generated by other employees for the organization or business unit (Chen et al., 2010:852).

STATISTICAL ANALYSIS

The Purpose and the Importance of the Research

It is a widely accepted view by both academic authorities and organizational theorists and practitioners that success in contemporary organizational structures is closely related to the ability to create and process knowledge and to produce economic and moral values by making maximum use of knowledge (Curado & Vieira, 2019:1452). Innovation can be fostered through knowledge sharing and ongoing learning. In other words, employees' knowledge sharing is a prerequisite for innovative service behavior; however, this requires practices for the development and preservation of individual and collective knowledge within the organization (Kmieciak & Michna, 2018:562). Innovation requires not only intellect and talent, but also knowledge. Therefore, knowledge needs to be shared among employees to increase innovation in enterprises. In other words, when employees share their knowledge, they can develop new ideas and new services that add value to the business (Leong & Rasli, 2014:595). For universities to increase their innovation capabilities, it is crucial that academics are willing to collect and share knowledge. Universities have to be knowledge producing institutions. therefore, the innovative behavior of academics is

Table 1. Research hypotheses

	Hypotheses
H_1	There is a significant relationship between knowledge sharing and innovative behavior
H_2	There is a significant relationship between knowledge sharing between employees and innovative behavior
H_3	There is a significant relationship between corporate knowledge sharing and innovative behavior
H_4	There is a significant relationship between individual knowledge sharing and innovative behavior

an important input to high performance. The aim of this study is to determine the effect of knowledge sharing on innovative behaviors for 890 academicians working in universities in Istanbul with the help of structural equation modeling.

Research Sample

The population of this study is academicians working at universities in Istanbul. According to the statistics of the number of academicians of the Higher Education Council (HEC) in 2021, the total number of academicians (research assistants, lecturers, doctor lecturers, associate professors, and professors) in state universities in Istanbul is 36,896. “Simple random sampling” was used as the sampling method. Surveys were conducted via e-mail between 01.05.2022 and 01.09.2022. The sample is 890 academicians who voluntarily participated in the research.

In the sample size table developed by Yazıcıoğlu and Erdoğan (2004), it was determined that 384 people should be studied for p=0.50 and q=0.50 with a sampling error of 0.05. In this paper, analyzes were carried out with 890 samples.

Research Hypotheses

The main hypothesis of the study is that knowledge sharing is significantly related to innovative behaviour. The developed hypotheses are presented in Table 1.

Data Collection Tool

Innovative Behavior Scale: In the study, Akkoç (2012) validated the Turkish version of the scale used by Scott and Bruce (1994). In the scale consisting of six statements and one dimension, the cronbach alpha value was found to be α=.891. In the scale, the answers were taken with a 5-point Likert scale (1=Strongly disagree, 5=Strongly agree).

Knowledge Sharing Scale: The knowledge sharing scale developed by Öneren et al. (2016) consists of a total of 18 items and 3 sub-dimensions: Inter-employee Knowledge Sharing, Organizational Knowledge Sharing and Individual Knowledge Sharing. A 5-point Likert scale was used for the answers (1=strongly disagree - 5=strongly agree). Items 14, 15, 16, 17 and 18 in the scale are reverse questions.

Statistical Methods

The data acquired from the scale utilized in the study were analyzed via IBM SPSS 27.0 and AMOS 25.0 programs. In the first stage, percentage distributions regarding demographic and general information were presented. In the second stage, exploratory factor analysis (EFA) was applied for the scales, and besides, confirmatory factor analysis (CFA) was performed to test construct validity. In the last stage, structural equation model estimation results were included in order to identify the relationships.

FINDINGS AND RESULTS

Survey Reliability

In this study, Cronbach Alpha, Split, Parallel, Absolute Precise Parallel (strict) tests were carried out as a reliability test. A Cronbach Alpha value above 70% indicates that the survey was successful. Some researchers take this value as 75% (Sezgin, 2016). The fact that other criteria are above 70% indicates that the questionnaire has internal consistency and that the results can be trusted (Sart et al., 2018). In this study, the results of the reliability analysis of the questionnaire are Cronbach-Alpha = 0.922, Parallel = 0.920, Split = 0.919-0.924, and Strict = 0.922.

Descriptive Statistics

47.5% of the participants were female, and 52.5% were male. 30.2% of the academician were aged 20-30, 29.6% were aged 31-40, 20.1% were aged 41-50, and 20.1% were aged 51 years and over. In terms of marital status, 56.3% of the individuals were married, while 43.7% were single. Looking at the participants' employment terms, 5.6% worked less than 1 year, 19.8% between 1-5 years, 29.4% between 6-10 years, 23.0% between 11-20 years, and 22.2% worked for 21 years or more. According to their job positions, 18.4% were research assistant, 22.1% assistant professor, 19.9% associate professor, 22.5% professor and 17.1% lecturer.

Exploratory Factor Analysis

Exploratory factor analysis attempts to identify underlying variables, or factors, that explain the pattern of correlations within a set of observed variables. Factor analysis is often used in data reduction to identify a small number of factors that explain most of the variance that is observed in a much larger number of manifest variables. Factor analysis can also be used to generate hypotheses regarding causal mechanisms or to screen variables for subsequent analysis. After the suitability of the data set was confirmed by the tests, the "Principal Components Analysis" approach was applied via the "Varimax" rotation technique as the factor retention method in order to reveal the factor structure.

As a result of the exploratory factor analysis (EFA): Regarding the factor structure, for the knowledge sharing scale, 3 factor structure was obtained, which explained 67.33% of the total variance; for the innovative behavior scale, 1 factor structure was obtained which explained 71.35% of the total variance. Following the EFA, there was no need to eliminate any questions since there was no item below 0.20 in the inference column and no item with anti-image matrix diagonal values below 0.50.

Table 2. Exploratory factor analysis results

Knowledge Management	PVE	CA	AVE	CR
Knowledge Acquisition	26.39%	0.920	0.796	0.918
Knowledge Conversation	21.09%	0.916		
Knowledge Application	19.85%	0.918		
KMO= 0.924; Bartlett ÷2= 6463.45 and p= 0.000; Percent of Variance Explained: 67.33%				
Organizational Learning	**PVE**	**CA**	**AVE**	**CR**
Organizational Learning	71.35%	0.919	0.801	0.915
KMO= 0.930; Bartlett ÷2= 5993.12 and p= 0.000; Percent of Variance Explained: 71.35%				

PVE: Percent of Variance Explained, CA: Cronbach's Alpha Coefficient; AVE: Average Variance Extracted; CR: Construct Reliability

Confirmatory Factor Analysis

Confirmatory factor analysis (CFA) is a multivariate statistical procedure that is used to test how well the measured variables represent the number of constructs. Confirmatory factor analysis (CFA) and exploratory factor analysis (EFA) are similar techniques, but in exploratory factor analysis (EFA), data is simply explored and provides information about the numbers of factors required to represent the data. In exploratory factor analysis, all measured variables are related to every latent variable. But in confirmatory factor analysis (CFA), researchers can specify the number of factors required in the data and which measured variable is related to which latent variable. Confirmatory factor analysis (CFA) is a tool that is used to confirm or reject the measurement theory. The objective of confirmatory factor analysis is to test whether the data fit a hypothesized measurement model.

In Table 3, X^2 /sd= 2.45, a "good fit" decision was made since it met the condition ≤3. NFI=0.930 resulted in the range of 0.94-0.90, thus an "acceptable fit"; other results were obtained as TLI (NNFI)= 0.922 was resulted in the range of 0.94-0.90, thus an "acceptable fit"; IFI= 0.987 was ≥0.95, a "good fit"; CFI= 0.984 was ≥0.97, an "good fit"; RMSEA= 0.024 was ≤0.05, a "good fit"; GFI= 0.938 was ≥0.90, a "good fit"; AGFI= 0.943 was ≥0.90, a "good fit"; and RMR= 0.011 was ≤0.05, thus it was a "good fit".

Structural Equation Modeling

Structural equation modeling (SEM) is a statistical technique used to test models in which causal relationships and correlation relationships between observed variables and latent variables coexist; it is a multivariate method that is formed by combining analyses such as variance and covariance analyses, factor analysis, and multiple regression in order to predict interdependency relationships. Considering the goodness of fit criteria for SEM, it was found that X^2 /sd= 2.31, a "good fit" decision was made because it met the condition £3. NFI=0.939 resulted in 0.94-0.90, thus an "acceptable fit"; other results were obtained as TLI (NNFI)= 0.928 was 0.94-0.90, a "acceptable fit"; IFI = 0.983 was ≥0.95, a "good fit"; CFI= 0.976 was ≥0.97, an "good fit"; RMSEA= 0.013 was ≤0.05, a "good fit"; GFI= 0.961 was ≥0.90, a "good fit"; AGFI= 0.970 was ≥0.90, a "good fit"; and RMR= 0.019 was ≤0.05, thus it was a "good fit". SEM estimates fit criteria yielded the result "acceptable" for only one criterion; others revealed that a "good fit" was attained and the model was appropriate for interpretation.

Table 3. CFA model fit indices

Measurement (Fit Statistic)	Good Fit	Acceptable Fit	Research Model Value	Fit Status
General Model Fit				
x^2 /sd	≤3	≤4-5	2.45	Good fit
Comparative Fit Statistics				
NFI	≥0.95	0.94-0.90	0.930	Acceptable fit
TLI (NNFI)	≥0.95	0.94-0.90	0.922	Acceptable fit
IFI	≥0.95	0.94-0.90	0.987	Good fit
CFI	≥0.97	≥0.95	0.984	Good fit
RMSEA	≤0.05	0.06-0.08	0.024	Good fit
Absolute Fit Indices				
GFI	≥0.90	0.89-0.85	0.938	Good fit
AGFI	≥0.90	0.89-0.85	0.943	Good fit
Residual-Based Fit Index				
RMR	≤0.05	0.06-0.08	0.011	Good fit

According to the SEM results, significant relationships were obtained among the variables. KS has a statistically significant positive effect on IB (β=0.593; p<0.01). KSBE has a statistically significant positive effect on IB (β= 0.585; p<0.01). CKS has a statistically significant positive effect on OL (β= 0.594; p<0.01). IKS has a statistically significant positive effect on OL (β= 0.578; p<0.01). According to the coefficient sizes for sub-dimension, the CKS variable is most effective on the IB variable.

CONCLUSION

Access to information and its sharing within the organization is one of the most important elements that an organization should attach importance to in order to achieve its goals and increase its performance. Within the scope of the formation of innovative work behavior, knowledge sharing among employees is an important driving force.

Table 4. SEM model estimation results

Structural Relation	Direction	Estimated Coefficient	Std. Error	t-Statistic	p	Result
KS → IB	+	0.593	0.085	6.976	0.000*	Significant relationship
KSBE → IB	+	0.585	0.093	6.290	0.006*	Significant relationship
CKS → IB	+	0.594	0.079	7.518	0.012*	Significant relationship
IKS → IB	+	0.578	0.066	8.757	0.000*	Significant relationship

*p<0.01, *KS: Knowledge Sharing, KSBE: Knowledge Sharing Between Employees*
CKS: Corporate Knowledge Sharing, IKS: Individual Knowledge Sharing IB: Innovative Behavior

It is reported in many empirical and theoretical studies that the potential for individual innovation behavior is furthered with knowledge sharing (Mura et al. 2013; Kim & Lee (2013) and Molose & Ezeuduji (2015), teams (Maccurtain et al., 2008; Hannemann-Weber et al., 2011) and organizations (Lin, 2007; Wangpipatwong, 2009; Mehrabani & Mohamad, 2011). While individual innovation behavior is explained as "exploring opportunities and generating new ideas (creativity-related behavior), this explanation can encompass actions aimed at implementing change, new knowledge, or furthering personal and/ or business performance by improving related processes (implementation-related behavior)" (De Jong & Hartog, 2008:5). Knowledge sharing and individual innovations were connected in many suggestions. It was suggested in the Mura et al. (2013) study that knowledge sharing improves personal desire and potential to promote and implement new ideas. The two phenomena are theoretically linked, however the suggested association was empirically examined in only a few studies.

Rapid changes and developments affect the dynamics of the education sector. It is not possible for the education sector, which trains human resources for Industry 4.0 and discusses the characteristics of people who can compete in the global market, to remain insensitive to innovation. Academics with innovative behavior are seen as "creative educators" who can create ideas and contribute to the progress of the university. By helping students to express their creativity in the learning process through innovative teaching, academics who exhibit innovative work behavior can respond to the different needs and characteristics of students by using new strategies.

It is very important to create the grounds for academics to produce all kinds of innovative ideas, and to share knowledge and information in order for creative ideas to be supported. Knowledge sharing also positively affects the trust between managers and employees in the organization. It is very important to create a knowledge sharing network in universities, to benefit from intellectual capital, to support the learning processes of employees, to give opportunities for new approaches and experiments, and to reward successful initiatives. In addition to the internalization of innovative behavior in academics, establishing and sustaining the phenomenon of trust within the institution is very important for university performance. In order to establish the concept of reputation in universities, it should be ensured that information sharing is implemented, a transparent management approach is adopted, and institutional ties fed by organizational culture are strengthened.

Universities seeking to increase innovative work behavior should examine their management processes, differentiate their bureaucratic structures and adopt effective leadership practices. Academics should be supported to experience innovative practices in the education and training process in line with their institutional and personal goals. Innovation-based human resource development is at the forefront of this support. Trainings to support the innovation process, practices that go beyond the routine, doing business with different methods, respecting different opinions, a culture of tolerance and giving importance to interdisciplinary studies will improve the innovative work behaviors of academics.

REFERENCES

Akhavan, P., Hosseini, M., Abbasi, M., & Manteghi, M. (2015). Knowledge sharing determinants, behaviours and innovative work behaviours: An integrated theoretical view and empirical examination. *Aslib Journal of Information Management*, *67*(5), 562–591. doi:10.1108/AJIM-02-2015-0018

Akkoç, I. (2012). Gelişim Kültürü ve Etik İklimin Yenilikçiliğe Etkisinde Dağıtım Adaletinin Rolü. *Uluslararası Alanya İşletme Fakültesi Dergisi*, *4*(3), 45–60.

Akram, T., Lei, S., Haider, M. J., & Hussain, S. T. (2018). Exploring the impact of knowledge sharing on the innovative work behavior of employees: A study in China. *International Business Research, 11*(3), 186–194. doi:10.5539/ibr.v11n3p186

Anser, M. K., Yousaf, Z., Khan, A., & Usman, M. (2020). Towards innovative work behavior through knowledge management infrastructure capabilities: Mediating role of functional flexibility and knowledge sharing. *European Journal of Innovation Management*, *24*(5), 1832–1859.

Casimir, G., Lee, K., & Loon, M. (2012). Knowledge sharing: Influences of trust, commitment and cost. *Journal of Knowledge Management, 16*(5), 740–753. doi:10.1108/13673271211262781

Chen, C.-J., Huang, J.-W., & Hsiao, Y.-C. (2010). Knowledge management and innovativeness. The role of organizational climate and structure. *International Journal of Manpower*, *31*(8), 848–870. doi:10.1108/01437721011088548

Choi, S. Y., Lee, H., & Yoo, Y. (2010). The impact of information technology and transactive memory systems on knowledge sharing, application, and team performance: A field study. *Management Information Systems Quarterly*, *34*(1), 855–870. doi:10.2307/25750708

Cummings, J. (2004). Work Groups, Structural Diversity and Knowledge Sharing in a Global Organisation. *Management Science*, *50*(3), 352–364. doi:10.1287/mnsc.1030.0134

Curado, C., & Vieira, S. (2019). Trust, knowledge sharing and organizational commitment in SMEs. *Personnel Review*, *48*(6), 1449–1468. doi:10.1108/PR-03-2018-0094

Davenport, T. H. (1997). Ten Principles of Knowledge Management and Four Case Studies. *Knowledge and Process Management*, *4*(3), 187–208. doi:10.1002/(SICI)1099-1441(199709)4:3<187::AID-KPM99>3.0.CO;2-A

De Jong, J. P. J., & Den Hartog, D. N. (2008). *Innovative Work Behavior: Measurement and Validation.* A working paper. EIM Business and Policy Research. Working Paper, 1-27.

Hanif, A., & Bukhari, I. (2015). Relationship between innovative work behavior and job involvement among the employees of telecom sector. *Pakistan Journal of Social and Clinical Psychology*, *13*(2), 23–29.

Hannemann-Weber, H., Kessel, M., Budych, K., & Schultz, C. (2011). Shared communication processes within healthcare teams for rare diseases and their influence on healthcare professionals' innovative behavior and patient satisfaction. *Implementation Science; IS*, *6*(40), 1–7. doi:10.1186/1748-5908-6-40 PMID:21510848

Hau, Y. S., Kim, B., Lee, H., & Kim, Y. G. (2013). The effects of individual motivations and social capital on employees' tacit and explicit knowledge sharing intentions. *International Journal of Information Management*, *33*(2), 256–266. doi:10.1016/j.ijinfomgt.2012.10.009

Higher Education Council (HEC). (2021). Retrieved from: https://istatistik.yok.gov.tr/, (12.09.2022).

Huang, C.-E., & Liu, C.-H. S. (2015). Employees and creativity: Social ties and access to heterogeneous knowledge. *Creativity Research Journal*, *27*(2), 206–213. doi:10.1080/10400419.2015.1030309

Hughes, M., Rigtering, J. P. C., Covin, J. G., Bouncken, R. B., & Kraus, S. (2018). Innovative behaviour, trust and perceived workplace performance. *British Journal of Management*, *29*(4), 750–768. doi:10.1111/1467-8551.12305

Kanter, R. M. (1988). When a Thousand Flowers Bloom: Structural, Collective and Social Conditions for Innovation in Organization. *Research in Organizational Behavior*, *10*, 169–211.

Khorakian, A., Mohammadi Shahroodi, H., Jahangir, M., & Nikkhah Farkhani, Z. (2019). Innovative work behavior in public organizations: The roles of ethical and knowledge sharing behaviors. *Creativity Research Journal*, *31*(2), 164–173. doi:10.1080/10400419.2019.1607444

Kim, T. T., & Lee, G. (2013). Hospitality Employee Knowledge-Sharing Behaviors in the Relationship between Goal Orientations and Service Innovative Behavior. *International Journal of Hospitality Management*, *34*, 324–337. doi:10.1016/j.ijhm.2013.04.009

Kim, W., & Park, J. (2017). Examining structural relationships between work engagement, organizational procedural justice, knowledge sharing, and innovative work behavior for sustainable organizations. *Sustainability*, *9*(2), 205–220. doi:10.3390u9020205

Kmieciak, R., & Michna, A. (2018). Knowledge management orientation, innovativeness, and competitive intensity: Evidence from Polish SMEs. *Knowledge Management Research and Practice*, *16*(4), 559–572. doi:10.1080/14778238.2018.1514997

Kuo, Y. K., Kuo, T. H., & Ho, L. A. (2014). Enabling innovative ability: Knowledge sharing as a mediator. *Industrial Management & Data Systems*, *114*(5), 696–710. doi:10.1108/IMDS-10-2013-0434

Leong, C. T., & Rasli, A. (2014). The Relationship between innovative work behavior on work role performance: An empirical study. *Procedia: Social and Behavioral Sciences*, *129*(1), 592–600. doi:10.1016/j.sbspro.2014.03.717

Liao, S.-H., Fei, W.-C., & Chen, C.-C. (2007). Knowledge Sharing, Absorptive Capacity, and Innovation Capability: An Empirical Study of Taiwan's Knowledge-Intensive Industries. *Journal of Information Science*, *33*(3), 340–359. doi:10.1177/0165551506070739

Lin, H. F. (2007). Knowledge sharing and firm innovation capability: An empirical study. *International Journal of Manpower*, *28*(3), 315–332. doi:10.1108/01437720710755272

Liu, Y., & Phillips, J. S. (2011). Examining the antecedents of knowledge sharing in facilitating team innovativeness from a multilevel perspective. *International Journal of Information Management*, *31*(1), 44–52. doi:10.1016/j.ijinfomgt.2010.05.002

Maccurtain, S., Flood, P. C., Ramamoorthy, N., West, M. A., & Dawson, J. F. (2008). *Top team trust, knowledge sharing and innovation*. Link Working Paper Series WP 05-08, 1-26.

Majchrzak, A., Cooper, L. P., & Neece, O. (2004). Knowledge reuse for innovation. *Management Science*, *50*(2), 174–188. doi:10.1287/mnsc.1030.0116

Mehrabani, E. S., & Mohamad, N. A. (2011). Identifying the important factors influencing the implementation of succession planning. *International Conference on Information and Finance*, *21*, 37-41.

Michna, A. (2018). The mediating role of firm innovativeness in the relationship between knowledge sharing and customer satisfaction in SMEs. *The Engineering Economist*, *29*(1), 93–103.

Mirzaee, S., & Ghaffari, A. (2018). Investigating the impact of information systems on knowledge sharing. *Journal of Knowledge Management*, *22*(3), 501–520. doi:10.1108/JKM-08-2017-0371

Molose, T., & Ezeuduji, I. O. (2015). Knowledge sharing, team culture and service innovation in the hospitality sector : The case of South Africa. *African Journal of Hospitality, Tourism and Leisure*, *4*(1), 1–16.

Mura, M., Lettieri, E., Radaelli, G., & Spiller, N. (2013). Promoting professionals' innovative behaviour through knowledge sharing: The moderating role of social capital. *Journal of Knowledge Management*, *17*(4), 527–544. doi:10.1108/JKM-03-2013-0105

Öneren, M., Çiftçi, G. E., & Harman, A. (2016). Bilgi Paylaşiminin Yenilikçi Davranişa Ve Örgütsel Güvene Etkisi Üzerine Bir Araştirma. *Akademik Bakiş Dergisi*, *58*(1), 127–157.

Ouakouak, M. L., & Ouedraogo, N. (2019). Fostering knowledge sharing and knowledge utilization: The impact of organizational commitment and trust. *Business Process Management Journal*, *25*(4), 757–779. doi:10.1108/BPMJ-05-2017-0107

Pittino, D., Martinez, A. B., Chirico, F., & Galvan, R. S. (2018). Psychological ownership knowledge sharing and entrepreneurial orientation in family firms: The moderating role of governance heterogeneity. *Journal of Business Research*, *84*(1), 312–326. doi:10.1016/j.jbusres.2017.08.014

Radaelli, G., Lettieri, E., Mura, M., & Spiller, N. (2014). Knowledge sharing and innovative work behaviorin healthcare: A micro-level investigation of direct and indirect effects. *Creativity and Innovation Management*, *23*(4), 400–414. doi:10.1111/caim.12084

Rao Jada, U., Mukhopadhyay, S., & Titiyal, R. (2019). Empowering leadership and innovative work behavior: A moderated mediation examination. *Journal of Knowledge Management*, *23*(5), 915–930. doi:10.1108/JKM-08-2018-0533

Sankowska, A. (2013). Relationships between organizational trust, knowledge transfer, knowledge creation, and firm's innovativeness. *The Learning Organization*, *20*(1), 85–100. doi:10.1108/09696471311288546

Sart, G., Sezgin, F. H., & Demir, N. (2018). Mobbingin Mesleki Tükenmişlik Algısı Üzerine Etkileri: Kadın Akademisyenler Örneği. *Beykoz Akademi Dergisi*, *6*(1), 118–135. doi:10.14514/BYK.m.21478082.2018.6/1.117-135

Scott, S. G., & Bruce, R. A. (1994). Determinants of innovative behavior: A path model of individual innovation in the workplace. *Academy of Management Journal*, *37*(3), 580–607. doi:10.2307/256701

Sezgin, F. H. (2016). Bayesci Faktör Analizi ve Maslach Tükenmişlik Envanteri Uygulaması. *International Conference on Scientific Cooperation for the Future in the Social Sciences (USAK) Bildiriler Kitabı*, 1283-1296.

Spanuth, T., & Wald, A. (2017). How to unleash the innovative work behavior of project staff? The role of affective and performance-based factors. *International Journal of Project Management, 35*(7), 1302–1311. doi:10.1016/j.ijproman.2017.07.002

Van de Ven, A. H. (1986). Central problems in the management of innovation. *Management Science, 32*(1), 590–607. doi:10.1287/mnsc.32.5.590

Wangpipatwong, S. (2009). Factors influencing knowledge sharing among university students. *Proceedings of the 17th International Conference on Computers in Education, ICCE 2009*, 800-807.

Yazıcıoğlu, Y., & Erdoğan, S. (2004). *SPSS Uygulamalı Bilimsel Araştırma Yöntemleri*. Detay Yayıncılık.

Yuan, F., & Woodman, R. W. (2010). Innovative behaviorin the workplace: The role of performance and image outcome expectations. *Academy of Management Journal, 53*(2), 323–342. doi:10.5465/amj.2010.49388995

Zhang, W., Sun, S. L., Jiang, Y., & Zhang, W. (2019). Openness to experience and team creativity: Effects of knowledge sharing and transformational leadership. *Creativity Research Journal, 31*(1), 62–73. doi:10.1080/10400419.2019.1577649

Zhao, S., Jiang, Y., Peng, X., & Hong, J. (2021). Knowledge sharing direction and innovation performance in organizations: Do absorptive capacity and individual creativity matter? *European Journal of Innovation Management, 24*(2), 371–394. doi:10.1108/EJIM-09-2019-0244

Zhou, J., & Shalley, C. E. (2003). Research on employee creativity: A critical review and directions for future research. *Research in Personnel and Human Resources Management, 22*, 165–218. doi:10.1016/S0742-7301(03)22004-1

ADDITIONAL READING

Bos-Nehles, A., Renkema, M., & Janssen, M. (2017). HRM and innovative work behaviour: A systematic literature review. *Personnel Review, 46*(7), 1228–1253. doi:10.1108/PR-09-2016-0257

Fernandez, S., & Moldogaziev, T. (2012). Using employee empowerment to encourage innovative behavior in the public sector. *Journal of Public Administration: Research and Theory, 23*(1), 155–187. doi:10.1093/jopart/mus008

Korzilius, H., Bücker, J. J. L. E., & Beerlage, S. (2017). Multiculturalism and innovative work behavior: The mediating role of cultural intelligence. *International Journal of Intercultural Relations, 56*(1), 13–24. doi:10.1016/j.ijintrel.2016.11.001

Li, X., & Zheng, Y. (2014). The Influential Factors of Employees' Innovative Behavior and the Management Advices. *Journal of Service Science and Management, 7*(2), 446–450. doi:10.4236/jssm.2014.76042

Pourkiani, M., Farahabadi, H. S., & Komak, M. D. (2013). Organizational innovation, barriers and factors. *European Online Journal of Natural and Social Sciences, 2*(3), 724–731.

Chapter 14
Assessment of University Students' Industry 4.0 Conceptual Awareness Levels

Gamze Sart
Istanbul University-Cerrahpasa, Turkey

ABSTRACT

The results of the intensive digital transformation in the world are manifest in all industries. Developments within the scope of Industry 4.0 in educational organizations change the structure and functioning of educational institutions on the one hand, while affecting and changing student competencies on the other. The aim of this study is to evaluate the Industry 4.0 conceptual awareness levels of 890 students from the Faculty of Engineering, Faculty of Economics, and Faculty of Education Sciences at Istanbul University-Cerrahpaşa in terms of gender, grade level, and faculty. As a result of the analysis, it was seen that there was a statistically significant difference in terms of gender, faculty, and grade level. It was determined that male students, engineering faculty students, and fourth year students had higher conceptual awareness of Industry 4.0.

INTRODUCTION

In particular, the rapid and unpredictable development of communication and information technologies enables the transfer of information in real time. In this context, the ability of developed countries or societies to maintain their competitive capacity requires that they develop their own strategies to encourage their industrial production and service sectors to stay relevant and compete in a dynamic and fluctuating global market (Morrar et al., 2017:13). It has also revealed a process that has been noticed especially with the beginning of the 21st century, where the future of changes and transformations has already been determined, and the speed of change and transformation continues to rise to higher levels. The transformation that emerged with this process is called digital transformation (Hirschi, 2018:195). Digital transformation and solving the problem of adaptation is one of the biggest challenges facing the industry. These challenges manifest themselves not only in terms of technology but also in terms of

DOI: 10.4018/978-1-6684-6620-9.ch014

management, human resources and education (Ballantyne et al., 2017:3). Researchers and all stakeholders continue to discuss the future impacts and consequences of digital transformation, both in terms of literature and practice.

The age we live in is leading to a process in which technological developments are trying to mutually transform societies in the world in an interactive process and determine their direction. In other words, technological transformation or digital transformation leads to a radical change that significantly affects individuals, businesses and societies in many aspects such as socio-cultural, economic and technology (Magruk, 2016:277). With the impact of digital transformation, also known as Industry 4.0, there is an era in industrial production that enables faster, more flexible and more efficient processes to produce higher quality products at lower cost (Collins & Halverson, 2010:20). The technological, sociological and industrial revolutions of this period will increase productivity, change all paradigms, stimulate industrial growth and, most importantly, transform the profile of human resources (Pfeiffer, 2017:109). As a result of these developments, countries and businesses need to understand Industry 4.0 and establish policies and processes to adapt. Awareness of Industry 4.0, the essence of digital transformation, is one of the priority issues for all countries, developed or developing. Universities are educational and training institutions that have to provide young generations with the requirements of the age (Wallner & Wagner, 2016:157). Although there are debates about whether universities provide these competencies to their students, there is a need to end these debates and to create and implement the education and training vision, strategies and goals of universities within a new understanding and discipline, and to create and implement plans and programs within the normative framework (Puncreobutr, 2016:94).

Measuring students' awareness of concepts related to Industry 4.0 is a necessity for higher education policies, universities, students and industrial organizations. The aim of this study is to evaluate the Industry 4.0 conceptual awareness levels of 890 students studying at Istanbul University-Cerrahpasa in terms of gender, class level and faculty.

CONCEPTUAL FRAMEWORK

In a rapidly digitalizing world, screens occupy more and more time in daily life. Data is constantly being generated from different sources. There is fierce competition for the information produced to attract the attention of consumers on different platforms (Bauer et al., 2015:419). In this context, Industry 4.0, in other words, the 4th Industrial Revolution affects the developments in the economic, social and political fields on a global scale. This issue is too complex and multidimensional to be evaluated only in terms of production systems (Kolberg & Zuhlke, 2015:1872; Gulicheva et al., 2017:132). Raising awareness of these different practices and technologies is extremely important. It is important for businesses to follow the changes and integrate them into their systems, and for educational institutions to implement them (Obitko & Jirkovský, 2015:219).

Another issue that started with Industry 1.0 and continues with Industry 4.0 today and on which studies are increasing day by day is the habit of using technology, in other words, technology addiction. With the development of technology, the number of people using technology increases, access to information becomes easier, it makes it possible to process and analyze big data and produce new information, and therefore it becomes an important element of economic and social change (Westerman et al.,2014:2). Technology addiction can be defined as the obsession with technology-related behaviors, which is increasing day by day, despite the fact that technology creates a number of negative consequences for individuals who use it (Vogel-Heuser & Hess, 2016: 412).

In today's social structure, young people have an important place in the use of technology. Young people, who put technology at the center of their lives, cannot accept a life without technology and try to maintain their lives as individuals who want to meet all their needs over the internet and see all the tools of technology as a part of their bodies (Sousa and Rocha, 2019:328). As in Industry 1.0, 2.0 and 3.0, it is thought that not only production but also many other factors such as social life, human needs, professional and commercial opportunities, individuals' leisure time, working hours, expectations from goods and services, and satisfaction with goods and services will change in Industry 4.0 (Posada et al., 2015:28). In today's world of Industry 4.0, the use of cloud technologies and the importance of big data are becoming more important. After this process, it is unpredictable how and in what direction technology will transform in an even faster structure. But a new revolution is coming, called Industry 5.0, which seeks ways for man and machine to work together to improve the efficiency of the means of production. It can be predicted that after Industry 5.0, there may be a period of revolutions and transformations in which even the concept of speed, which will be difficult to name numerically, will not be explained in itself (Matt et al., 2015:341).

With the concept of Industry 4.0, the importance of information infrastructure has increased even more and new concepts and applications have emerged that are used and applied in daily life. Concepts with unknown meanings and only known names, which usually find application in engineering fields, have entered daily life with the Industry 4.0 movement. Knowing and naturally understanding these concepts will expand the field of use (Özsoylu, 2017).

The development of technology has led to the formation of a large mass of people who use and develop technological systems such as technological literacy. However, concepts such as technology addiction have emerged (Anggraeni, 2018:15). There is a need for more studies on Industry 4.0 in the academic field together with the studies on technology that have been studied in recent years (Ciolacu et al.,2017:440). When evaluated in this context, the starting point of this research is to determine the Industry 4.0 awareness levels and technology usage habits of university students who use technology the most and follow it closely (Benesova & Tupa, 2017:2197). It is thought that the suggestions developed based on the results of the research will contribute to the education-training activities of educational institutions, the education-training policies of state governing bodies and the related literature.

LITERATURE REVIEW

Motyl et al (2017) conducted a two-semester survey of 463 students at three different universities in Italy to determine what skills and expertise young engineers need to be ready for Industry 4.0. According to the results of the analysis, it was determined that the conceptual awareness levels of the students increased as their grade levels increased, especially in fundamental concepts such as virtual and augmented reality, three-dimensional printers, rapid prototyping. The researchers emphasized that not only university students but every citizen should be digitally literate. Akgul et al., (2018) aimed to determine the perceptions of university students about the advantages and disadvantages of Industry 4.0 in their study conducted at a state university in Turkey. In particular, it was found that the perceptions of students studying in social programs were lower than the perceptions of students in engineering programs. It was found that concepts such as cloud and Big Data were well known by social program students and engineering students were more familiar with the Internet of Things and simulation concepts. Kaygisiz & Sipahi (2018) focused on Industry 4.0 knowledge levels of undergraduate students in Turkey. The results of the survey, which was intended to measure their level of knowledge, show that students were introduced to the concept of Industry 4.0 on the internet or in courses. A study to determine the Industry 4.0 awareness of undergraduate students studying at the Faculty of Economics and Administrative Sciences and the Vocational School of Social Sciences was conducted by Temel (2019). The study compared both results in the context of Industry 4.0 awareness.

Similarly, Ozkoc & Karalar (2019), used metaphors to determine K12 and undergraduate students' perceptions' on Industry 4.0. K12 students were found to explain the Industry 4.0 with 48 metaphors with generally positive metaphors. Undergraduate students used 89 metaphors to explain the concept with generally negative views. Halili (2019) argued that the use of almost all dimensions of technological developments in learning and teaching processes has a direct impact on productivity and creativity, especially developments such as artificial intelligence, augmented reality, cloud computing and hologram technologies, which are Industry 4.0 technologies, will have a positive impact on every stage of education. A study conducted by Torun & Cengiz (2019), attempted to measure the perspectives of FEAS students on Industry 4.0 within the scope of a theoretical basis called Technology Acceptance Model (TAM). The researcher determined the technology acceptance levels of the students with the help of a four-dimensional scale with a total of 15 questions named as Perceived Usefulness, Perceived Ease of Use, Intention to Use and Usage Behavior. The results of the research revealed that the students have positive perceptions towards the concept of Industry 4.0, the awareness of the students of the Department of Management Information Systems (MIS) regarding the concept of Industry 4.0 is higher than the students of other departments in the faculty, and these results obtained do not differ significantly according to the gender variable of the students.

STATISTICAL ANALYSIS

The Purpose and the Importance of the Research

Digitalization is an important fact of life. It is widely accepted and expressed that the new age is the digital age, and large-scale research and development work is being carried out in both the public and private sectors. The results of the digital transformation, which is experienced intensively in almost every sector in the world, show its impact in all industries. Developments within the scope of Industry 4.0 in educational organizations change the structure and functioning of educational institutions on the one hand, while affecting and changing student competencies on the other. This research will attempt to determine the state of conceptual awareness of Industry 4.0 in university students. The results of the research are of great importance in this respect. It is expected that the results to be obtained in the research will provide important contributions to higher education institutions, higher institutions that are in a position to determine policies for higher education institutions, university administration and students. In addition, it is thought that this study will give direction to the ongoing Industry 4.0 research in different fields.

Table 1. Research hypotheses

	Hypotheses
H_1	There is a statistically significant difference for Industry 4.0 Conceptual Awareness in terms of gender
H_2	There is a statistically significant difference for Industry 4.0 Conceptual Awareness in terms of faculty
H_3	There is a statistically significant difference for Industry 4.0 Conceptual Awareness in terms of grade level

Research Sample

The universe of the study consists of students in Istanbul University-Cerrahpasa. According to the statistics of the number of students of the Higher Education Council (HEC) in 2021, the total number of students in Istanbul University-Cerrahpasa is 31.439. "Simple random sampling" was used as the sampling method. Surveys were conducted via e-mail between 01.03.2022 and 01.05.2022. In the sample size table developed by Yazıcıoğlu and Erdoğan (2004), it was determined that 384 people should be studied for p=0.50 and q=0.50 with a sampling error of 0.05. In this application, analyzes were carried out with 890 samples.

Research Hypotheses

The developed hypotheses are presented in Table 1.

Data Collection Tool

Industry 4.0 Conceptual Awareness Scale: In the first part of the questionnaire, demographic characteristics of the students were used, and in the second part, the "Industry 4.0 Conceptual Awareness Scale" consisting of 39 statements and one sub-dimension, the validity and reliability of which was conducted by Doğan & Baloglu (2020), was used. The lowest score value to be obtained from the Industry 4.0 Conceptual Awareness scale was calculated as "39" and the highest score as "195". The Cronbach's alpha coefficient value of the scale was found to be 0.96. Awareness level responses were in Likert format as "1=none", "2=some", "3=medium", "4=much" and "5=full".

Statistical Methods

The data acquired from the scale utilized in the study were analyzed via IBM SPSS 27.0 program. In the first stage, percentage distributions regarding demographic and general information were presented. In the second stage, independent sample t test and ANOVA test were applied for the scale score.

Assumptions and Limitations of the Study

It was assumed that the people participating in the research acted with their true feelings and thoughts while answering the propositions in the scales. It was accepted that each participant answered the survey study voluntarily and answered each statement in the survey correctly and completely. It was also assumed that each participant understood the real meanings of all the words while answering the propositions.

FINDINGS AND RESULTS

Of the students participating in the study, 53.3% were female and 46.7% were male. When the grade variable is examined, the percentage of participants studying in the first grade is 28.3%. While the percentage of second grade participants was 16.5%, the percentage of third grade participants was 20.3%. The percentage of fourth grade participants was 22.9% and those with more than 5 years of education was 12.0%. Of the students who participated in the survey, 29.8% were studying at the Faculty of Economics, 31.1% at the Faculty of Engineering and 39.1% at the Faculty of Education.

According to the Kolmogorov-Simirnov test, the score value obtained from the scale shows a normal distribution since p>0.05. In group difference analyses, independent sample t test will be applied for 2 groups and ANOVA test will be applied for more than three groups.

When the independent sample t test results of the Industry 4.0 conceptual awareness levels of university students were examined according to the gender variable, the assumption of equality of variances was accepted as "p>0.05" in the Levene test results applied for the interpretation of the t test. A statistically significant difference was obtained for Industry 4.0 Conceptual Awareness in terms of gender. When the mean values of the answers were examined, it was determined that men's awareness was significantly higher.

As a result of the homogeneity of variances test for the groups formed according to the faculty variable of university students' Industry 4.0 conceptual awareness levels, the p value was found as "0.381>0.05". This statistical result shows that the assumption of homogeneity of variances required for the use of ANOVA test is fulfilled. According to the faculties in which the students are studying, there is a statistically significant difference (F=9.564). Engineering faculty students have significantly higher awareness.

As a result of the homogeneity of variances test for the groups formed according to the grade level variable of the university students' Industry 4.0 conceptual awareness levels, the p value was found as "0.415>0.05". The assumption of homogeneity of variances is met. It was determined that there was a statistically significant difference according to the grade levels of the students (F=12.923). 4th grade and 5th and above students had significantly higher levels of awareness.

CONCLUSION

In this context, it is important to increase awareness of Industry 4.0, which is one of the important stages of professional development in changing and rapidly developing technology. In particular, it is important to include current technologies that can be evaluated within the scope of Industry 4.0, which are directly related to the departments in the education and training curricula of higher education institutions, which are the first and important stages of professional development, and to teach the working principles of these technologies at a fundamental level (Hussin, 2018:93).

The aim of this study is to evaluate the Industry 4.0 conceptual awareness levels of 890 students from the Faculty of Engineering, Faculty of Economics and Faculty of Education Sciences at Istanbul University-Cerrahpasa in terms of gender, class level and faculty. As a result of the analysis, it was seen that there was a statistically significant difference in terms of gender, faculty and grade level. It was determined that male students, engineering faculty students and fourth grade and above students had higher conceptual awareness of Industry 4.0.

Table 2. The average Industry 4.0 conceptual awareness scores of university students

Industry 4.0 Concepts	$\bar{x}$
1. Internet of Things	2.79
2. Artificial intelligence	3.24
3. Learning (Smart) Robots	3.01
4. 3-D Printers	3.19
5. Advanced Automation	2.67
6. Cyber Security	2.59
7. Cyber Physical Systems	2.09
8. Cloud Computing Technology	2.31
9. Big Data and Data Analytics	2.24
10. Virtual Reality	2.90
11. Augmented Reality	2.40
12. Mixed Reality	2.02
13. Smart Manufacturing Technologies	2.72
14. Dark Factories	2.23
15. Embedded Systems	2.47
16. Machine-Machine Cooperation	2.77
17. Sensor Technologies	3.01
18. Computer Vision	2.54
19. Personal Product Development	2.64
20. Deep Learning	2.22
21. Data-Driven Service	2.25
22. Energy 4.0	2.62
23. Digital Supply Chain	2.36
24. Unmanned Systems	2.91
25. Agile and Flexible Production-Service	2.24
26. Hologram Technologies	2.23
27. Wearable Technologies	2.51
28. Digital Diagnostics, Diagnosis, Treatment	2.52
29. Nano Technology	2.70
30. Industrial Internet	2.42
31. Advanced Production Techniques	2.49
32. Technological Innovation	2.40
33. Rapid Prototype Production	2.33
34. Micro Factories	2.17
35. Factories Producing Their Own Energy	2.65
36. Artificial Neural Networks	2.39
37. Smart Storage and Transfer Technologies	2.36
38. Simulation Technologies	2.54
39. Additive Manufacturing	2.06
Total	**2.51**

Table 3. t-Test results of awareness levels according to gender variable

	Gender	x̄	SS	Levene's Test (p)	t Statistic	p
Industry 4.0 Conceptual Awareness	Male	3.95	0.99	0.261	5.742	0.001*
	Female	2.78	0.87			

*p<0.01

Table 4. ANOVA results of awareness levels according to faculty variable

	Faculty	x̄	SS	Homogeneity Test (p)	ANOVA F Value	p	Post-hoc
Industry 4.0 Conceptual Awareness	(1) engineering	3.23	0.93	0.381	9.564	0.001*	1>2 1>3 3>2
	(2) education	2.89	0.88				
	(3) economics	3.01	0.74				

*p<0.01

It may be recommended to work on an educational model that closely follows digital developments and can teach students basic working principles as well as awareness. In addition, with the determination of sectoral needs, studies can be carried out to improve the technical infrastructure with updates in the education-training curriculum for sectoral expectations. Interdisciplinary studies can be carried out within universities to increase the level of knowledge on the issues.

Based on the results obtained in the research, it can be recommended to determine the level of conceptual awareness of Industry 4.0 comparatively with different sample groups and especially with technical university students in Turkey and abroad. The results will mirror both the participants and university administrators. In addition, determining the level of Industry 4.0 conceptual awareness of university lecturers and administrators should also be addressed. The course contents of higher education institutions should be revised in line with the developments in this field and necessary measures should be taken to increase the awareness of students in this context.

Table 5. ANOVA results of awareness levels according to grade level variable

	Class Level	x̄	SS	Homogeneity Test (p)	ANOVA F Value	p	Post-hoc
Industry 4.0 Conceptual Awareness	(1) 1st class	2.45	0.68	0.415	12.923	0.001*	4>1.4>2.4>3 5>1.5>2.5>3.5>4
	(2) 2nd class	2.84	0.87				
	(3) 3rd grade	3.12	0.90				
	(4) 4th grade	3.43	0.93				
	(5) 5 and above	3.59	0.96				

*p<0.01

REFERENCES

Akgul, A., Akbas, H. E., & Taskin Gumus, A. (2018). A Survey of Students' Perceptions on Industry 4.0 in a Large Public University in Turkey. In F. B. Candan & H. Kapucu (Eds.), *Current Debates In Busıness Studies* (pp. 237–247). IJOPEC Publication Limited.

Anggraeni, C. W. (2018). Promoting Education 4.0 in English for survival class: What are the challenges? Metathesis. *Journal of English Language, Literature, and Teaching*, *2*(1), 12–24. doi:10.31002/metathesis.v2i1.676

Bauer, W., Hammerle, M., Schlund, S., & Vocke, C. (2015). Transforming to a hyperconnected society and economy-towards an "Industry 4.0". *Procedia Manufacturing*, *3*, 417–424. doi:10.1016/j.promfg.2015.07.200

Benesova, A., & Tupa, J. (2017). Requirements for education and qualification of people in Industry 4.0. *Procedia Manufacturing*, *11*, 2195–2202. doi:10.1016/j.promfg.2017.07.366

Ciolacu, M., Tehrani, A. F., Beer, R., & Popp, H. (2017). Education 4.0—Fostering student's performance with machine learning methods. In *2017 IEEE 23rd International Symposium for Design and Technology in Electronic Packaging (SIITME)* (pp. 438-443). IEEE.

Collins, A., & Halverson, R. (2010). The second educational revolution: Rethinking education in the age of technology. *Journal of Computer Assisted Learning*, *26*(1), 18–27. doi:10.1111/j.1365-2729.2009.00339.x

Dogan, O., & Baloglu, N. (2020). Industrial 4.0 Conceptual Awareness Levels of University Students. *Turkish Science Research Foundation*, *13*(1), 126–142.

Gulicheva, E., Lisin, E., Osipova, M., & Khabdullin, A. (2017). Leading factors in the formation of innovative education environment. *Journal of International Studies*, *10*(2), 129–137. doi:10.14254/2071-8330.2017/10-2/9

Halili, S, H. (2019). Technological advancements in education 4.0. *The Online Journal of Distance Education and e-Learning*, *7*(1), 63-69.

Hirschi, A. (2018). The fourth industrial revolution: Issues and implications for career research and practice. *The Career Development Quarterly*, *66*(3), 192–204. doi:10.1002/cdq.12142

Hussin, A. A. (2018). Education 4.0 made simple: Ideas for teaching. *International Journal of Education and Literacy Studies*, *6*(3), 92–98. doi:10.7575/aiac.ijels.v.6n.3p.92

Kaygısız, E., & Sipahi, H. (2018). Y kuşağı üniversite öğrencilerinin bireysel yenilik ve Endüstri 4.0 bilgi düzeyleri arasındaki ilişkinin incelenmesi. *Gaziantep University Journal of Social Sciences*, *18*(2), 922–936.

Kolberg, D., & Zuhlke, D. (2015). Lean automation enabled by industry 4.0 technologies. *IFAC-PapersOnLine*, *48*(3), 1870–1875. doi:10.1016/j.ifacol.2015.06.359

Magruk, A. (2016). Uncertainty In The Sphere Of The Industry 4.0-Potential Areas To Research. Business, Management & Education / Verslas. *Vadyba Ir Studijos*, *14*(2), 275–291.

Matt, C., Hess, T., & Benlian, A. (2015). Digital transformation strategies. *Business & Information Systems Engineering*, *57*(5), 339–343. doi:10.100712599-015-0401-5

Morrar, R., Arman, H., & Mousa, S. (2017). The fourth industrial revolution (Industry 4.0): A social innovation perspective. *Technology Innovation Management Review*, *7*(11), 12–20. doi:10.22215/tim-review/1117

Motyl, B., Baronio, G., Uberti, S., Speranza, D., & Filippi, S. (2017). How will change the future engineers' skills in the Industry 4.0 framework? A questionnaire survey. *Procedia Manufacturing*, *11*, 1501–1509. doi:10.1016/j.promfg.2017.07.282

Obitko, M., & Jirkovský, V. (2015). Big data semantics in industry 4.0. In *International conference on industrial applications of holonic and multi-agent systems* (pp. 217-229). Springer.

Ozkoc, H. H., & Karalar, H. (2019). K12 ve lisans öğrencilerinin Endüstri 4.0 kavramına ilişkin algıları. *Eskişehir Osmangazi Üniversitesi Sosyal Bilimler Dergisi*, *20*, 243–258.

Pfeiffer, S. (2017). The vision of 'Industrie 4.0' in the making a case of future told, tamed, and traded. *NanoEthics*, *11*(1), 107–121. doi:10.100711569-016-0280-3 PMID:28435474

Posada, J., Toro, C., Barandiaran, I., Oyarzun, D., Stricker, D., de Amicis, R., & Vallarino, I. (2015). Visual computing as a key enabling technology for industrie 4.0 and industrial internet. *IEEE Computer Graphics and Applications*, *35*(2), 26–40. doi:10.1109/MCG.2015.45 PMID:25807506

Puncreobutr, V. (2016). Education 4.0: New challenge of learning. St. Theresa. *Journal of the Humanities and Social Sciences*, *2*(2), 92–97.

Sousa, M. J., & Rocha, A. (2019). Digital learning: Developing skills for digital transformation of organizations Futur. *Gener. Comput. Syst.*, *91*(1), 327–334. doi:10.1016/j.future.2018.08.048

Temel, K. (2019). Endüstri 4.0 farkındalığının belirlenmesi: Çanakkale Onsekiz Mart Üniversitesi örneği. *Girişimcilik ve Kalkınma Dergisi*, *14*(1), 31–44.

Torun, N., & Cengiz, E. (2019). Endüstri 4.0 bakış açısının öğrenciler gözünden teknoloji kabul modeli ile ölçümü. *Uluslararası İktisadi ve İdari İncelemeler Dergisi*, *22*, 235–250.

Vogel-Heuser, B., & Hess, D. (2016). Guest editorial Industry 4.0-Prerequisites and visions. *IEEE Transactions on Automation Science and Engineering*, *13*(2), 411–413. doi:10.1109/TASE.2016.2523639

Wallner, T., & Wagner, G. (2016). Academic education 4.0. *Proceedings of the international conference on education and new developments END 2016 icinde*, 155-159.

Westerman, G., Bonnet, D., & McAfee, A. (2014). The nine elements of digital transformation. *MIT Sloan Management Review*, *55*(3), 1–6.

ADDITIONAL READING

Ballantyne, N., Wong, Y. C., & Morgan, G. (2017). Human services and the fourth industrial revolution: From husITa 1987 to husITa 2016. *Journal of Technology in Human Services*, *35*(1), 1–7. doi:10.1080/15228835.2017.1277900

Berman, S. J. (2012). Digital transformation: Opportunities to create new business models. *Strategy and Leadership*, *40*(2), 16–24. doi:10.1108/10878571211209314

Drath, R., & Horch, A. (2014). Industrie 4.0: Hit or hype? [industry forum]. *IEEE Industrial Electronics Magazine*, *8*(2), 56–58. doi:10.1109/MIE.2014.2312079

Hariharasudan, A., & Kot, S. (2018). A scoping review on Digital English and Education 4.0 for Industry 4.0. *Social Sciences, 7*(11), 1-13.

Lee, J., Kao, H. A., & Yang, S. (2014). Service innovation and smart analytics for industry 4.0 and big data environment. *Procedia CIRP*, *16*, 3–8. doi:10.1016/j.procir.2014.02.001

Chapter 15
Business Transformation Projects:
The Role of a Transcendent Software Engineering Concept (RoTSEC)

Antoine Toni Trad
https://orcid.org/0000-0002-4199-6970
IBISTM, France

ABSTRACT

The hyper evolution of technologies is a major problem for transformation projects because such projects take a long time to terminate. That is why there is a need to find transcendent technological artefacts for all technology generations. The role of a transcendent software engineering concept (RoTSEC) is central for implementation projects in general and is also especially crucial for business transformation projects (or simply projects) because transformation phases incur major changes in the existing sets of the archaically defined requirements, development, and platform models. Software engineering (SE) is probably the riskiest part of a project because it consists of many related complex factors and dependencies, for example, the need for SE's artefacts to interact between various actors like project management, business users, business architects, implementation developers, and other project actors.

INTRODUCTION

This chapter keywords clearly show the complexity and the need of having a holistic *Project* approach in order to scope its strategic goals and objectives. That is achieved by defining a generic and standardized Enterprise Architecture (EA) based RoTSEC that can be used in any final application (or business) domain, like finance, education, industry, finance, or any other. Where RP's main concerns are reversing common Source Code Components (SCC), business models, algorithmic codebases, and data structures refinements. These operations are carried out mainly in phases B and C, as shown in Figure 1. An RP can be conducted using a polytechnical mathematical (or simply a Polymathic) model or the AHMM4SE,

DOI: 10.4018/978-1-6684-6620-9.ch015

Figure 1. The architecture development cycle
Source: The Open Group (2011a)

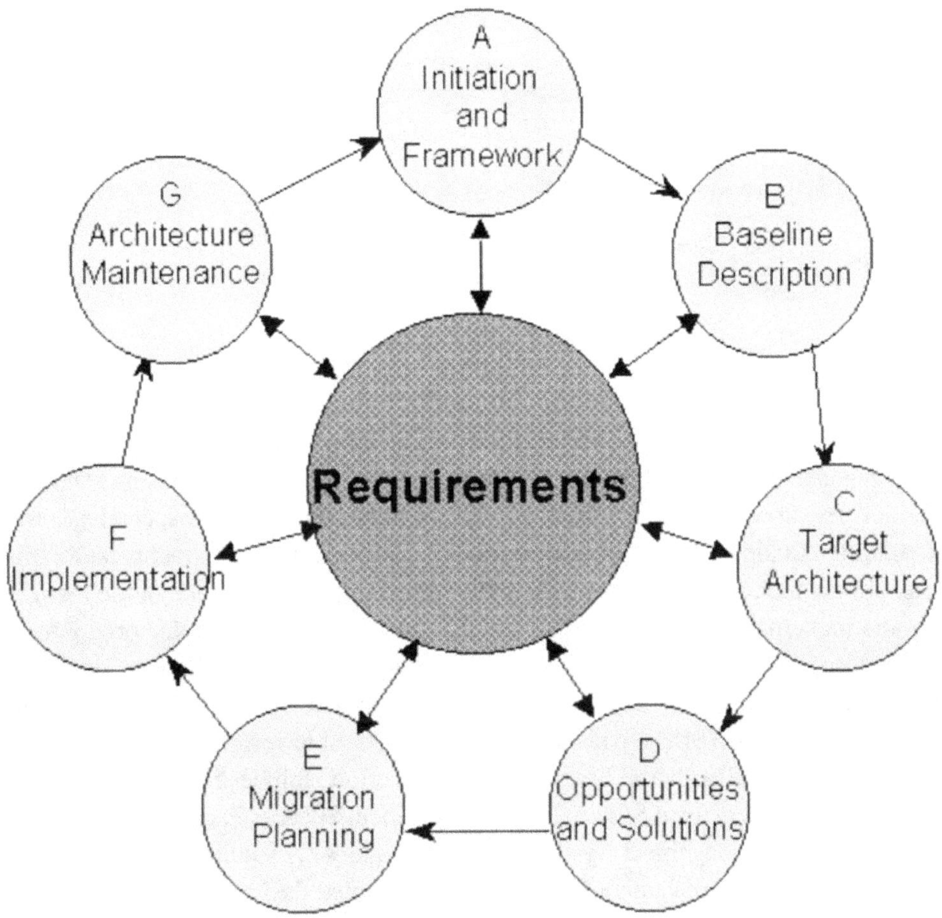

which uses a transcendent approach, that refers to surpassing the complexity of heterogenous approaches and ensures its integrity.

The proposed Polymathic AHMM4SE, supports the iterative transformation of a legacy system, using standard methodologies, like The Open Group's (TOG) Architecture Framework's (TOGAF) Architecture Development Method (ADM) as shown in Figure 1.

Like in all Information and Communications Systems' (ICS) related development works, the recommended approach is a cyclic one, which is based on *Project's* implementation phases, that includes the SEP and RP. RPs are performed mainly for SCCs that include: 1) Refinement; 2) Development and Operations (DevOps); 3) Automated tests and qualifications; and 4) Modelling and design activities. The RoTSEC proposes an efficient use of RP, which might face complexities due to: 1) The implementation of complex and heterogenous software components; and 2) Their maintenance (Koenig, Rustan, & Leino, 2016). In this chapter the RoTSEC was applied on a concrete *Project* in the form of a Proof of Concept (PoC), that is related to a leading *European Bank*. The mentioned *Project* was mainly used to support an SEP/RP transformation process of the Bank's legacy framework which is based on an International Business Machine (IBM) Mainframe and Java Extended Edition (JEE) 1st tier concept and architecture.

Figure 2. The project's construct

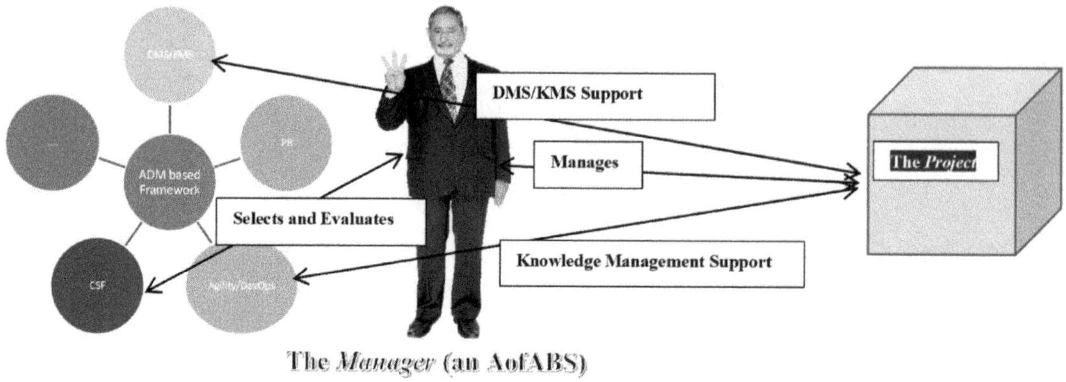

The used EA methodology was TOGAF and its main iterative dynamic cycle manager the ADM; where the modelling language was ArchiMate. The ADM manages the underlying design, DevOps and governance activities. As shown in Figure 2, such a *Project* needs a qualified EA specialists, *Managers* (or Architect of Adaptive Business Information System-AofABIS)) and an RP capable team; this team was the main weak point which proves that the SEP is the most critical one. Although the main focus of the ADM was on the implementation of the organization's (simply the *Entity*) refactored EA patterns and SCCs, the ADM also was responsible of storing and managing the *Entity's Enterprise Continuum* with the transformed Building Blocks (BB); knowing that there are two types of BBs: 1) Architecture BBs (ABB); and 2) Software BBs (SBB), which englobe the SCC libraries (The Open Group, 2011a). As shown in Figure 2, The RoTSEC based Framework's Interaction includes: 1) Decision Making System (DMS); 2) Knowledge Management System (KMS); 3) Critical Success Factors (CSF) (and areas Critical Success Areas-CSA) Management System (CSFMS); and 4) An inhouse implemented RoTSEC. To prove RoTSEC's feasibility the author uses his unique adapted PoC and Research and Development Process (RDP).

THE RESEARCH AND DEVELOPMENT PROCESS

RP based legacy ICS transformation initiatives have generated create a paradigmatic shift in the domains of SE innovation, delivering agile implementation models. Legacy systems distributed across various ICS nodes, use a whole set of heterogenous technologies and methodologies. The transformation of ICS artefacts in the form of *Project* BBs is a major paradigm shift in the ICS industry, that is supported by RoTSEC's Evolution using RP (PERP). As shown in Figure 3, the PERP focuses on the refactoring of SCCs.

Transformed legacy SCCs, objects, resources, and artefacts are classified in specialized repositories and independent components that can interact together and roam across ICS' nodes, using a unique and flexible Global Unique Identifier (GUID); which can be used to persist resultant BBs. This RDP proposes RoTSEC's PERP, to help *Managers* and *Project* members in refining legacy systems. The PERP main activity is to extract business scenarios and relate them to SCCs.

Figure 3. Project's polymathic approach

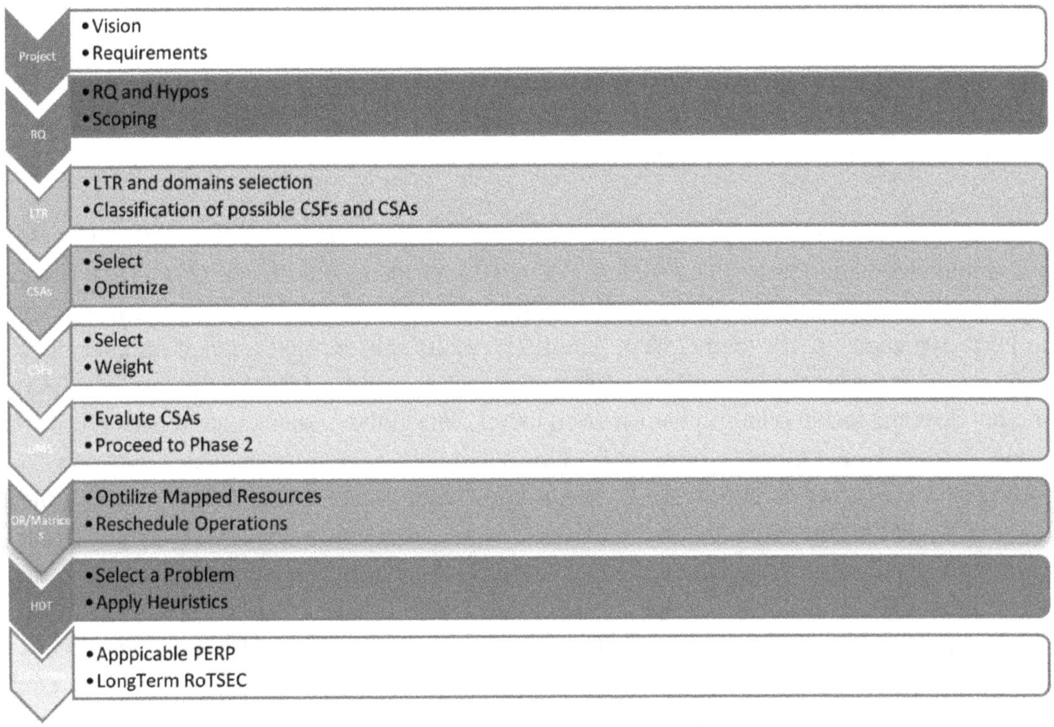

Figure 4. RoTSEC-based project's holistic approach

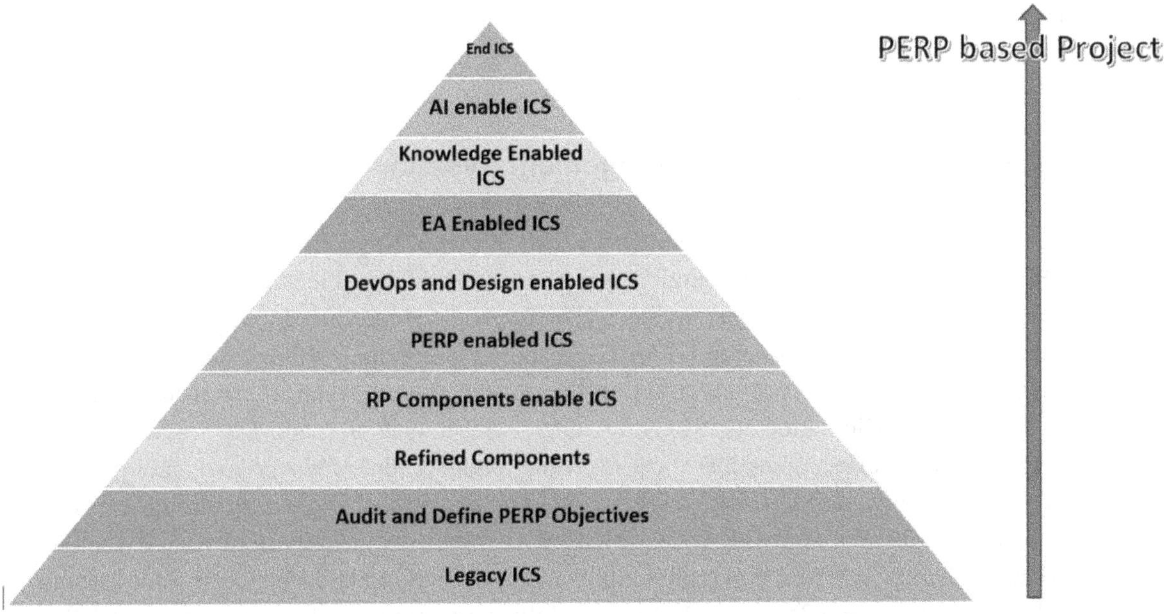

The RDP presents the methodology behind the implementation and an adapted PoC; which mainly based on a concrete professional *Project* done for a major Bank (simply called zBank). Figure 4 shows the holistic approach to be used by RoTSEC based *Project* for the zBank. For these mentioned topics the first RDP step was to establish the Research Question (RQ) and the Literature Review Process (LRP).

The Research Question and Literature Review

This research's global research topic is related to *Projects* and this chapter's RQ is: "Can an RoTSEC support the implementation phase of a *Project*?" Where this chapter's auxiliary RQ is: "How can a PREP support a business transformation project?". Where the kernel of this RDP is based on EA, AHMM4SE, and the DMS which use CSFs (and CSAs). The LRP results showed that a small number of relevant scholar resources exist on the RQ and its related subjects, except for TOGAF, which delivers the EA structure but does not tackle minor transformation topics. The author considers that his RDP works, are pioneering, innovative and the most relevant finding is that there is an important gap, between *Projects'* methodologies and existing solutions. An immense knowledge gap exists, mainly because of the existing LRP on *Project's* failure rates and on the RoTSEC, where there is practically no insight on a holistic and Polymathic approach to design the RoTSEC; Added to these facts, all RP related activities are done manually. In this LRP the following resources and topics were used: 1) Various articles and resources related to RP and transformation projects; 2) The author's general works and his genuine framework; 3) RoTSEC's integration capacities; 4) An initial set of CSFs and CSAs; and 5) The RDP uses empirical engineering research methods. This chapter and all other author's works are based on his Framework's and RDP's maturity, in the domains of SE, risk management, and transformation initiatives. The RDP is based on the Literature Review Process (LRP) which has logical structure. This chapter is a part of many years research cluster that has produced a large set of articles and research concept/framework, LRPEs, and usable items. The RDP proved the existence of an immense knowledge gap, as well as the necessity its recommendations, as a complement to existing holistic approach to the mentioned subjects. The main gap is mainly due to the fact that there isn't something to the proposed RoTSEC; but there are some RP initiatives that concern heterogenous code sources, and which use mainly manual processes. As shown in Figure 4, the next step is to select and classify the sets of CSFs and CSAs in the CSFMS.

Critical Success Areas, Factors Management System

A CSA is a category (or set) of CSFs where in turn a CSF is a set of Key Performance Indicators (KPI), where a KPI maps (or corresponds) to a single requirement and/or *Project* feature, known as a Microartefact. For a given requirement or *Project* problem, the *Manager* (or enterprise architect; or even a business analyst) can identify the initial set of CSAs, CSFs and KPIs, for the DMS and map them to the sets of requirements. Hence the CSFs are important for the mapping between the requirements, knowledge constructs, SE modules/Microartefacts, SCCs, organizational items/units, and DMS/KMS (Peterson, 2011). Therefore, CSFs reflect areas that must meet the main strategic *Project* goals and predefined constraints. Measurement's technics, which are provided by the author's Transformation Research Architecture Development framework (TRADf), which can be used to evaluate performance in each CSA, where CSFs can one of: 1) RoTSEC's overall status evaluation; 2) Mapping levels of BBs and resources; 3) RoTSEC's gap analysis; and 2) DMS/KMS system calls in real time, as shown in Figure 5. KPIs can

Figure 5. The CSAs and CSFs integration using the RDP
Source: Trad and Kalpić (2020a)

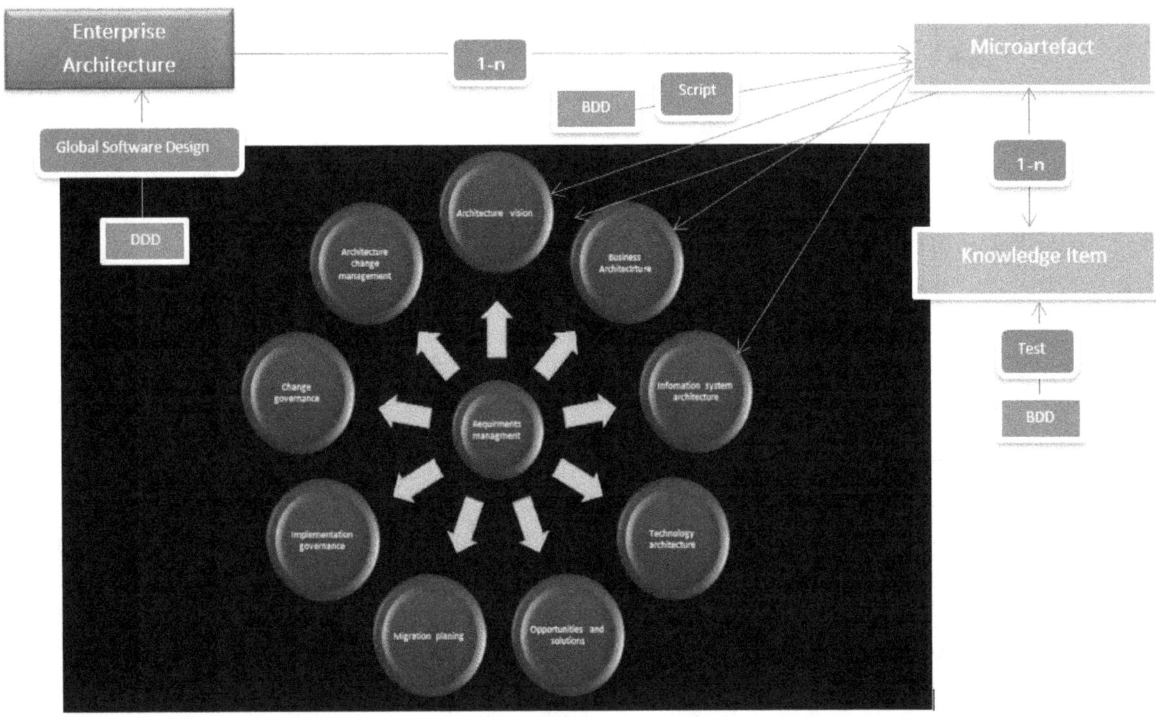

integrated in SCCs, so the Heuristics Decision Tree (HDT) based evaluation processes can automatically estimate the values of CSAs, and CSFs.

As shown in Figure 6, CSFs and risks estimations have the following characteristics (Ylimäki, 2006): 1) Understanding the transformation activities related to *Projects*, like the PERP; 2) CSFs correct EA and implementation fallouts; 3) The *Project* team has also to assign a mitigation strategy for each *Project* risk, which is in turn related to a specific CSFs; 4) CSFs for *Projects* are a set of potential key domains KPIs selected by the transformed SCC to be the most significant; and 5) These CSAs, CSFs, and KPIs, are carefully chosen by the *Project team* using the RoTSEC.

As shown in Figure 5, once the initial sets of CSFs and CSAs have been identified and weighted, then the *Project* team can use the DMS/KMS to generate corresponding sets of solutions for a given RoTSEC problem. The HDT-based RDP would use the DMS in all the *TRAD's* modules. RDP's phase 1 (represented in automated tables), form the empirical checking part of the RDP. This phase checks the following set of CSAs: 1) The RDP, which is synthesized in Table 1; 2) The ICS, which is synthesized in Table 2; 3) The ADM, which is synthesized in Table 3; 4) The KMS, which is synthesized in Table 4; 5) The DMS, which is synthesized in Table 5; 6) The RoTSEC; which is synthesized in Table 6; and 7) This chapter's RDP outcome, which is synthesised in Table 7. The RoTSEC delivers a set of (managerial and technical) recommendations and solutions for a *Project,* and a concept to integrate an in-house transformation framework, like TRADf.

Figure 6. Initial set of potential CSFs for enterprise architecture
Source: Ylimäki (2006)

The Proposed Framework: TRADf

TRADf and its new module, the RoTSEC supports the transformation of legacy SE blocks into agile SCCs or Microartefacts, which are implemented by means of TCDM for various types of technologies and formalisms. The TCDM ensures that the *Project* and future transformations are not handicaped by the hyper evolution of technologies.

Figure 7. The implementation environment

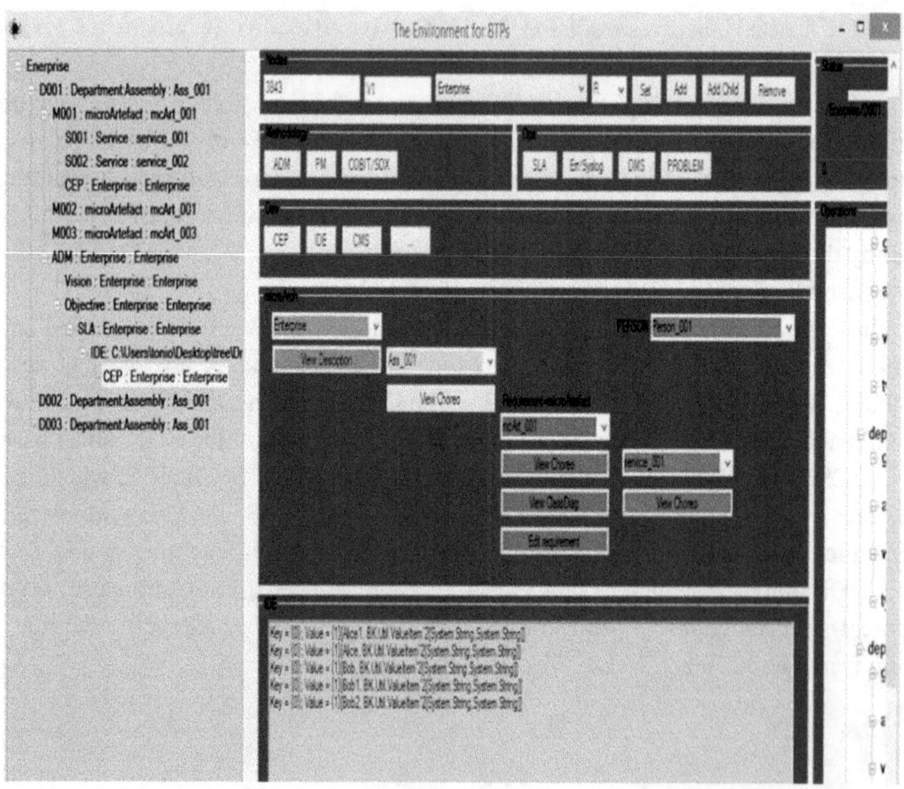

The RoTSEC is a complex concept due to the unviable, heterogenous, and archaic ICS' components formalisms are mammothlike; that makes the PERP very hard to predict resultant artefacts and SCCs. A holistic, systemic, and Polymathic approach for the RoTSEC's integration is optimal for such complex activities (Daellenbach, & McNickle, 2005). The RoTSEC and its PERP can be applied to any type of *Project* and other versions of automated refactoring activities and are a part of *TRADF's* Software engineering or the Implementation module (Im), and the Architecture module (Am). In this chapter the author proposes a set of RoTSEC managerial and technical recommendations and a reusable real-world concept of an in-house framework. The RoTSEC module is supported by ADM's phases and DevOps, where each RoTSEC transformed SCC, BB or Microartefact, circulates through their phases. *Project* and RoTSEC Microartefacts contain their private sets of CSFs and KPIs. This chapter is a part of research work that has took many years and has produced a large set of articles, chapters, a research *TRADf*, LRP, reusable items, Microartefacts, and research artefacts. In this RDP, parts of previous works are reused for the better understanding of this complex subject and RQ. So, it can be considered as an iterative work research; and if everything that is needed should be referenced it would be tedious to follow and understand. This research work can be considered as a non-conventional and pioneering one, in the field of global complex transformation and EA change initiatives. The RoTSEC is business-driven and is founded on a genuine and empirical engineering research model that in turn is based on HDT, DMS, EA and ICS concepts (The Open Group, 2011a).

Empirical Engineering Research Model

This RDP is based on an empirical engineering research model which is optimal for transformational engineering projects (Easterbrook, Singer, Storey, & Damian, 2008). It applies *TRADf* (where mixed research is a very simplistic synonym) that can be considered as a natural complement to conventional research models, like: 1) Basic reasoning; 2) Quantitative Analysis for AI (QNA4AI); and 3) Qualitative Analysis for AI (QLA4AI) research methodologies, to deliver empirical pragmatism concepts as a possible approach for complex tuned mixed methods research. The author and his *TRADf* consider that are all methods are compatible, and the difference is just in the scope and depth of the research process. Empirical research validity checks if the research work, like this RDP, is acceptable as a contribution to existing scientific (and engineering) knowledge, and the author wants in this RDP, to convince the valuable reader(s) that the offered recommendations and the related PoC, or engineering experiment, are valid and applicable for various types of PERP based transformations. In engineering, a controlled experiment or a PoC is a design and software prototype of a testable RQ (and hypothesis) where one or more CSFs and KPIs (or independent variables, in theoretical research) are processed to evaluate their influence on the research model's dependent variables. Controlled experiments or PoCs permit to evaluate with precision the CSFs and if they are related, whether the cause–effect relationship exists between these CSFs and CSAs. The RoTSEC is transformation centric and is founded on *TRADf* that is in this chapter and uses RP, EA, DMS standards and basics (The Open Group, 2011a). But the *Project* should not underestimate the role of complexity, which is mainly due to the change of technologies.

The Role of Complexity and Transformation Readiness

Projects are very complex and are even chaotic, with a very failure rate. Failures are mainly due to its SEP and a so-called supper agile design and development cycles and methods, which generate many collateral problems. *Projects* face failures and many facts confirm these facts (O'Riordan, 2021; Standish, 2011): 1) It is the *Entity's* executive *Managers* who stop *Projects* with a failure rate of 70%; 2) In general people are blamed to justify the financial aspects, but the main reasons are that technology evolves very fast and the business schools *Managers* are submerged by this very complex mainly engineering domain; 3) In an Oxford study, 90% of *Projects* were stopped because of budgets overruns, especially ICS budgets which have a 200% overruns rate; 4) Failure rates are also due to the excessive demands of stakeholders to make excessive gains; 5) *...these failure facts and numbers represent a downtick in the success rates from the previous study, as well as a significant increase in the number of failures...*; 6) *They are low point in the last five study periods. This year's results represent the highest failure rate in over a decade...*; 7) Business transformation initiatives for change is a critical subject for business organizations; where various research show that the failure rates of such initiatives are around 70-80%, while other business organizations are struggling for their projects' and business survival; 8) The Chaos Reports, produced by the Standish Group over the last fifteen years; they assert that: ... *only about 29% of transformations come in on time and budget...*; and 9) It is hard to define the profile and to find an the needed skills. Using the references (O'Riordan, 2021; Standish, 2011) shows that in fact that the failure rates are dramatically increasing... So why continuing such *Projects*? *TRADf* and RoTSEC enable PERP based *Project's,* where refactoring tasks are the main activities; but needs to include the selection of needed and experienced skills, inhouse RP tools, synchronized refactoring processes, and EA/ADM capabilities. *Projects* with a recent successful execution of such similar strategic objective(s), where the legacy ICS had a similar size and complexity, the used appropriate PERP processes, structure/discipline, skills, and a rationale model for deciding what skills and activities to be externally sourced. Such a procedure is called the *Enterprise Capacity to Execute*, which is the ability of the *Entity* to perform all the transformation tasks required by the *Project's* strategic objective, in areas outside of the ICS, including the ability to make real-time DMS based decisions within limited time constraints, which is typical to *Projects* based on the most recent successful execution of a similar strategic objective; and which has at least half the size and complexity. In such *Projects* there are also non-ICS (or more precisely domain) processes, disciplines, and skills to support these types of *Project's* strategic objectives. The *Entity* must acquire the ability and capacities to deal with the various types of PERP based applications conversions, to restructure its portfolio, and has to manage *Project* management issues and users' defined requirements. The RoTSEC needs the following types of activities (The Open Group, 2011a):

- A *Project* team must have in-depth knowledge of EA (and eventually PERP) environments, like the TOGAF framework and its *Business Transformation Readiness Assessment* capacities; which means that the *Project* must have the capacity to execute all business and ICS tasks required by the *Project*, including: the holistic management skills, modeling background, tools management, business processes, PERP activities, and hands-on management capability for the implementation phase.
- In the last decade, there has been a limited number of successful execution of such complex undertakings, and that is due to use of appropriate standardized processes, methods, modeling skills, and a heuristics-based model (like the HDT) for deciding what skills and activities are needed.
- The *Project* team must be capable also to design the *Entity's* capacity to execute, which is the ability to perform all the *Project's* transformation tasks, which includes also sections and domains not directly related to ICS technologies, and the capacity to integrate an DMS/KMS, by using a built-in tree reasoning model, like the HDT, within the limited time constraints.

Table 1. CSFs that have the average of 9.0

Critical Success Factors	KPIs	Weightings
CSF_RDP_ RoTSEC_PERP_Standards	Complex	From 1 to 10. **08 Selected**
CSF_RDP_CSA_CSF_KPI_Integration	Proven	From 1 to 10. **10 Selected**
CSF_RDP_Complexity	High	From 1 to 10. **08 Selected**
CSF_RDP_Empirical_Engineering	Proven	From 1 to 10. **10 Selected**
CSF_RDP_RoTSEC_Manager_Profile	Complex	From 1 to 10. **09 Selected**
CSF_RDP_RoTSEC_Components	Feasible	From 1 to 10. **09 Selected**
CSF_RDP_TRADf_Transformation_Framework	Possible	From 1 to 10. **09 Selected**
CSF_RDP_LRP	Feasible	From 1 to 10. **09 Selected**

valuation

- The team must have the ability to manage such a *Project* DMS based learning concept, and to relate acquire knowledge to its business requirements. Such a tasks needs KMS skills, interoperability architecture concepts, and to use the TCDM's level of methodologies.

RDP's CSFs

Based on the LRP and DMS, the most important RDP's CSFs were extracted, weight and used, and the resulted are shown in Table 1.

To implement RoTSEC based *Projects* the author will propose a Polymathic model and the first step is to present the role and evolution of TCDM compatible methodologies.

THE ROLE OF METHODOLOGIES

Legacy, Structured Concepts, and Methodologies

A Project must use a TCDM compatible methodology, which in fact can a mix of existing methodologies. The RoTSEC tries to focus on the methodologies that resulted from known methodologies, like Object Oriented (OO) evolution. Even though the roles of legacy of traditional methodologies, like the Structure Analysis and Structured Design (SA/SD) are still very present, and from SA/SD the *structure* artefact can be used. Refining 2nd generation (and mapping) resources or mainly software code components, must be done using the following steps: 1) To use the PERP to transform code into SA/SD modelled components; 2) Move to a TCDM compatible OO Methodology (OOM) modelled component(s); 3) Move to TCDM compatible Unified Modelling Language (UML); 4) Move to a TCDM compatible TOGAF and ADM; and 5) Move to a TCDM compatible Decision Making Notation (DMN). The RoTSEC recommends avoiding the costly and risky transformation adventure from the 2nd generation to integrate all TOGAF and DMN features, which is a failure in case of zBank. But it proposes to choose a non-locked in approach that is assisted by TCDM that includes the following steps: 1) To transform the existing Mainframe legacy system to a well-designed and mapped SA/SD model, where for example a structure corresponds to an OO class; 2) To transform the existing OOM/UML diagrams based system to a well-designed and mapped UML/BPM modelled components, using mainly class, sequence, communication, Entity Relationship Diagram (ERM), and BPM diagrams; 3) Move to Spiraled, TOGAF and ADM/DevOps development cycles, in a light version; and 4) Adopt basic DMN artefacts, like the requirements diagrams and tables evaluations for the DMS. But what is common to all these mentioned methodologies and central to the TCDM is OOM.

Object-Oriented Methodologies

The *Project* uses various types of standard methodologies which are based on OO approach; and all these are considered to inherit the features of OO, which has its roots in three OOM techniques namely Rumbaugh, Booch and Jacobson Methodologies, which are the fundaments of what has become the most known ICS standard, the UML (Liu, 2022); knowing that all following methodologies are developed using an UML metamodel. Where that was the first major step to transformation paradigms.

Rumbaugh's Methodology

Rumbaugh's Object Modeling Technique (OMT) is a methodology used to develop manageable OO based ICS and to support OO oriented Integrated Development Environments (IDE). OMT's main aim was to allow class attributes, methods, inheritance, and association to be coherently and simply exposed to developers and analysts. OMT was used in important real-world *Projects*, where OMT's main advantage is to simulate classes (or types) before starting the *Project* and another one was to support, and enable the requirements engineering process and to reduce complexity through the use of diagrams. The OMT consists of four phases: 1) Analysis, which assigns an object, dynamic, and a functional model to the design process, and determines the important properties and main domain issues; 2) Systems' design, depicts the ICS' basic structure and its components, including data access, and concurrency topics; 3) Object design, classifies OO artefacts, and data structures; and 4) Implementation, translates of OO artefacts to implement code. It also delivers the following types of models: 1) The OO, which is responsible mainly of classes and their associations and attributes; 2) The dynamic, which is responsible for the interaction between OO model's artefacts, by using events and objects' states; and 3)

The functional, which is responsible for the data flows, storage, constraints, and related processes (Liu, 2022).

Booch Methodology

This methodology focuses on OO Analysis (OOA) and OO Design (OOD) phases, having major five activities: conceptualization, analysis, design, evolution, and maintenance of requirements and their related SCCs. It is cyclical (in fact spiral) model which uses the concept of incremental implementation processes, and that is the origin of the ADM and DevOps. Booch's OOA and OOD phases incorporate the following processes: 1) Conceptualization, which establishes functional requirements; 2) Analysis, develops a model by using objects/classes, attributes, methods, and relations; 2) Design, develops a structure and architecture where logical and physical features are modelled; 3) Evolution, relates to the implementation process; and 4) Maintenance activities that follow the deployment process. OOA and OOD phases, use six different models/diagrams: class diagrams, state transition diagrams, object diagrams, process diagrams, module diagrams, and interaction diagrams; which all of them are TCDM artefacts. Class and module diagrams are static while state transition diagrams are dynamic (Liu, 2022).

Jacobson Methodology (OOSE)

Jacobson's methodology, which is known as the OO Software Engineering (OOSE) (and even as Objectory). OOSE is a methodology that can be used to plan, design, and implement OO ICS components; and has five types of models: 1) Requirements' model, which delivers problem domain OO diagram(s), that satisfy requirements and specifies Use cases (UC) diagrams; 2) Analysis models, use OOA and OOD diagrams; 3) Design models, use transition and interaction diagrams like the ones from OOD diagrams; 4) Implementation models, used in the SEP; and 5) Testing models, which test the implemented models. UC are one of the fundamental TCDM diagram type.

Use Cases

UC helps the *Project* understand, analyze and design the target system; in fact, they are scenarios for detailed domain analysis of the *Project's* requirements. UCs are defined as an interaction between a user and the transformed end-system. Each UC has a *goal* and a responsibility, where a goal is for the user to be achieved and a responsibility is for the end-system to execute. UCs include: 1) OO diagrams; 2) Non-formal code; 3) Events flow; 3) Pseudo code; 4) They are related using actors supporting the scenarios; and 5) Free style low charts. However, they may contain the following elements: 1) Conditions for starting and ending points; 2) Manages interactions between the various actors; 3) The management of used data; 4) Exceptions' management; 5) The management of constraints for a set of selected problems; and 6) They depict features and elements of the end ICS components. OOM, UC are the basis of the actual EA modelling language, ArchiMate.

ArchiMate

ArchiMate has many artefacts, diagram types and views, that is why only the UC View (UCV) will be presented to show how the TCDM can include common ArchiMate artefacts and diagrams.

Use Case View

ArchiMate can be used for analyzing business scenarios from the functional perspective of an application, by using the UCV. UCs (as known in UML) can be mapped to *Application Services* in the form of Microartefacts. A UC can be divided into following sections: 1) Business UCs; and 2) System UCs or non-functional UCs. A UC can be modelled with *Business Services*, and a subsequent set of system UCs, like *Application Services*. When the *Project's* UCs are identified as application services, which can be afterwards used in other modelling diagrams (like in EA's layered view). These diagrams are elements of functionalities of the target application or SCC. Which means that application services (or Microartefacts) represent the behavior (the functionalities) of an application or SCC (Hosiaisluoma, 2022).

Table 2. CSFs that have the rounded average of 8.6

Critical Success Factors	AHMM4SE enhances: KPIs	Weightings
CSF_Method_StructuredConstructs	Stable	From 1 to 10. **09 Selected**
CSF_Method_OOM	Stable	From 1 to 10. **09 Selected**
CSF_Method_EA	Complex	From 1 to 10. **08 Selected**
CSF_Method_Archimate	Stable	From 1 to 10. **09 Selected**
CSF_Method_TCDM	Complex	From 1 to 10. **08 Selected**
CSF_Method_DevOps	Complex	From 1 to 10. **08 Selected**
CSF_Method_PERP	Complex	From 1 to 10. **08 Selected**
CSF_Method_ADM	Proven	From 1 to 10. **10 Selected**

valuation

Methodologies' CSFs

Based on the LRP and DMS, the most important methodologies' CSFs were extracted, weight and used, and the resulted are shown in Table 2.

To organize various types of refined Microartefacts there is a need to adopt SEP based on a patter-driven approach.

A PATTERN-DRIVEN APPROACH

Categories of Patterns

The TCDM needs the minimum and coherent set of patterns; and the RoTSEC needs a strategy for the use of design of patterns, which can be used to enforce the *Project*. This TCDM strategy suggests that implementers must be able to reuse proven components that emerge from *the best architecture & modeling practices*, in order to solve transcendent generic business implementations. Without the use of transcendent design patterns, team members like *Managers*, business analysts, designers, and SCC implementers, are badly applying *architecture & modeling techniques*, which can result in that the target business solution: 1) Has bad performance; 2) Lacks scalability; and to 3) SCCs become un-usable and un-maintainable. Added to that and for practical reasons, many EA specialist have the tendency to *reinvent the wheel* when attempting to implement an architecture or SCC template, to solve various types of business transformations issues. The RoTSEC set of patterns are BBs for modeling, designs and implementation constructs for *Projects*, need PoC to present the feasibility of patterns that can be used to create generic types of EA and SCC models. These patterns are suitable for implementing various architectures including the RoTSEC, which in turn uses TOGAF. While composing TCDM patterns to create reusable BBs that can be mapped to different types of *Projects* and PERP, to develop solutions for the *Project* (The Open Group, 2011a). For that goal there is a need for the reduction of complexity and the adaption of a bottom-up approach that based on a *1:1* mapping architecture. This section illustrates existing patterns in various domains and tries to propose a TCDM for existing patterns (Fowler, 2003; The Open Group, 2021). The existing categories of patterns are:

- EA or TOGAF patterns are a quick aide-memoire and are handy links which include a set of model patterns that can be used to generate separate models in a *Project,* like the following (The Open Group, 2021): Starter model, ADM, Enterprise continuum, Technical Reference Model (TRM), Catalogs, ….
- Gang of Four (GoF) patterns offers three categories of patterns and deal with the creation of objects and lay the foundation of core design patterns for SEP.
- Domain logic patterns include: Transaction script, Domain model, Table module, and Service layer.
- Data source architectural patterns include: Table data gateway, Row data gateway, Active record, and Data mapper.
- Object-relational behavioral patterns include: Unit of work, Identity map, and Lazy load.
- Object-relational structural patterns include: Identity field, Foreign key mapping, Association table mapping, …
- Object-relational metadata mapping patterns include: Metadata mapping, Query object, and repository.
- Web presentation patterns include: Model View Controller (MVC), Page controller, Front controller, Template view, …
- Distribution patterns include: Remote facade, and Data transfer object.
- Offline concurrency patterns; Optimistic offline lock, Pessimistic offline lock, ...
- Enterprise Service Bus (ESB) patterns include: Event stream processing, API,…
- Enterprise Application Integration (EAI) patterns include: Messaging systems, Messaging channels, Message construction, …

There are many other categories of patterns, where many overlaps and creates EA models redundancy which makes *Projects* difficult to manage. That is why the TCDM offers a set of transcendent patterns like the MVC. The main goal of these basic mainly ICS patterns is to support complex intelligent data and knowledge artefacts and patterns.

Intelligent Data and Knowledge Artefacts

The RoTSEC and PERP are specialized in designing and extracting various types of more complex patterns; and this section will take some of these patterns, because there is a large set of such patterns.

Business Data or Interaction Modeling Patterns

The complex description, design and eventual extraction of business data and interaction models (and assioated modelling patterns), does not depend on the types of databases or SE environments, that are used in the *Project*; but on the diversity and chaotic management of data-sources, IDEs and SE generates major problems in *Projects*. Especially in its implementation SEP, which its complexity is undermined. Atomic business data services for such business environments focuses primarily on the encapsulation of the data and behavior schema (Pavel, 2011). The "1:1" mapping concept is to applied for business artefacts management and access, where the business requirement corresponds to a *data entitty, business data view or a business interaction*, if the artefact can be encapsulated in a single class. Such classes are the basis of the *Business Knowledge Management Pattern*.

Business Knowledge Management Pattern

Business Processes (BP) and their Models (BPM), can be persisted to form knowledge classes and the following artefacts exist:

- The Business Process Oriented Knowledge Management (BPOKM) framework can be applied for the *Project's* knowledge management that will help in the all types of just-in-time knowledge assistance.
- The Delivery of Context -Sensitive Organizational Knowledge (DECOR) supports BPOKM and can be used by the RoTSEC. DÉCOR that implements, integrates, and maintains *Entity* solution for BPOKM, including a BP analysis method and tool, a BP oriented intelligent knowledge archive system, and a workflow enactment (Abecker, Papavassiliou, Mentzas, & Müller, 2014). The DÉCOR approach to BPOKM is based upon the extension of BPM paradigm, that includes the automation of knowledge access mechanisms, complementary knowledge management sub-processes, and business rules processes (Papavassiliou, Ntioudis, Mentzas, & Abecker, 2002).
- BPOKM services are offered by the DECOR framework, which supports the *Project* team a just-in-time business knowledge and information services (Abecker, Papavassiliou, Mentzas, & Müller, 2014).

All the mentioned patterns are complex construct that need a specfic Microartefacts' assembling model.

Microartefacts Assembling Model

The Requirements Integration Pattern

PERP is a part of *TRADf*, where PERP extracts types of Microartefacts to be used and mapped to the *Project's* requirements and ICS' components. The extracted Microartefacts are orchestrated by the AHMM4SE actions that process the refinement processes. PERP actions map requirements to the various Microartefacts and application processes, located in repository. The assembly processes are responsible for the implementation of mechanisms needed to deliver the mapped relations and synchronization actions. The PERP concept is implemented in all of the *Project's* processes and the implementation of Microartefacts to deliver a final business system; such a set of actions can be modelled and orchestrated by the AHMM4SE (The Open Group, 2011a).

The Code Blocks Integration Pattern

Even if SE related methodologies and RP technics have made many major advances, the use of standard Microartefacts and coding blocks patterns generally result in an inefficient ICS and the *Project's* failure. To improve the RP for code blocks, it is recommended to use the PERP approach that uses TCDM artefacts and requires minimal transformation effort; and minimizes code blocks overhead. The PERP, proposes that the *Project* team profiles the code blocks in order to determine critical regions, and then then to apply basic refinement guidelines to convert those regions and make them amenable to transformation tools. The extracted critical regions correspond to a limited set of artefacts, which can transformed easily, resulting in more efficient and transformable code blocks (Stitt, Stitt, Vahid, & Najjar, 2006). The RoTSEC and PERP propose a set of refinement models.

Refinement Models

Refinement processes or the PERP's main goal is to extract standard models and the TCDM artefacts. As there are many standards and types of artefacts, this section will focus on the latest Object Management Group's (OMG) DMN.

Refining and Mapping DMN Pattern

DMN is a standard that is used for modeling operational decisions; and DMN decision models are shared between DMN-compliant systems, so that *Project* team members are unified in designing and implementing DMN decision services. The DMN standard is similar to and can be used together with the BPM and Notation (BPMN) standard for designing and modeling BPs (RedHat, 2022). DMN main diagrams and artefacts that can be used in the TCDM- are: 1) Decision Requirement Diagrams (DRD), to cover requirements' engineering; 2) BPM; 3) Business Knowledge Model (BKM); and 4) Decision Tables, similar to the ones used in this chapter. PERP process results in a set of TCDM artefacts and models that include DMN artefacts, and their mapping. These resultant artefacts are persisted in components. PERP can use the following steps: 1) Define TCDM main artefacts and models; 2) Transform the current code base to deliver well-designed and mapped UML/BPM modelled components; 2) Move to TOGAF and ADM; and 3) Include DMN. TCDM helps to avoid that PERP delivers a patterns hairball.

Refining the Patterns Hairball

As shown in Figure 8 the RoTSEC that be applied in a bottom up approach to propose a TCDM that also offers the minimal set of patterns (The Open Group, 2021).

Pattern's CSFs

Based on the LRP and DMS, the most important pattern's CSFs were extracted, weight and used, and the resulted are shown in Table 3.

To implement RoTSEC based *Projects* the author will propose a Polymathic model.

Figure 8. The RoTSEC-based project pyramid

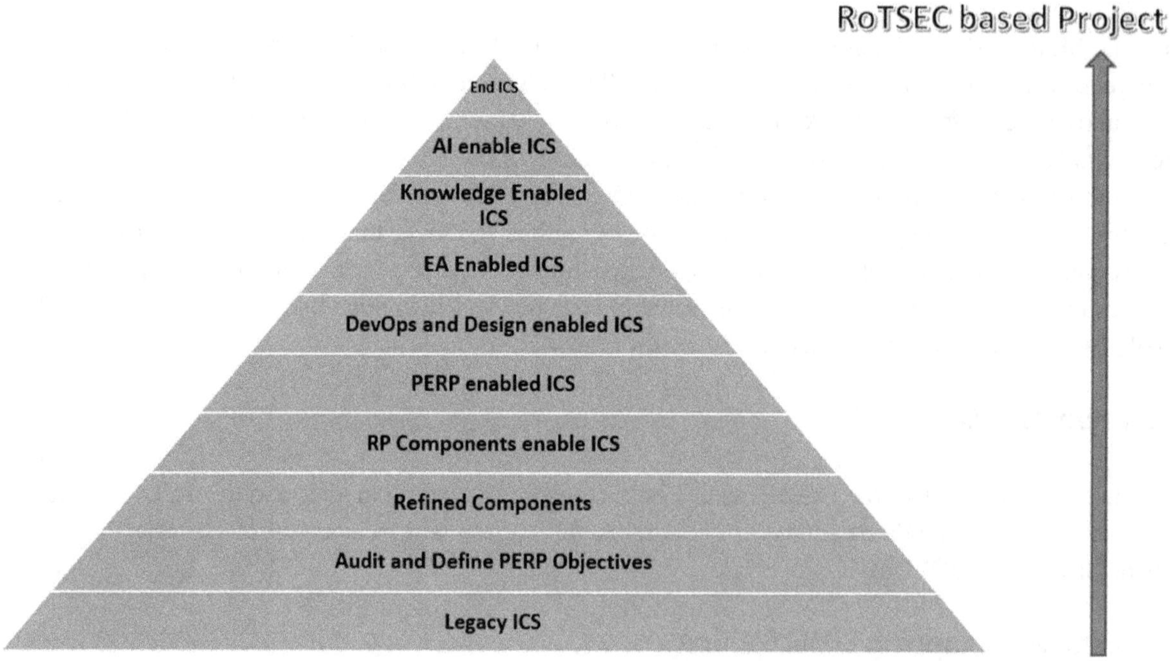

A Polymathic Transformation Model

Evolution and a Major Paradigm Transformation

This section presents a model that follows the evolution of various *Project* components and tries to extract common artefacts, because a *Project* takes ten to twenty years, but ICS related technologies, components, and methodologies which evolute must faster, hence there is a need to find a transcendent Polymathic concept and to be backward compatible… The RoTSEC can be applied in ICS fields, like

Table 3. CSFs that have the rounded average of 8.8

Critical Success Factors	HMM enhances: KPIs	Weightings
CSF_Patterns_Categories_Integration	Feasible	From 1 to 10. **09 Selected**
CSF_Patterns_Intelligent_Microartefacts	Complex	From 1 to 10. **09 Selected**
CSF_Patterns_TCDM_Usage	Complex	From 1 to 10. **08 Selected**
CSF_Patterns_DMS_KMS	Feasible	From 1 to 10. **09 Selected**
CSF_Patterns_PERP_Refinement	Feasible	From 1 to 10. **09 Selected**

valuation

services' technologies, methodologies … The AHMM4SE is RoTSEC's basic structure and is based on various mathematical disciplines, development methods, engineering standards, and risk response strategies (Trad, & Kalpić, 2020a).

Risk Response Strategies for RoTSEC

As shown in Figure 9, major ICS players state that *Projects'* team(s), select the optimal RoTSEC risk strategy, which is based on the following types of risks: 1) Risk avoidance; 2) Risk reduction, where mitigation of the severity of losses; 3) Alternative actions to offer possible RP actions to reduce risks; 4) Sharing actions of transferring risks to third parties; and 5) Risk acceptance, is the willingness to accept their consequences. RoTSEC critical risk estimation, capabilities include (Pratap, & Predovich, 2020): 1) Risk analysis; 2) Risk remediation; 3) Compliance content mapping; 4) Coherent workflow design; 5) User experience; 5) Board and senior *Managers* executive reporting; 6) Basic and advanced integrations; 7) Digital asset discovery; and 8) Real-time assessment. Risk mitigation artefacts are linked to the Polymathic AHMM4SE basic elements.

Polymathic Model's Basic Elements

The RoTSEC identifies, includes, and assesses various strategic and main risks to guaranty PERP coherency; and needs the following elements: 1) A clear strategy setting; 2) *Project* related risk's identification, assessment, and response; and 2) Communication and monitoring.

AHMM4SE's nomenclature is presented in a basic form to be understandable by the readers. The RoTSEC uses the AHMM4SE that is formalized as shown in Figure 10; and the main artefacts and characteristics are:

- Basic SE actions = support basic SE operations and DevOps activities with the goal of finalizing the SEP.
- PERP = A set of basic refinement actions.
- *Project* Environment = Includes the ICS, components, and its networks.
- RoTSEC = transformation of *Project* Environment + the defined goals of *Project* operations.
- Organizational RoTSEC = includes *Project* Environment + RoTSEC + PERP.
  Entity RoTSEC = $\sum$ Organizational RoTSEC.

The Entity Model

As shown in Figure 11, the symbol å indicates summation of all the relevant named set RoTSEC related members, while the indices and the set cardinality have been omitted. The summation should be understood in a generic sense, more like a set. The AHMM4SE uses services model to support the RoTSEC based RoTSEC and is in this RDP, represented in a simplified form. The RoTSEC's interfaces are based on the ADM and uses services to enable the polymathic transformation model. The EA based AHMM4SE is the combination of EA and AHMM4SE looks as follows:

EAHMM4SE = EA + AHMM4SE (16).

Figure 9. The magic quadrant for risk management
Source: Pratap and Predovich (2020)

The Polymathic Transformation Model

The RoTSEC transformation model is the combination of an EAHMM4SE and *IterationGap* that can be modelled using the following formula:

Project = EAHMM4SE + *IterationGap* (17).

Figure 10. The AHMM4SE nomenclature

Basic Mathematical Model's (BMM) Nomenclature

Iteration	= An integer variable "*i*" that denotes a *Project*/*ADM iteration*	
microRequirement	= KPI	(B1)
CSF	= Σ KPI	(B2)
Requirement	= CSF = $\underline{\cup}$ microRequirement	(B3)
CSA	= Σ CSF	(B4)
microKnowledgeArtefact	= $\underline{\cup}$ knowledgeItem(s)	(B1)
neuron	= action->data + microKnowledgeArtefact	(B5)
microArtefact / neural network	= $\underline{\cup}$ neurons	(B6)
microArtefactScenario	= $\underline{\cup}$ microartefact	(B9)
AI/Decision Making	= $\underline{\cup}$ microArtefactScenario	(B10)
microEntity	= $\underline{\cup}$ microArtefact	(B7)
Entity or Enterprise	= $\underline{\cup}$ microEntity	(B8)
EnityIntelligence	= $\underline{\cup}$ AI/Decision Making	(B11)
BMM(*Iteration*) as an instance	= EnityIntelligence(*Iteration*)	(B12)

The *Project's* model is based on a concurrent and synchronized choreography activities.

The Role of Choreography

A *Project* and RoTSEC are related to: transformations, *Project* related RP operations, BPM/choreography, and EA. All these mentioned fields need choreography capability that in turn is based on the HDT that uses Microartefacts. The AHMM4SE is composed of a large number of interconnected neurons, to solve a specific RoTSEC (or other *Project*) types of problems. RoTSEC units are connected to each other, like nodes of the HDT and there is a real number associated with each connection, which are called weightings, and CSFs.

Figure 11. The AHMM4SE main formulas

The Generic AHMM's Formulation

AHMM	= $\underline{\cup}$ ADMs + BMMs	(B13)

AHMM's Application and RoTSEC

Domain	= RoTSEC	(B14)
AHMM4(*Domain*)	= $\underline{\cup}$ ADMs + BMMs(*Domain*)	(B15)

Table 4. CSFs that have the average of 9.6

Critical Success Factors	HMM enhances: KPIs	Weightings
CSF_AHMM4SE_Project_Support	Proven	From 1 to 10. **10 Selected**
CSF_AHMM4SE_Risk_Response	Possible	From 1 to 10. **09 Selected**
CSF_AHMM4SE_Basic_Elements	Proven	From 1 to 10. **10 Selected**
CSF_AHMM4SE_TCDM_Interface	Possible	From 1 to 10. **09 Selected**
CSF_AHMM4SE_Entity_Model	Proven	From 1 to 10. **10 Selected**

valuation

The Model's CSFs

Based on the LRP and DMS, the most important model's CSFs were extracted, weight and used, and the resulted are shown in Table 4. The AHMM4SE and all the presented artefacts are used to support the RoTSEC based *Projects.*

RoTSEC-Based Project

The RoTSEC processing, is supported by a predictive DMS that depends on the selected CSFs, like the types of PERP activities, types of *Project* related risks, budget's situation, types of BPM models, skills, … A *Project* should be adapted to handle complex RoTSEC that can produce complex designs and eventual problems, which can be the source of *Project* risks... RoTSEC problems can be measured and weighted; where the *Project's* related risks are not easy to measure. This explains the difficulty of estimating *Project's* risks of a consequential set of PERPS operations; and the need for a decision holistic model.

A Decision Holistic Model

DMS is used to deliver a set of weightings for possible PERP actions (AI Microartefacts). Weightings' DMS concept support the RoTSEC's to deliver solutions in the form of SE recommendations. The DMS is responsible for the HDT based approach for RoTSEC problem solving, as shown in Figure 12. The RoTSEC adopts a holistic systemic approach which makes the *Project* robust and the Microartefacts management subsystem is the basis of a successful *Project.* RoTSEC's Microartefacts are managed .by the ADM.

This chapter proposes an alignment between the RoTSEC, ICS, PERP, and EA activities; by the mean of a TCDM. The TCDM provides a long-term systematic support to manage refined *Project* Microartefacts, which are compatible EA models. Developing an RoTSEC strategy for an *Entity,* ensured the control of the *Project's* planning and needs guidance that is supported by EA's ADM. EA is used to discover deficiencies, show complex and risky interactions between strategies, global processes, services

Figure 12. The entity's risk management concept

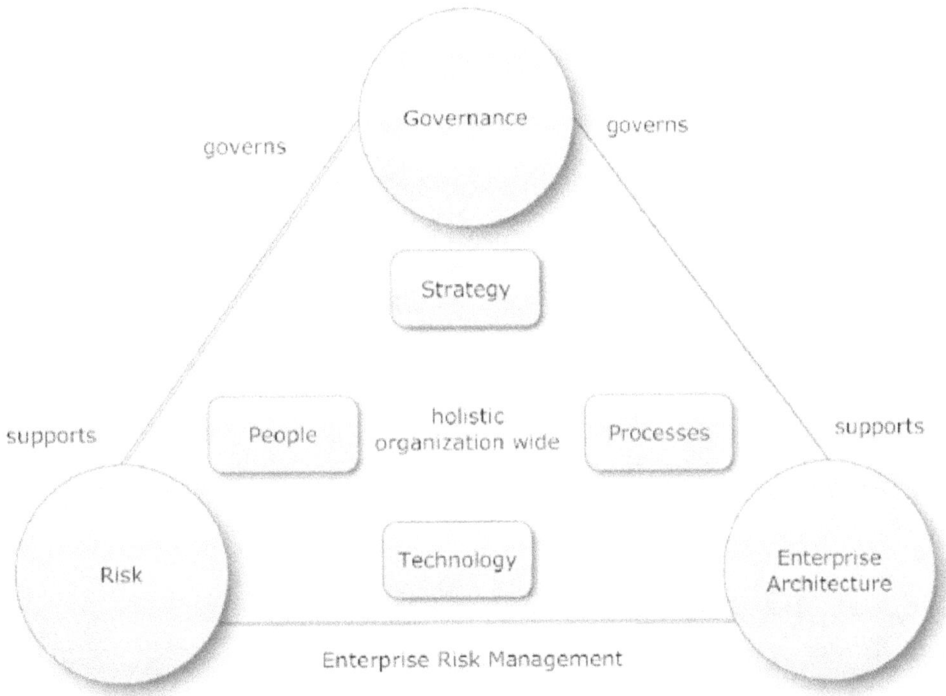

and the platform. The DMS controls PERP risks optimally. To implement RoTSEC based *Project* the author will propose a *Project* vision and strategy.

The Strategy and Evolution

Concerning RoTSEC the following concepts can be integrated: 1) DevOps/Agility, and hence PERP components extraction; 2) Establish a TCDM set of artefacts; 3) RoTSEC and its ADM interfacing; 4) Mapping resultant TCDM artefacts to *Project's* resources; 5) Microartefacts' granularity and unbundling; and 6) Requirements mapping to SE services and Microartefacts. *Projects*' PERP capacities, and transformations capabilities, are the main reasons for *Entities'* need for integrating emerging technologies, such AI, EA, and scalable platforms, according to Gartner's Hype Cycle for Emerging Technologies 2021. As shown in Figure 13, technologies to explore to accelerate growth include multi-experience, RP based innovation, digital transformations, and the use of major standards (Sargent, 2021).

THE ROLE OF STANDARDS AND AVANT-GARDE TECHNOLOGIES

Today there are many SE and methodologies standards and they are stable. These standards and the corresponding environments, support the *Project's* PERP. An important CSF for PERP is the capacity of refinement of existing components into dynamic ones (The Open Group, 2011a). The refined BPs, Microartefacts, components, and services can be used with existing standards by the means of the TCDM. The PERP contains an architectural pattern to be used as a template to deliver ABBs and SBBs.

Figure 13. Hype cycle for emerging technologies 2021
Source: Sargent (2021)

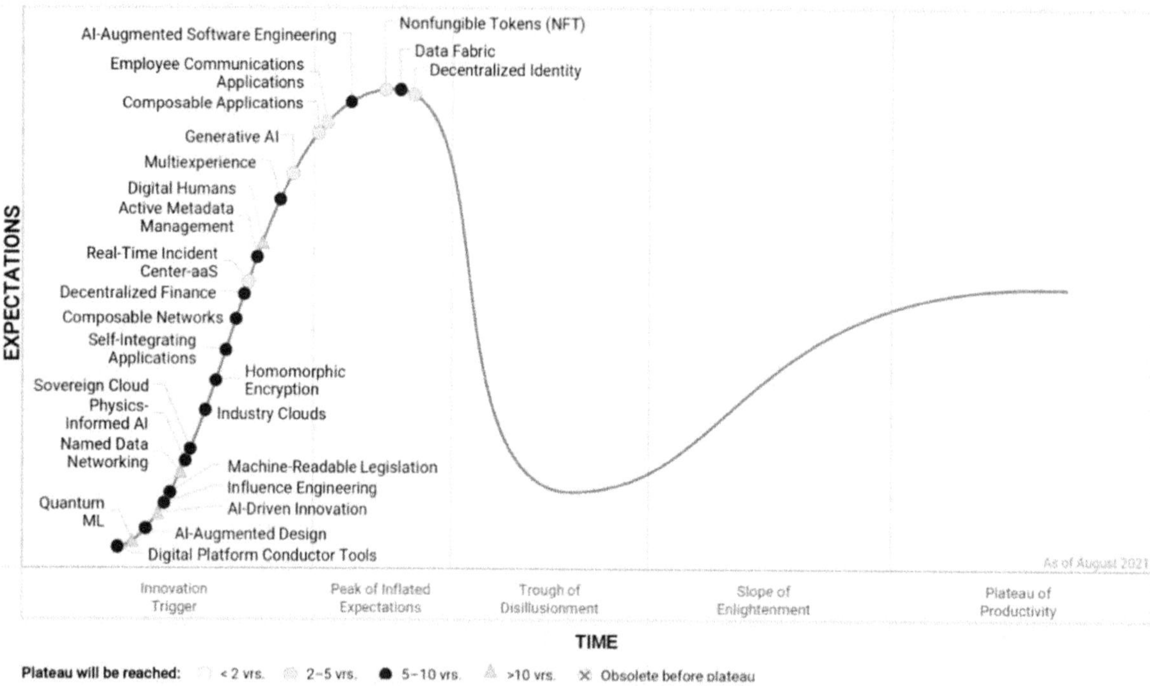

The RoTSEC is used to enforce SE and EA standards and the *Project's* implementers must be able to reuse refined components that result from the PREP. The PERP and standards are not integrated properly that results in that the target solution: 1) Has bad performance; 2) Lacks scalability; 3) Generates many errors; and to 4) Becomes un-usable and un-maintainable. Most SE initiatives have the tendency to *reinvent the wheel* when creating new components. The RoTSEC delivers refined BBs for architecture/ modeling, designs, and implementation constructs for *Projects*. Mixing Microartefacts to create robust BBs that can be mapped to different types of EA standardized frameworks, the *Project* team has to implement a PERP that is based on standards like TOGAF (The Open Group, 2011a). Using the TCDM enables the reduction of complexity and the adaption of a bottom-up cycle based on a "1:1" mapping approach. RoTSEC is to apply: 1) Standardized tools for modeling methodology, like the UML; 2) Standardized EA framework like TOGAF; 3) Standardized service architecture like the Service Oriented Architectures (SOA) Markup Language (SOAML); 4) Standardized BP modelling methodology, like the BPM; and 5) Apply the "1:1" mapping model. Applying the mentioned standards and the classification of the behavior and interoperation of Microartefacts, enables the negative impacts of frequent changes; and it also facilitates the understanding process of solving *Project* problems. The RoTSEC relies on the mentioned standards to deliver an adequate TCDM which is based on:

- **The BPM Standard:** The evolution and stability of BPM imposes a de facto standard for *Projects*. It also changed the nature of *Project* management from pure ICS domain to more of an agile business management activity. That enables *Projects* to become a business-driven activity. The 2012 year's hype cycle for BPM highlights a range of emerging, standardized ICS technologies and methodologies that enable BPM to deliver significant, tangible business value to *Entities*.

- **The BP Integration (BPI) Standard:** Focuses on the capabilities of the integration of refined BPs by the use of advanced EAI infrastructure. BPI supports inter-business environments, by using BP interfaces.
- **BPOKM Standards:** BPI, knowledge services, and SOA, are the basis of the *Entity's* knowledge system. This type of adaptive knowledge management model can be used also dynamically, to enhance KMS access. The KMS supports the use of CSFs to integrate the DMS.
- **Business/Domain Documents Standards:** Business/domain environments use standards like the eXtensible Markup Language (XML). This standard is results from business and ICS standards that have been boosted by Fortune 500 companies.
- **Governance Standards:** Using governance standards in *Projects* is crucial for its control and eventual success. One of the main CSFs for *Project* is to apply governance, tracing, and monitoring to of its refined components. Financial governance and control of the transformed *Entity* has to be automated.
- **Avant-Garde Applications and Web Technology Standards:** The *Project* team must capable of using standard technologies, like dynamic web pages, style sheets, MVC, and other. Which are the fundament of the EA and SEP of the *Project*. These technologies are to be used using PERP aspects for future changes. The team must be capable of refining existing components to meet the defined standards.
- **The SE Stack Standard:** Includes various levels of business and ICS resources and components needed for the *Project*. The SE stack standards includes: 1) A lower layer, which is the business infrastructure; 2) Microartefact catalog; 3) An upper layer that contains the enterprise's business modeling and EA.
- **The EA Standard:** The initial step is to generalize the use of the ADM and to extract the TCDM, to support the RoTSEC and to support the following operations: 1) Use EA (like TOGAF) to coordinate the transformation and PERP initiatives; 2) Defined ICS/technical architecture phases and the needed interfaces. 3) Platform for integration activities and PERP operations; 4) Use DevOps standards to interact with the ADM; 5) Use *Project* management standards to interact with the ADM; 6) Use client design standards standards to interact with the ADM; and 7) Use the TCDM to interact with the ADM and the RoTSEC.

The TCDM

As ICS technology evolves faster than one major iteration of a *Project*, it is difficult to finalize the *Project* with the initial goals and defined ICS structure. That is why it is important to define TCDM artefacts that are transcendent for all the *Project's* iteration. For this chapter a sample of TCDM includes: 1) OO basics; 2) The class and object diagrams (which can map to an ERD); 3) The BPM and collaboration diagrams; 4) Delimiters, actors and interfaces; 5) Circular implementation methods, like DevOps or ADM; 6) Services technologies, abstracted by Microartefacts; 7) Use Case diagram; and 8) DMN artefacts; … The ADM's integration with the RoTSEC, enables the automation and auto-generation of the *Project's* mainly TCDM Microartefacts, where these Microartefacts go through all the ADM phases and its various iterations. The ADM encloses cyclic iterations, where information about all its phases' RoTSEC activities are logged. RoTSEC is not dedicated to any specific business or technology platform.

The ADM is controlled and monitored in real-time and supports the Microartefacts' refinement, mapping, and interactions using RoTSEC in various types of tests and integration-driven developments. That all enables the business integration and inter-operability.

Business Integration and Inter-Operability

Business integration and inter-operability capacities have the following requirements and characteristics:

- Business integration ensures that the *Entity* can easily executes concrete integration of refined business scenarios in its eco-system, and in turn that ensures its long-term business sustainability. The following RoTSEC modelling artefacts patterns enable such a business integration process: 1) Services inter-operability pattern; 2) EA methodology pattern; 3) An anti-locked-in strategy; 4) TCDM artefacts, like the BPM or services choreography; 5) A generic and inter-operable business communication layer that enables the exchange of modelling patterns and integration interfaces.
- Business inter-resources operability is supported by the XML based interchange format (XMI), that makes the *Entity* resources and Microartefacts generic, standardized, independent of implementation tools, methods, and above all consulting companies locked-in model.
- *Project* management tools support to serialize and synchronize BPMs and other TCDM artefacts. The serialized and standardized format file contains: model semantics, and the diagram-interchange information. This standardized interchange format, supports the business integration process using BPI, which facilitates the use of the *business interaction matrix*; which is as shown in Figure 14. The *business interaction matrix* shows the mapping between the business services and other business functional domains.

Business integration and inter-operability capacities depend mainly on critical issues like the business infrastructure and platform.

Business Infrastructure and Business Development Environments

Managing the *Entity's* infrastructure, by the *Project* team implies that to transform it into an agile business environment's platform that is based on: 1) Resources management and sharing, by the means of the Cloud; 2) High business availability without downtime; 3) Massive data management; and 4) Load-balancing activities. *Entity's* security is also important, where the *Project* team manages security controls and standards; which are related to the design and implement security concepts for BPM, BPI, services, and Microartefacts patterns. This is needed for the protection of the *Entity's* business logic and know-how from being copied, which can avoid that the *Entity* erodes. All the presented CSAs can be verified in the PoC's implementation.

RoTSEC Project's CSFs

Based on the LRP and DMS, the most important RoTSEC *Project's* CSFs were extracted, weight and used, and the resulted are shown in Table 5.

To implement RoTSEC PoC the author will propose his *TRADf*.

Figure 14. TOGAF's business interaction matrix
Source: TOGAF (2011)

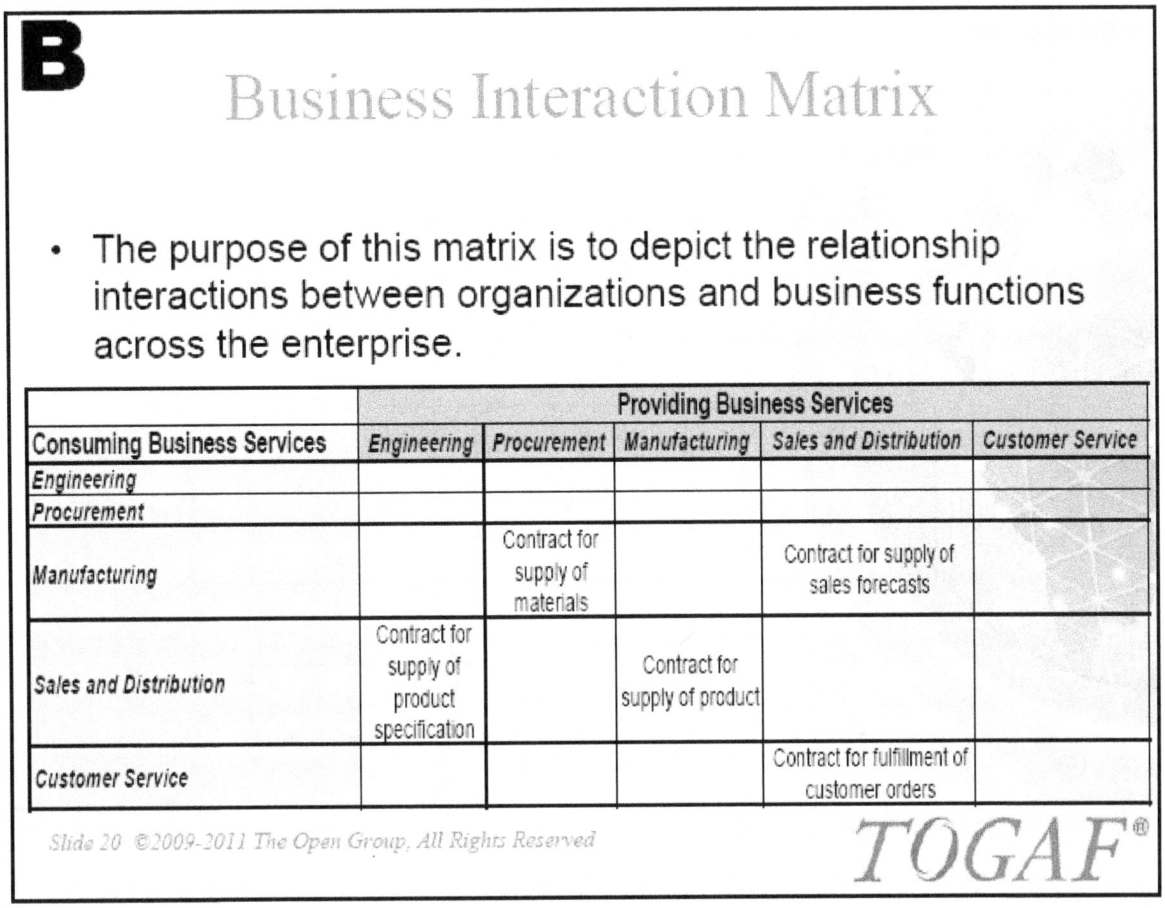

Table 5. CSFs that have the rounded average of 8.90

Critical Success Factors	KPIs	Weightings
CSF_RoTSEC_Project_DMS_Integration	Possible	From 1 to 10. **09 Selected**
CSF_RoTSEC_Project_Strategy_Evolution	Unstable	From 1 to 10. **07 Selected**
CSF_RoTSEC_Project_Standards_ICS	Complex	From 1 to 10. **08 Selected**
CSF_RoTSEC_Project_TCDM	Feasible	From 1 to 10. **09 Selected**
CSF_RoTSEC_Project_BPI_Inter-operability	Possible	From 1 to 10. **09 Selected**

valuation

Figure 15. RDP's flow
Source: Phu and Thao (2017)

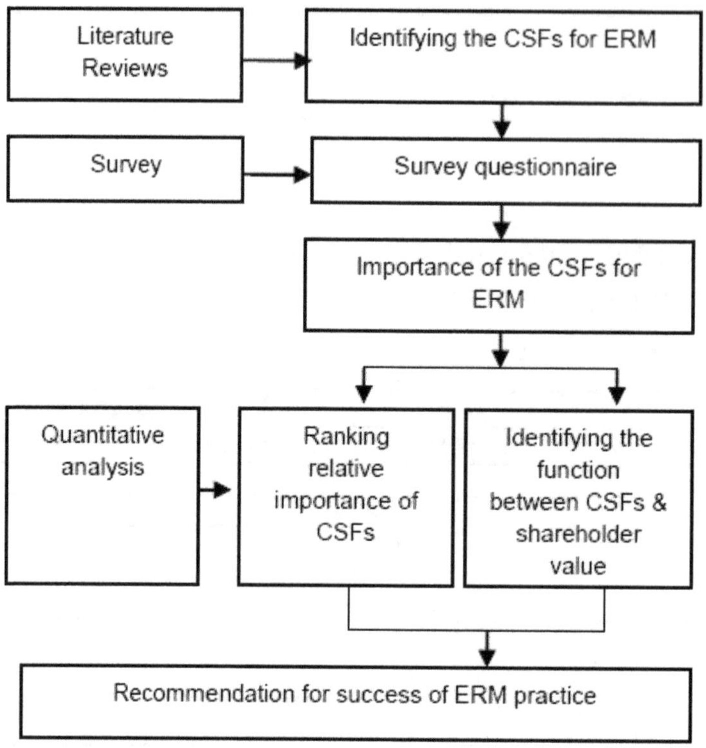

THE PoC's IMPLEMENTATION

RoTSEC Design and Implementation

An essential constraint for RoTSEC's PoC is to use of existing standards in a reduced form, what corresponds to the TCDM. In this case TOGAF, BPM and UML transcendent artefacts and diagrams are used. These standards include Microartefacts to be used to integrate RoTSEC modules in the existing *Project* BPMs. To identify the initial sets of CSAs' CSFs and test whether the RQ's of CSFs affect RoTSEC's integration. The PoC uses the HDT based mixed qualitative and quantitative method. The CSF's analytical process is illustrated in Figure 15. The PoC in the beginning uses phase 1 that is mainly based on the HDT tables, which weights using CSFs. Phase 1 is used to weight the relative importance of CSAs and CSFs for the usage of RoTSEC and that is done using a decision table (Quang Phu, & Thi Yen Thao, 2017).

The Proof of Concept: Phase 1

LRP's Outcome proved the existence of an important knowledge gap and it's (or Phase 1's) outcome supports the PoC's background, by the use of an archive of an important set of references and links. After selecting the CSA/CSFs tag is linked to various PERP microartefacts scenarios. The PoC is based on the CSFs' binding to a specific research resources, where the RoTSEC was prototyped using *TRADf.*

Table 6. The RoTSEC PoC's Phase 1 outcome is 9.15

CSA Category of CSFs/KPIs	Transformation Capability	Average Result	Table
The RDP's Integration	Usable-Mature	From 1 to 10 9.0	1
The Methodology ADM usage	Transformable-Possible-Complex	From 1 to 10 8.60	2
The Patterns' usage	Transformable-Possible-Mature	From 1 to 10 9.60	3
The Polymathic Model's Appliance	Transformable-Possible-Mature	From 1 to 10 9.00	4
The RoTSEC's Integration	Implementable-Complex	From 1 to 10 8.40	5
Evaluate First Phase			

The HDT represents the relationships between this RDP's requirements, Microartefacts and selected CSFs. PoC's interfaces were achieved using Microsoft Visual Studio .NET environment and *TRADf*. The RoTSEC uses calls to refined Microartefacts, to execute various actions related to *Project* requests. CSFs were selected and evaluated (using the HDT and DMS) in this chapter's tables; and the results are illustrated in Table 6. Table 6 shows that the RoTSEC is not an independent component and in fact it is strongly related to the *Project's* and *Entity's* overall transformation risk concept. HDT's main constraint is that CSAs having an average result below 7.5 will be ignored. This fact leaves the RoTSEC's CSAs (marked in green) to make this work's conclusions; and drops the CSAs marked in red, which means that the *Project* and RoTSEC transformation is feasible, but will face major difficulties; and is a complex. As Phase 1 is feasible then the PoC can proceed to Phase 2.

The Proof of Concept: Phase 2

EA's Setup and Related CSFs

The Phase's 2 implementation major settings look as follows:

- Sub-phase A or the Architecture Vision phase's goals, establishes a PERP approach and main goals.
- Sub-phase B or the Business Architecture phase shows how the RoTSEC target architecture realizes the key SE actions.
- Sub-phase C shows and uses the Application Communication Diagram to describe main SE activities.
- Sub-phase D or the Target Technology Architecture shows the end RoTSEC optimal infrastructure.
- Sub-phases E and F, Implementation and Migration Planning; the transition architecture, proposing possible intermediate situation and evaluates the RoTSEC status.

The HDT based Microartefacts have mappings to specific *Project* resources and where the RoTSEC defines relationships between the code components, TCDM artefacts, and Microartefacts.

Processing Control on a Concrete Node

The DMS solves RoTSEC requests, where CSFs link to a specific SE problem type and a related set of actions that starts to be processed at a specific HDT node. For this goal, the CSF_RoTSEC_Control_Procedure was selected from the defined CSAs and delivered solutions related to the selected problem. Solving the given RoTSEC problem involves the determination of actions and related solutions for multiple activities for the *Project* team. The applied RFP mixed models are based on quantitative analysis, beam search and grounded HDT; that is in fact a dual-objective of the DMS solves the problem using the following steps:

- Relating the PoC's resources to CSF_RoTSEC_Control_Procedure that is done in Phase 1.
- Link the selected node to the mixed modules and deliver the root node of the HDT.
- The DMS engine is configured, weighted, and tuned using configuration information.
- The HDT starts with the initial CSF_RoTSEC_Control_Procedure and then finds the set of solutions in the form of possible and successfull transformations and eventual improvements.
- Once *TRADf's* interface is activated to implement the needed Microartefact scripts to process the defined six CSAs.

Node Solution

These scripts support AHMM4SE's instance that are processed in the background to deliver *Project* related risk value. The AHMM4SE uses data services that are called by the DMS actions, which deliver *solutions and recommendations*; which are a set of PERP actions.

SOLUTION AND RECOMMENDATIONS

In this chapter, the author proposes the following set of RoTSEC's architecture, refinement, technical and managerial recommendations:

- A *Project* must build a holistic EA concept to ensure an efficient PERP based RoTSEC.
- Efforts are applied to integrate the ICS to support *Project* related risk operations and the problem is the alignment of various refined modules. Each *Entity* constructs its own refinement strategy.
- The RoTSEC replace traditional *Project* related component conversion concepts in order to improve *Project's* success.
- RoTSEC should fit in the *Entity's* EA framework and deliver a TCDM set of artefacts and diagrams.
- The ADM's and DevOps' integration in an RoTSEC enables the automation of all its activities.
- *Entity's* codebase stability is crucial for its future and is refined continuously.
- It is recommended to avoid consulting firms and do PERP activities internally.

FUTURE RESEARCH DIRECTIONS

Based on the conclusions *TRADf's* future research will focus on the evolution intelligence-based refinement concepts.

CONCLUSION

In this chapter the author presented the PERP and RoTSEC in the form of a concept which was verified by the PoC, which can be used as a BB for RPs and automated tests & qualification activities of the *Project's* future transformed components. The LRP outcome presented an important knowledge gap which is mainly due to the lack of transcendent holistic approach which is recommended by the RoTSEC. Actually, but there are primitive RP initiatives that concern heterogenous software components and which use manual refinement. The main aim is to show how an inhouse developed RoTSEC can inspire *Projects* and other innovation activities to refine and transform *Entity's* sources/resources (Koenig, Rustan, & Leino, 2016). This RDP phase is part of a series of publications related to *Projects*, EP/PERP, SE, EA, Polymathic models, RoTSEC and EAs; and is based on mixed HDT model; where CSFs and CSAs are offered to help *Project* teams to improve refinement SE activities. In this chapter, the focus is on evaluating and supporting the very complex SEP, in which RoTSEC and SE alignment are RP driven. The RoTSEC describes a structured relationship between: RP, SE, risks, EA, technical constraints, and HDT based solutions which can be used to support the long-term needs of the *Project*. RoTSEC's integration in the end system and the most important recommendation that was generated by this RDP, is that the *Manager* must be an architect of adaptive business systems and an experienced SEP specialist. In this chapter, the PoC is based on the CSFs' binding to a specific RDP resources, the DMS, KMS, RQ, and selected CSAs/CSFs; and the result is that RoTSEC is feasible and the main complexity lies in RoTSEC's integration and transformation refinement operations. Because of the score, 9.15, Table 6 shows that RoTSEC's implementation is possible but is a very risky transformation process. The RoTSEC can be an inhouse only refinement concept, methodology and framework. In this chapter, the author proposes the following set of managerial recommendations:

- The LRP has identified an important gap.
- The PERP supports an RoTSEC in order to ensure an efficient global RP and SE concepts.
- The TCDM and RoTSEC should fit in the *Entity's* global EA framework.
- ADM's integration in the RoTSEC enables the automation of all its RP activities.
- SEP and PERP constraints are controlled and monitored by the ICS.
- *Entities'* business sustainability is orthogonal to its transformation capacities.
- To avoid any form of locked-in scenario the *Entity* must build its own RoTSEC.

REFERENCES

Abecker, A., Papavassiliou, G., Mentzas, G., & Müller, S. (2014). *Methods and tools for business-process oriented knowledge management: Experiences from case studies*. Available online at: https://www.researchgate.net/publication/228769241_Methods_and_Tools_for_Business-Process_Oriented_Knowledge_Management_Experiences_from_Case_Studies

Daellenbach, H., McNickle, D., & Dye, Sh. (2012). *Management Science. Decision-making through systems thinking* (2nd ed.). Palgrave Macmillan.

Easterbrook, S., Singer, J., Storey, M., & Damian, D. (2008). *Guide to Advanced Empirical Software Engineering-Selecting Empirical Methods for Software Engineering Research*. Springer.

Fowler, M. (2003). *Catalog of Patterns of Enterprise Application Architecture*. Fowler. https://martinfowler.com/eaaCatalog/

Hosiaisluoma. (2022). *Holistic Enterprise Development*. https://www.hosiaisluoma.fi/blog/archimate-examples

Koenig, J., Rustan, K., & Leino, M. (2016). *Programming Language Features for Refinement*. Stanford University. doi:10.4204/EPTCS.209.7

Liu, A. (2022). *Rumbaugh, Booch and Jacobson Methodologies*. Opengenus. https://iq.opengenus.org/rumbaugh-booch-and-jacobson-methodologies/

O'Riordan, B. (2021). INNOVATION-Why Transformations Fail And How They Can Succeed With People Power. *Forbes*. https://www.forbes.com/sites/unit4/2021/10/11/why-transformations-fail-and-how-they-can-succeed-with-people-power/#:~:text=It%20is%20a%20bold%20decision,all%20major%20transformation%20projects%20fail

Papavassiliou, G., Ntioudis, S., Mentzas, G., & Abecker, A. (2001). The DECOR approach to Business Process Oriented Knowledge Management (BPOKM). In *DEXA '01 Proceedings of the 12th International Workshop on Database and Expert Systems Applications*. IEEE Computer Society. https://astimen.wordpress.com/2009/10/28/the-decor-approach-to-business-process-oriented-knowledge-management-bpokm/

Pavel, F. (2011). Grid Database—Management, OGSA and Integration. Academy of Economic Studies Romania, Bucharest. Database Systems Journal, 2(2).

Peterson, S. (2011). *Why it Worked: Critical Success Factors of a Financial Reform Project in Africa*. Faculty Research Working Paper Series. Harvard Kennedy School.

Pratap & Predovich. (2020). *The 2020 Gartner Magic Quadrant for IT Risk Management*. Gartner.

RedHat. (2022). *Decision Model and Notation (DMN)*. RedHat. https://access.redhat.com/documentation/en-us/red_hat_process_automation_manager/7.1/html/designing_a_decision_service_using_dmn_models/dmn-elements-example-con

Standish. (2011). *The Chaos Reports*. http://www.standish.com

Stitt, G., Stitt, F., Vahid, W., & Najjar, W. (2006). A Code Refinement Methodology for Performance-Improved Synthesis from C. *International Conference on Computer-Aided Design (ICCAD'06)*, San Jose, CA.

The Open Group. (2011a). *Introduction to the Architecture Development Method (ADM)*. The Open Group.

The Open Group. (2021). *TOGAF Model Patterns*. The Open Group. https://sparxsystems.com/enterprise_architect_user_guide/15.2/model_domains/togaf_model_patterns.html

Trad, A., & Kalpić, D. (2020a). *Using Applied Mathematical Models for Business Transformation*. IGI Global. doi:10.4018/978-1-7998-1009-4

Ylimäki, T. (2008). *Potential Critical Success Factors for Enterprise Architecture*. University of Jyväskylä, Information Technology Research Institute.

Chapter 16
In Search of Cooperative Compliance:
International Compliance Assurance Programme (ICAP) and Turkey

Hakkı Odabaş
Yıldırım Beyazıt Üniversitesi, Turkey

İmran Arıtı Erdem
Yıldırım Beyazıt Üniversitesi, Turkey

ABSTRACT

International Compliance Assurance Programme (ICAP) of OECD, a brand-new application, is the international reflection of the tax compliance expectations rising over cooperative compliance. ICAP is an application in which the tax risks of MNE groups are evaluated by bringing together the tax administrations of more than one country in which they operate and the relevant MNE group, and it has the capacity to create positive outputs at the point of ensuring voluntary tax compliance based on the establishment of tax certainty and consistency. In this study, the charactetistics, composition, and function of ICAP is revealed with the concept of cooperative compliance by comparing similar concepts with ICAP. Based on these elements, the chapter discusses Turkey's future participation in the programme.

Taxpayers' decisions to pay or avoid/evade taxes usually emerge as a result of a cost-benefit analysis. The risk of encountering tax audits or tax penalties has always play a leading role in decisions of taxpayers. Consequently, it is known that tax systems respond in various ways to the efforts of taxpayers to reduce their tax burdens, and the majority of these responses involve punitive mechanisms. On the other hand, the main factor affecting taxpayers' decisions is not always the existence of the state's power to impose punishments or the fear of this authority. At this point, the concept of cooperative compliance emerges as a policy tool. It is thought that tax compliance can be achieved by establishing more communication, cooperation and trust between taxpayers and tax administrations. On the other hand, tax compliance is a multidimensional concept and it should be ensured in the international manner. International Compli-

DOI: 10.4018/978-1-6684-6620-9.ch016

ance Assurance Programme (ICAP) of OECD, a brand-new application that we have now encountered, is the international reflection of the tax compliance expectations rising over cooperative compliance. ICAP is an application in which the tax risks of multinational enterprise (MNE) groups are evaluated by bringing together the tax administrations of more than one country in which they operate and the relevant MNE group, and it has the capacity to create positive outputs at the point of ensuring voluntary tax compliance based on the establishment of tax certainty and consistency. In this study, the characte-tistics, composition and function of ICAP is revealed, besides, the concept of cooperative compliance by comparing similar concepts with ICAP. Based on these elements, it is aimed to discuss Turkey's future participation in the programme.

INTRODUCTION

The taxation relationship, which positions the state and the taxpayer as opposite sides, has always included a tension between these two parties throughout history. States always consider how taxpayers perceive taxes and how they react to them, and in light of this, ongoing efforts are made to avoid incompatibilities. In general, such efforts have typically taken the form of penalties and performing audits. By acting this regard, the goal is to make taxpayers feel the weight of the state's deterrence power. However, a means of convergence between the state and taxpayers has been sought over time, particularly in light of the public interest in tax collection. At this stage, initiatives that prioritize the idea of assuring tax compliance come to the fore rather than those that aim to eliminate tax incompatibility. In this context, states developed a set of practices to ensure cooperation between tax administrations and taxpayers, and ICAP is one of them.

As a risk assessment programme on a voluntary basis, ICAP constitutes one of the sound policy tools to improve communication between tax administrations and tax compliance. From this point of view, this programme can be considered as an example of cooperative compliance effort. Cooperative compliance, which is a relatively new concept, poses an attitude that aims to increase trust between taxpayers and the government in order to ensure tax compliance. This concept entails taking various actions to collect taxes within the context of a relationship based on communication and guidance, away from the pursuit of "thieves and cops" and ICAP has served as the representation of this idea in the international arena.

When seen from this angle, it becomes clear that in order to comprehend this matter, one must first evaluate the historical stages that the issue of tax compliance has undergone. Following this assessment, the subject of cooperative compliance needs to be looked at. From this vantage point, it is possible to evaluate the ICAP framework, which is the multilateral representation of domestic cooperative compliance procedures. Put it differently, examining the application of ICAP requires addressing taxpayers' reactions to taxes and the issue of tax compliance, as well as the term cooperative compliance, which can be observed in country practices.

TAXPAYERS' ATTITUDES TOWARDS TAXES: COMPLIANCE OR NON-COMPLIANCE

Taxes can be approached as a financial burden on taxpayers, so it is observed that some taxpayers can resist taxes. It is possible for taxpayers to choose strategies like avoiding the taxable event or acting in non-taxable areas, as well as to avoid paying taxes in ways that the legislator does not intend, or to evade taxes illegally. This broad range of taxpayer behaviours or attidutes brought the issue of tax compliance out. Before examining the framework of cooperative compliance and ICAP, it would be appropriate to look into the models we encounter in the evaluation of taxpayers' attitudes towards taxes. Since ICAP, as a concept, is based on the enhanced relationship between taxpayers and revenue authorities, we need to focus on this relationship. Due to the fact that interaction between revenue administrations and taxpayers is determined by their basic assumptions about the others' goals and strategies (Kirchler, E., 2007: 168) compliance behaviour of taxpayers are also influenced this perceptions.

Individual interests conflict with collective interests under the tax system, posing a societal conundrum. Within this perspective, the taxation relationship contains a conflict of interest which poses the non-compliance problem. This problem is as old as taxes themselves (Kirchler, E., 2007: 103) and the literature has looked at the non-compliance issue brought on by the above-mentioned conflict of interest from a number of angles. Within this context, there are two basic approach to the issue of (non) compliance in the literature which are based on economic rationality and wider behavioral concepts (Baisalbayeva, K., 2014).

The Traditional Approach

The formal theory, which reflects the traditional approach, was shaped by the adaptation of Becker's (1968) economics-of-crime model to the issue of tax non-compliance by Allingham and Sandmo (1972) and Srinivasan (1973). In Becker's (1968) model, criminal behavior is typically regarded as a logical personal choice that is influenced by the likelihoods of detection, conviction, and the severity of the punishment and it is claimed that this approach may be applicable to combat tax evasion and avoidance (Graetz, M. J., Reinganum, J. F., & Wilde, L. L., 1986) (namely, non-compliance behaviour). Accordingly, Allingham and Sandmo (1972) suggest that an increase in the penalty rate or the probability of imposition of sanctions increases the reported taxable income of a rational maximizing taxpayer.

Based on these points in the literature on tax compliance, in formal theory which can be seen as the traditional theory, the taxpayer is typically viewed as an independent decision-maker in the literature on tax compliance (Klepper, S., & Nagin, D., 1989: 167) and it is accepted that the taxpayer makes strategic decisions (Kirchler, E., 2007: 167). In this sense, the attitude to adopt toward taxes is assessed as the outcome of this strategic thinking. According to this, people weigh the expected gains of evasion against the risk of getting caught and the penalties imposed when evasion is being detected (OECD, 2013). If starting from this point, one can deduce that certain obligations are effective in ensuring tax compliance. In other words, when more obligatory transparency and disclosure are mandated, it is crucial in modifying the real and perceived possibility of aggressive tax preparation from the perspective of an economically sensible model of taxpayer behavior (OECD, 2013). So we can conclude that, based on the formal theory, rational individuals are expected to comply with norms when the expected utility of compliance is higher than that of non-compliance; therefore, the likelihood and severity of sanctions and punishments for violators are crucial elements in explaining the levels of norm adherence that have been found (Baird, D., Gertner, R., & Picker, R., 1994).

It should be noted that this formal theory is criticized for having unrealistic assumptions (Boudon, R. 2009). According to Mehta (1998), there are no functional links in the realm of human action, and it is debatable whether preference relations even exist from a philosophical standpoint. The fact that the rational choice approach predicts a much higher level of tax evasion than is typically observed or estimated is actually the most common criticism leveled against it in this area. Since audit probabilities and fine amounts are typically low, most taxpayers should be expected to rationally conceal the majority of their income, but they do not (Bergman, M. & Nevârez, A. 2005). To conclude, one big failure of this theory is portraying taxpayers as socially isolated decision-makers who are simply affected from deterrence. Revenue bodies also recognized the limitations to the explanatory power of this model (OECD, 2013).

The Economic Psychology Approach

At this point, another approach to the issue of non-compliance should be highlighted. Since the formal theory is not sufficient to explain the issue of non-compliance, another hypothesis that emphasizes the social characteristics of taxpayers has been sought after. Although the importance of personal and social norms as determinants of compliance behaviour has also long been acknowledged, however, the factors such as legitimacy, fairness (Wenzel, M. 2002) and trust (Murphy, K. 2004) or ethics (Aim, J., & Torgler, B. 2011) have only recently been recognized another factors affecting compliance behaviour and this changed the opinions on compliance. Consequently, it is clear that there are circumstances in which taxpayers comply with tax laws for reasons other than deterrence mechanisms and it would seem that a sizeable portion of taxpayers would just pay their taxes regardless of the presence of authority or power (Martini, M. 2020).

If the traditional approach is followed, taxpayers as profit maximizers may be perceived as "robbers" from the pointview of the revenue authority, while such authorities see themselves as "hunters" or "cops" (Kirchler, E., 2007: 168). Contrasted with, taxpayers perceiving themselves as being persecuted may adopt the respective role and try to escape by using creative constructions of non-compliance and making use of loopholes in the tax law or by taking the risk of evading if it pays. However, tax authorities may establish the role of advisors with the primary objective of assisting in preparing accurate tax reports and supervising taxpayers, though, if they believe that taxpayers are willing to follow the spirit and letter of the law and accept that the complexity of tax law may lead to mistakes in tax filings, both to the advantage and disadvantage of taxpayers (Kirchler, E., 2007: 168). This strategy is likely to foster feelings of empathy and fairness among taxpayers, which is likely to be reciprocated by compliant behavior.

As is evident, a variety of circumstances influence whether taxpayers comply with the law or not. However, avoiding compliance expenses associated with various sanctions for tax evasion as well as cooperating with the tax authority might enhance taxpayers' compliance capacities. Nations and international organizations who are aware of addressed the issue of collaboration. As a result of these efforts, the idea of cooperative compliance has been established in this case.

BETWEEN THE PARTIES TO THE TAXATION RELATIONSHIP: COOPERATIVE COMPLIANCE

In the course of history, it has become clear that in addition to removing the causes of non-compliance, it is also possible to change taxpayers' attitudes regarding taxes by offering the assistance and guidance required to assure compliance. Revenue authorities realized that authority is not the only factor which affects tax compliance behaviours and for the sake benefiting limited resources in a more efficient way, cooperative taxpayers began to be treated differently from others (Martini, M., 2020). In this regard, new policy objectives and tools have been established. The influence of some incidents on this change cannot also be ignored. Especially after the major corporate scandals of the late 1990s and early 2000s, the need for new approaches to promote a shift away from traditional adversarial interaction and toward a more collaborative approach grew.

Actually, during the start of the 1990s, cooperation between tax authorities and taxpayers hit a low point, there was little to no comprehension of one another's views, and taxpayers lacked the motivation to fulfill their commitments (Bronżewska, K., 2016). They had to deal with time-consuming audits from numerous departments, and there was also inadequate communication between the parties and as a result, the taxpayers were unable to determine what their proper tax position should be (Bronżewska, K., 2016). However, as the initial result of these efforts, the early 2000s witnessed a reevaluation of the conventional manner in which taxpayers and revenue authorities have been communicating with one another, solidifying the idea of more cooperative communication (Martini, M., 2020).

The assumption that tax compliance can increase when there are changes in the attitudes of revenue administrations toward taxpayers was supported by country practices during this time. In this period, it can be observed that three nations—the Netherlands, Ireland, and the United States—started particular programs intended to gain the participation of large corporate taxpayers at a higher level, as well as the UK started a number of programs (de Widt, D., & Oats, L., 2019). Collectively, all of these actions could be viewed as being under the umbrella of the term cooperative compliance.

Before moving on to the definition of the concept of cooperative compliance, it is necessary to mention the efforts of the OECD in the historical sequence of the background of this definition. Because even while different nations had the ability to implement local cooperative compliance examples in the 2000s, it wasn't until the OECD's efforts that the issue was treated broadly and brought to the international stage.

It should be noted that the OECD's research of the experiences of the three aforementioned nations that pioneered cooperative compliance began in 2008. In this year, the Forum on Tax Administration of the OECD prepared "Study on the Role of Tax Intermediaries" and suggested creating a more cooperative, trust-based relationship between revenue authorities and substantial business taxpayers who abide by the law and go beyond statutory duties to cooperate. The OECD did not yet use the word cooperative compliance at the time this early step was taken; instead, the term "enhanced relationship", which reflects the change in the obligation based traditional relationship, is choosen.

Enhanced Relationship Era

Enhanced relationship, which entails a shift in the tax compliance process and in the behavior of the tax authorities and taxpayers, is one that results in open communication, real-time working, and ultimately, higher tax certainty and efficiency gains. According to the OECD's 2008 study's findings, there was a huge opportunity to affect the "demand side" of aggressive tax planning strategies with regard to major business taxpayers. The establishment of a cooperative and trustworthy relationship between these taxpayers and revenue authorities was fostered. The study created the term "enhanced relationship" to describe these interactions and developed a conceptual framework for them. This concept was established as a solution to three primary problems: 1) aggressive tax planning, 2) the outdated methods of tax collection and compliance, and 3) the necessity for governments to adapt to society's changes and evolution (Bronżewska, K., 2016).

In addition, the approach used in the OECD's 2008 study is based on a total of seven pillars, some of which are related to the revenue administration and others to the taxpayers. It can be separated into two categories; (a) in dealings with taxpayers, revenue authorities demonstrating understanding based on commercial awareness, impartiality, proportionality, opennes through disclosure and transparency, responsiveness and (b) in dealings with revenue authorities, taxpayers providing disclosure and transparency (OECD, 2013).

First among these characteristics, commercial awareness is important that revenue authority should be a good understanding of the commercial drivers behind the transactions and activities (OECD, 2013). Without knowing the commercial drivers, revenue authorities might not comprehend the full context of a transaction and might react in a way that leads to uncertainty and perhaps costly disputes (Huiskers-Stoop, E., & Gribnau, H. 2019). On the other hand, according to the impartiality concept, revenue authorities must approach issue resolution with a high degree of consistency and objectivity. Proportionality is all about choices. When allocating resources, revenue authorities must prioritize which taxpayers, whose tax returns, and which tax issues will have the most impact on overall revenue. Openness and responsiveness is another pillar which create interaction in building early certainty. On the other side, disclosure and transparency come into play in the taxpayer dimension of the issue.

Following the 2008 study, which was created using the expertise of three nations, numerous revenue authorities have put into practice compliance measures built upon these pillars, such as the creation of compliance risk management strategies. The OECD (2013) has tried to advance the topic by analyzing each of these new country experiences, and as a result of these research, a new study named "Cooperative Compliance: A Framework" was released in 2013. It is the main study of the OECD which the term "cooperative compliance" introduced instead of "enhanced relationship". This report includes the term "cooperative compliance" and this makes it apparent that the strategy is built on cooperation but with the aim of ensuring compliance, or paying the appropriate amount of tax at the appropriate time. This phrase refers to a brand-new, purportedly mutually advantageous method of cooperation between taxpayers and revenue authorities (Bronżewska, K. 2016).

The Framework of the Term Cooperative Compliance

Cooperative compliance is the development of a trust-based working relationship between revenue authorities and taxpayers that is based on voluntary tax compliance and results in the timely payment of the correct amount (Huiskers-Stoop, E., & Gribnau, H., 2019). In order to encourage voluntary compliance and counteract non-compliance among large taxpayers, cooperative compliance tries to create the necessary conditions. Within this framework, "cooperative" emphasizes the reciprocal nature of the relationship which is aimed at enhancing compliance and "compliance" includes transparency and disclosure on the part of both parties (Huiskers-Stoop, E., & Gribnau, H., 2019).

Cooperative compliance may be seen as the solution to issues that tax authorities in many jurisdictions have faced for a long time, including the inability to thoroughly investigate all cases that called for a closer look, to audit taxpayers as frequently and thoroughly as they deemed necessary, to react quickly and effectively to the market's shifting conditions and new, more sophisticated tax planning opportunities brought on by globalization (Bronżewska, K., 2016). Perhaps for this role of cooperative compliance, nations have started to concentrate on actions that can be assessed within the framework of this concept, in addition to instruments like BEPS that result from non-compliance.

Cooperative compliance has several advantages for both sides of cooperation. From the pointview of revenue authorities, these benefits include improved relationships, a better understanding of business, risk management, certainty in advance, a decrease in administrative burdens, increased compliance, accurate and timely filing of tax returns and timely payment, better real-time information about commercial developments, better use of revenue body sources, and increased confidence in the tax system and revenue bodies. On the other hand, from the perspective of the taxpayer, the benefits include the following: improved connection, reputation, risk management, the chance to draw attention to issues with the tax code or its administration, certainty in advance, and a decrease in administrative costs.

The phrase "cooperative compliance" has become commonly used to describe any strategy that encourages a connection between taxpayers and revenue authorities based on trust, collaboration, and cooperation, especially since the 2013 OECD publication, however, because there is no common definition of cooperative compliance, the pillars and practices implemented by each nation may differ significantly (Martini, M., 2020). However, in this report, the experiences of 24 countries at the time that put the idea of cooperative compliance into practice, were analyzed and certain overarching themes were identified. Pursuant to these assessments, it was observed that all cooperative compliance approaches functioned within the confines of the law and regulations, however the majority of countries formalized cooperative compliance in some sort of agreement with taxpayers rather than changing existing laws and regulations. On the other hand, the cooperative compliance strategy was generally only made available to a specific class of taxpayers in most countries, as determined by a number of objective factors like size and complexity and it was observed that some countries explicitly exclude high-risk taxpayers from the practice of cooperative compliance arrangements. In most countries, the adoption of cooperative compliance starts with the application of taxpayers who meet the requirements. In other words, it is clear that the norm is for taxpayers to apply voluntarily (OECD, 2013).

After outlining the broad contours of domestic procedures, the OECD questioned in this report if a multilateral cooperative compliance model might be used based on the growing number of country practices. Multilateral cooperative compliance, according to the OECD, is an agreement between a taxpayer and two or more revenue agencies in which all parties concur to use cooperative compliance principles to manage tax issues that are in everyone's best interests. Practical clarity and a reduction in the compliance requirements related to cross-border transactions are the goals of a multilateral cooperative compliance relationship (OECD, 2013).

Based on the analysis of country experiences in the 2013 report of the OECD, a multilateral cooperative compliance approach is viewed favorably in theory in a setting of trust between the parties, with a shared understanding of the factual position leading to a closer reflection of the reality of the operations of the large business. Additionally, multilateral cooperative compliance programmes may help reduce tax uncertainty, according to the 2017 IMF/OECD report on tax certainty (Majdańska, A., 2021: 3).

Breaking the traditional approach and establishing a communication based on trust and cooperation between taxpayers and revenue administrations is a farsighted step in the conditions of that day, but it has started to prove insufficient for problem-solving in an increasingly globalized world where taxpayer mobility is rising. The expansion of collaboration to a global level has been made necessary, particularly by the emergence of multinational corporations. In order to ensure tax conformity, the OECD underlined the importance of a global strategy. This multilateral approach has been a very novel step, particularly during the time when the BEPS Action Plan was introduced and it became clear that the individual efforts of the countries were insufficient. Only a few years later, with ICAP, was this strategy's concretization, or reality, accomplished. So now that the background of ICAP has been made known, it is possible to study this application, which is the end product of the quest for multilateral compliance.

The Multilateral Cooperative Compliance Dimension: International Compliance Assurance Programme

Even though the notion of transferring the cooperative compliance practice from the domestic to the international level has been debated for about ten years, the first significant action on this topic wasn't made until 2018. The ICAP, a multilateral risk assessment program that gave the foundations of cooperative compliance an international scope, was introduced in 2018 by a group of eight OECD countries. Actually, development on ICAP started in 2016 as part of the Large Business and International Programme of the OECD Forum on Tax Administrations, which was sponsored by the Canada Revenue Agency and the Internal Revenue Service of the United States (Pross, A. & Johnson, M., 2020: 15). So, the step taken in 2018 can be seen as a result of this 2016 endeavor.

ICAP is a voluntary programme for a multilateral cooperative risk assessment and assurance procedure that aims to give multinational enterprice (MNE) groups wishing to participate actively, openly, and fully transparent manner with improved tax certainty with respect to some of their activities and transactions (OECD, 2021: 6). In this process, various tax risks are evaluated in active cooperation of the participating revenue administrations of countries in which multinational enterprises operate and MNE groups and determined whether these covered risks mean low risk.

This programme supports revenue authorities in coordinating their assessments of the main international tax risks facing MNE groups and thus, the program's goal is to increase the efficiency of these risk assessments and possibly give MNE groups long-term comfort over the issues covered in the various jurisdictions where they operate. However, this comfort should not be seen as a legal certainty. The OECD (2021: 15) highligted that the outcome of ICAP risk assessment does not mean the legal certainty, it only indicates that the program's covered risks are considered to be low risk.

Beyond the tax authorities' experiences in engaging with taxpayers through domestic cooperative compliance programmes, the G20 tax certainty agenda, the jurisdictions' commitment under BEPS Action 14 to improve the effectiveness and efficiency of dispute resolution through the Mutual Agreement Procedure, the availability of better or more standardized information for transfer pricing risk assessment as a result of BEPS Action 13, were among the factors that led to the need for such a program at the time (Pross, A. & Johnson, M., 2020: 16).

In such a period, in which multilateral agreements on BEPS as well as international exchange of information and cooperation of tax administrations come to the fore, it has become inevitable for domestic cooperative compliance programs to become multilateral. ICAP's introduction can be viewed in this light as a significant move taken by revenue administrations in the international arena, which is the setting where tax non-compliance is most actively encouraged. In this regard, a thorough evaluation of the program in question is crucial.

Although the issue of tax compliance and the concept of cooperative compliance in the background of this program are important to understand ICAP, which is a relatively new practice, this effort is not sufficient. In this respect, it is necessary to take two pilots of ICAP into consideration and then put up against similar applications. A deeper understanding of the programme can be possible only after these are explained. Therefore, the functioning of the ICAP process should be discussed at first place and then the motives of this program as well as it's goal and advantages shlould be evaluated.

UNDERSTANDING THE PROCESS: HOW DOES THE ICAP FUNCTION?

ICAP, a programme where taxpayers and tax administrations collaborate to assess certain tax risks, has only been used as two pilot applications up to this point. The application's design allows for the evaluation of the tax risks that specific taxpayers may face. But at this point, it's important to note that not all taxpayers can apply for the ICAP process; only MNE groups can. ICAP, in this context, essentially entails the activity of assessing the tax risks of an MNE group through the examination of documents submitted by the MNE group by the country where the ultimate parent entitiy (UPE) of the group is located and other countries in which the group has significant operations and that are connected to the tax risks being discussed. As a result of this action, each country determines the level of tax risks evaluated.

ICAP's core component is risk assessment and this risk assessment is conducted in a number of stages. The roles of the MNE group and the country where the UPE resides alter in each of these stages, which each make up a separate part of the risk assessment process. In this regard, these stages can be made concrete by looking at the two ICAP pilot applications.

ICAP 1.0

The country practices that the OECD considered in its 2013 report on cooperative compliance served as the basis for the first pilot application. Within this context, together with the OECD Secretariat, tax authorities from Australia, Italy, Japan, The Netherlands, Spain, and the UK joined Canada and the US in 2016 and 2017 to prepare a handbook that served as the foundation for the first ICAP pilot (Pross, A. & Johnson, M., 2020: 17). The eight member tax administrations and MNE groups with headquarters in their jurisdictions who had volunteered to participate in the pilot were brought together at an event sponsored by the Canada Revenue Agency and Internal Revenue Agency in January 2018 to officially launch it. The size of the pilot was kept to a minimum, and it was decided that each jurisdiction would include one MNE in the risk assessment process (Rijsbergen, H., 2020: 26). These MNEs work in a variety of industries and have a wide range of assets, revenues, operational complexity, footprints in the covered jurisdictions, and PE risk and materiality of intercompany transactions.

There were a total of four steps to the ICAP 1.0 procedure. Stage I constituted "Pre-entry", while Stage II was named as "Scoping" and Stage III was named as "Risk assessment and issue resolution". The last stage (Stage IV) was "Outcomes". MNE groups are required to prepare various documents in some of these stages, each of which has a varied scope, and then tax authorities write a result letter for the MNE group. Each of these stages are explaining as follows in more detail according the the handbook about ICAP 1.0.

The Stage I (OECD, 2019: 12), "Pre-entry", serves the purpose of giving MNEs a straightforward, unbureaucratic, and affordable means to determine whether their involvement in ICAP will be viewed favorably and, if so, by which tax administrations. An MNE may express interest in taking part in ICAP to the tax authority overseeing its ultimate parent corporation (the UPE tax administration) or may be contacted by the UPE tax administraion to inquire about the possibility of taking part in the program. The UPE tax administration is willing to serve as the lead tax administration, and it may, as necessary, get in touch with other FTA member tax administrations to ask if they would be interested in participating in an ICAP risk assessment of the MNE as covered tax administratio. By this way, it is established that the foundation for the countries that will conduct the risk assessment.

Following this pre-entry, Stage II, "Scoping", begins with submission of the scoping documentation package by the MNE via a secured virtual data room. It is important to highlight that the scoping documentation package includes documents such as Country-by-country-reports (CbCR), master file, a summary of the MNE's current global group structure, a list of all unilateral, bilateral and multilateral advance pricing agreements (APA). The desired timescale for the risk assessment stage, any adjustments that may be necessary to the usual main documentation package, and whether any covered transactions should be omitted from the scope of the MNE's ICAP risk assessment are all discussed further during this stage (OECD, 2019: 13). So this stage can be evaluated a period, in which risk assessment has not started yet and further preparations continue.

"The risk assessment ans issue resolution" stage (Stage III), which is the most crucial stage, is started after the scoping stage is completed. The main documentation package from the MNE is submitted to start this stage. This package is broader than the scoping documentation package. It includes any necessary updates to the documents included in the scoping documentation package, taking into account any modifications to the ICAP risk assessment's scope that were agreed upon during the scoping stage, also a schedule containing details of covered transactions, CbCR, local file, a schedule containing details of covered transactions etc (OECD, 2019: 47-48). Typically, at least one multilateral call or meeting will be held during the risk assessment stage with the MNE, the lead tax administration, and other covered tax administrations (Pross & Johnson, 2020). In order for each covered tax administration to be certain that the covered risks constitute a minimal risk or that such a finding is not achievable, the lead tax administration and other covered tax administrations will share their findings.

Based on the shared opinions, the risk assessment and issue resolution stage is ended and then Stage IV, "Outcomes", begins. In here, the lead tax administration will provide a completion letter to the MNE stating that the ICAP risk assessment and assurance process has been successfully completed. If the lead tax administration determines after conducting an ICAP risk assessment that an MNE group offers a low risk either generally or in specific areas, this will have an impact on the outcome letter it delivers to the MNE group. The outcomes of each covered tax administration's risk assessment and assurance of the covered risks for the covered periods are also included in separate outcome letters that each covered tax administration sends to the MNE.

The process that was intended to occur in this way was carried out 2018 and 2019, according to the handbook. Meetings with the lead and covered tax administrators were held in this case between March and October of 2018, and the procedure was finished in 2019 with the last MNE group receiving the outcome letter.

The ICAP 1.0 was unprecedented, since tax authorities had never before tested if and how they could jointly carry out a risk assessment and provide tax assurance to MNE groups. On the other hand, it was also unique that MNE groups chose to take part in this pilot, which had an unproven process and an unknown result.

In general, participants gave the pilot's essential components favorable reviews. MNE groups, in particular, appreciated the pivotal role played by the lead tax administration as well as the precise deadlines set for covered tax administrations to submit questions and for responding to those questions. On the other hand, there are some procedural issues that needn to be mentioned. It became evident during the ICAP 1.0 process that the intended process in fact performed differently from what was anticipated. For instance, it was discovered that it was challenging to distinguish between Stage I and Stage II, so participating the tax authorities to decide to drop Stage II.

The multidimensional transparency and understanding between all involved parties was ICAP 1.0's most significant accomplishment and it is also important that multilateral exchanges of information between tax authorities' methods to large corporation compliance were conducted.

ICAP 2.0

Compared to the eight in the first pilot, more tax administrations have expressed a desire to take part in ICAP 2.0, increasing the chance that an MNE will be able to acquire comfort over a larger amount of its operations. However, the number of covered tax administrations for a certain MNE's ICAP risk assessment will still continue to be capped. On the other hand, ICAP 2.0 differs from ICAP 1.0 in that there are a total of three stages, and Stage II, known as "Scoping," has been eliminated. As another difference, Stage I's name is changed from "Pre-entry" to "Selection".

According to the OECD's ICAP Handbook, the first stage of ICAP 2.0 is "Selection" stage. This method is essentially the same as Stage I in ICAP 1.0, with the exception of the name change. The application of the MNE group to the tax administration of the country where UPE resides and the participation of other administrations where the group conducts business are both under consideration in this context.

The suitability of an MNE group for ICAP is evaluated on a case-by-case basis. An MNE group should consider (1) whether its UPE is a resident of a Forum on Tax Administration member tax administration participating in the program, (2) the footprint of the MNE group and the materiality of the MNE group's covered transactions in jurisdictions of Forum on Tax Administration member tax administrations participating in the program, (3) whether a CbCR filing requirement exists in the MNE group's UPE jurisdiction, (4) whether the MNE group has ever taken part in ICAP in the past, (5) whether the MNE group has internal processes to establish and manage its tax policies, as well as a well written group tax strategy that is controlled by senior management at the board level, and (6) whether the MNE group is willing to participate in the ICAP procedure and accept the requirements, such as documentation when determining whether it is suitable for an ICAP risk assessment (OECD, 2021: 14).

Here, the procedure starts with the MNE group submitting the selection documentation package, unlike Stage I in ICAP 1.0. The MNE group's information form, which includes a list of proposed covered tax administrations and periods, information about the group's UPE etc. and a covered risk overview, which includes a list of categories of transactions or arrangements falling under the proposed covered risks etc., are both included in the selection documentation package. Also documents such as the CbCR, master file, a summary of th MNE group's current global structure clearly take place in the pacakge. After the submission of this package, tax administrations agree on the timeframe of the risk assessment stage.

Stage II of ICAP 2.0 includes "Risk assessment and issue resolution," which is Stage III in ICAP 1.0, however this is the most significant stage in both pilot programmes. It involves the multilateral risk assessment and assurance of covered risks by the tax administrations. This stage starts with the submission of the main documentation package by the MNE group, which includes any updates to the documents in the selection documentation package, a schedule with information on covered transactions, the value of the transaction, information on how the transfer pricing policy was applied to this transaction, as well as the CbCR, local files, information on the MNE group's tax strategy, etc. The lead administration and covered tax administrations discuss the tax risks in this stage and this process resembles Stage III of the ICAP 1.0.

In the Stage III, which is the "Outcomes" stage, the lead tax administration sends an outcome letter to the MNE group to certify that the ICAP risk assessment and assurance procedure is complete. Each covered tax administration sends the MNE group an outcome letter with the outcome of the covered risks for the covered periods.

International Compliance Assurance Programme and Other Tools to Provide Tax Certainty

Tax administrations have a number of tool to provide tax certainty both in domestic and international aspects. From the pointview of tax certainty, such tools include tax audits and tax rulings as well as APA's and Mutual Agreement Procedures (MAP) (Pross & Johnson, 2020: 15). These instruments may be used at different phases of the relationship between the taxpayer and the tax authority and may be carried out entirely locally or only in an international manner.

According to the OECD's Glossary of Tax Terms, advance rulings are "letter ruling(s), which is a written statement, issued to a taxpayer by tax authorities, that interprets and applies the tax law to a specific set of facts". Advance rulings are fully domestic tools to provide tax certainty. Both full and limited taxpayers may request an advance ruling for the tax administration and this ruling may only be given in situations which included in the taxation authority of this country.

ICAP can be regarded as a different tool from advanced rulings from based on a few sides. First of all, advanced rulings cannot reflect the other tax administrations views. Therefore, the taxpayer cannot receive such views on the subject. This affects negatively the tax certainty when more than one country is related with the concerned subject. On the other hand, ICAP can be regarded as an improvement over advanced rulings, because ICAP tries to give tax authorities a fuller picture by giving them information about the entire business as well as its surroundings, while avanced rulings can be thought of as a single image of a particular business. Thus, more information and trust is presented on ICAP.

On the other hand, ICAP constitutes a different manner from tax audits. Except from the joint audit issue, tax audits have only a domestic side, while it is broader than ICAP considering its content. Besides, there is no other tax administration actively involved in the audits which is a matter of tax sovereignty. However, some tools such as exchange of information can be benefited in the audit process but the information obtained from abroad means only data, the view of the tax administraion on the subject of the audit cannot be obtained. Thus, tax audit has only a one dimension, while ICAP has different dimensions. As another difference, tax audits create legal consequences, such as sanctions or fiscal penalties. However, as mentioned above, ICAP only provides a comfort for the covered transactions and this cannot be seen as a legal outcome.

As another tool to provide upfront legal certainty, APA can be compared with ICAP. According to the OECD's Glossary of Tax Terms, APA is identified as "an arrangement that determines, in advance of controlled transactions, an appropriate set of criteria (e.g. method, comparables and appropriate adjustments thereto, critical assumptions as to future events) for the determination of the transfer pricing for those transactions over a fixed period of time". This concept differs from ICAP in that a tax administration may provide comfort for a covered tax risk, even though the other covered tax administration did not conclude a low risk outcome. This makes the ICAP's scope broader than APA's.

A Closer Look to International Compliance Assurance Programme

As a voluntary compliance tool, ICAP constitutes one of the best international practices in providing an effective and efficient collaboration between the both sides of the taxation relationship. In this situation, it is important to look at the circumstances that led to the relationship in question in order to comprehend it. This analysis will also help us to see the benefits of this relationship.

Addressing the factors that lead to the emergence of a relationship is important in terms of understanding why such a relationship arises and its limits. On the other hand, the understanding of the benefits, when evaluated together with the factors behind the relationship, provides a holistic picture of the implementation. From this point of view, the emergence of the application of ICAP can be attributed to a variety of factors. This issue is especially focused on since these factors will also clarify the anticipated benefits from this initiative. As another difference, the level of documentation may be mentioned. For an ICAP risk assessment, less documentation is needed than is generally needed to reach an APA agreement. On the other hand, an MNE group's ICAP risk assessment will normally be finished and outcome letters delivered within 24 to 28 weeks of the completion of the main documentation package and thus ICAP has a clear timframe. Compared to APA's, this timeframe is shorter.

The last tool that can be evaluated is MAP. According to the OECD's Glossary of Tax Terms, MAP is expressed as "a means through which tax administrations consult to resolve disputes regarding the application of double tax conventions. This procedure, described and authorized by Article 25 of the OECD Model Tax Convention, can be used to eliminate double taxation that could arise from a transfer pricing adjustment". These two tools are different based on their extent to which early certainty is required. A recent completed fiscal quarters for which information, including the MNE group's CbCR, is available are the topic of an ICAP risk assessment. This will frequently be done sooner after the end of the covered period than a tax audit would be able to, and it will undoubtedly be sooner than under the MAP system (OECD, 2021, 9).

As it can be understood, all of the mechanisms have their own characteristics and one can make a list of the differences between them. However, for the sake of providing tax certainty and compliance, all of these mechanisms may be regarded as complementary policy tools. If circumstances permit, combining these technologies will produce quicker and more precise outcomes.

The Derivers for International Compliance Assurance Programme

The OECD (2021) listed six key derivers behind the development of ICAP in the *International Compliance Assurance Programme Handbook*. Some of these refer to the necessity of developing mechanisms that will give MNE organizations and revenue authorities more assurance while building on the achievements of the OECD BEPS Project and the creation of the OECD (2021) Inclusive Framework on BEPS, while other factors that contributed to the creation of such systems include the trend toward more coordination and cooperation between various tax authorities and MNE groups.

The first key driver is "providing a pathway to improved tax certainty for MNE groups". This means that initiatives like ICAP, which are based on more transparency, collaboration, and coordination between revenue administrations and MNE groups, have the potential to promote tax compliance and have a favorable impact on taxpayer behavior while also giving MNE groups more tax certainty. The MNE groups are specifically mentioned in the context of program participation because it is frequently these taxpayers who are exposed to significant foreign tax risks. If the concept is described from the other side; in terms of tax administrations, MNE groups are the most probable actors of tax non-compliance in the global sense, and collaboration is crucial in assuring their compliance. Therefore, it is important to evaluate certain risks and provide tax certainty in terms of cooperation with these taxpayers.

"More effective dispute resolution" is the second driver which is mentioned in the OECD's Handbook for ICAP. As a result of BEPS Action 14, more efficient MAP's are put into practice and these endeavors for dispute resolution can be aided by a multilateral risk assessment and assurance program that prevents disputes from occurring and restricts the increase of MAP inventories. In this way, the conditions for the emergence of the dispute can be eliminated before the dispute arises. Through methods to stop these disputes from forming, one of the most efficient ways to reduce the risk of disputes between tax administrations. The phrase "dispute prevention" can also be used to describe this circumstance.

The third deriver is "well-established MNE group compliance frameworks" and the OECD's Forum on Tax Administrations has identified early resolution programmes amongst the best practices in the areas of cooperative compliance. Early resolution programmes prevent stances from getting entrenched and can be implemented by close to real-time while there is current documentation, access to key decision makers, and staff who are knowledgeable about recent transactions. In light of current data and as a result of a process where multiple tax administrations communicate and share their perspectives, output is thus obtained.

As a fourth deriver, the OECD mentions "advances in international collaboration". Collaboration lies in the hearth of the all cooperative compliance programmes and ICAP brings this Notion to the international arena. The depth and frequency of cooperation among the tax administrations that are members of the OECD Forum on Tax Administrations have significantly risen in recent yearsa and well-developed platforms can be used to support multilateral contacts with MNE groups. ICAP can be regarded as one of these instruments.

"Better and more standardised interactions for transfer pricing risk assessment" is included by the OECD as the fifth driver of ICAP. This issue is related with the Action 13 of the OECD's BEPS Project. This action calls for the introduction of regulations and the requirement that certain taxpayers submit transfer pricing documents. In this regard, the tax administration receives documents containing diverse information on a regular basis from taxpayers by obtaining CbCR, local files and master files. Action 13 of the BEPS Project has given tax authorities access to a more widespread set of data that they may use to evaluate the risks associated with BEPS 13 and other BEPS-related risks across numerous jurisdictions. Tax administrations may jointly de-select compliance issues and taxpayers who pose a lower risk of non-compliance while selecting those issues and taxpayers for further review and this allows tax administrations to use the information in an MNE group's CbCR, master file, and local files more effectively.

"Capitalising on greater opportunities for multilateral engagement to provide improved assurances for tax administrations" is regarded as a final deriver. As it can be understood from the explanation on the fifth deriver, tax authorities now have access to a larger number of data that may be utilized to more effectively identify and address compliance risks. Thus, multilateral programs can give participating tax administrations more assurance by supplying a more solid and well-thought-out basis for risk assessment using this data, allowing decisions to be taken in conjunction with other tax administrations rather than unilaterally.

The Benefits of International Compliance Assurance Programme

Documents which are suggested in ICAP can provide data about current operations and global chains of MNE groups. Besides this, the MNE group's itself, which carried on these operations and created these structures, participates actively in the assessment process. Additionally, analyses of other participating countries where the MNE group conducts business can also be enlightening. Thus, such a method of risk assessment in the ICAP process includes number of advantages.

The OECD addresses the benefits of ICAP in the ICAP Handbook in the five items. The first benefit is stated as the "fully informed and targeted use of CbCR and other information held for risk assessment". This benefit has two side. On the first side, this means that MNE groups can talk through their CbCR and other information and provide additional clarity to aid understanding their cross-border activities. Thus, the active engagement of MNE group in the assessment may serve a better understanding and in terms of putting the economic approach concept into practice in tax law, this outcome is advantageous to taxpayers. On the other side, tax administrations have the ability to communicate their findings to one another, enhancing the uniformity of data interpretation and this should assist tax administrations in making an early determination regarding the severity of tax risks. This issue is crucial for achieving fair taxation and the ideals of certainty and predictability of taxation.

"The efficient use of resources" is mentioned as another benefit of ICAP. As it can be understood, targeted documents and an atmosphere based on cooperation can help tax administrations to reach a most cost-effective outcome. Instead of having to deal with numerous distinct inquiries, an MNE group can interact with various tax administrations at once, either through multilateral engagements or through the lead tax administration. By doing this, the tax administration has the opportunity to view the event through the eyes of a different tax administration, giving it a better understanding of the situation and allowing it to make decisions with lower administrative costs.

ICAP also serves as "a faster, clearer route to multilateral tax certainty" and this can be regarded as another benefit of this programme. This issue has also two sides. As tax authorities point of view, they are able to obtain a complete picture of an MNE group's international operations and may be confident that either the tax status is satisfactory or that any potential tax risks have been recognized. On the other hand, MNE groups may benefit from the reduction of uncertainties in reporting and revenue audits and in the long run, it may be result in lower compliance costs for taxpayers .The procedure also gives an MNE group the chance to use ICAP findings more strategically when operating its business across its international operations. As a result, it significantly affects how MNE groups perceive the perspective of the tax administrations.

"Cooperative relationships between MNE groups and tax administrations" constitutes one of the benefits that has been most frequently highlighted in this study and may also be the most visible so far. ICAP involves a commitment by MNE organizations and tax authorities to collaborate in an open, transparent, and cooperative manner throughout the risk assessment and assurance process which developed a relationship of mutual trust based on reciprocal openness and good faith.

As a last benefit addressed by the OECD in the ICAP Handbook is "fewer disputes into entering MAP". As it is mentioned above with the term "dispute prevention", ICAP gives tax administrations the chance to discuss transactions at the risk assessment stage before opposing viewpoints are embraced and cemented in place. By ensuring greater uniformity among tax administrations, this can lessen the frequency of disputes and MAP. Such an advantage is also related with the second one, namely "the efficient use of resources".

INTERNATIONAL ASSURANCE PROGRAMME AND TURKEY

As a country which takes place in the OECD and G20, and also a member of Forum on Tax Administration, Turkey has been involved in studies on international tax issues. It is possible to assess Turkey's status in these areas given that the Forum on Tax Administration concentrates on tax certainty and tax cooperation concerns. However, Turkey is not yet a participating country of ICAP. However, Turkey draws investment since it is an emerging economy. In this regard, it could be useful for our tax administration to assume the roles of both the lead and the covered tax authority in the taxation of MNE organizations and conduct a risk assessment.

In this regard, it is essential to assess whether Turkey's participation in ICAP is feasible or advantageous. Therefore, the notion of cooperative compliance should be first explored in relation to Turkish tax policies and then it should be determined if Turkey satisfies the requirements outlined in the ICAP Handbook of the OECD.

Cooperative Compliance and the Turkish Tax System

As it is mentioned above, since the idea of cooperative compliance has recently taken on a global dimension, it is evident in many national practices. Applications that may be assessed within this idea have started to emerge in this regard, particularly since the start of the 2000s, in the tax policies of many nations. This concept, which has become more well-known thanks to the OECD's perspective on the matter, may take the form of adjustments and improvements to the organizational layout of revenue bodies or to the interactions between taxpayers and the administration.

From Turkey's standpoint, addressing cooperative compliance entails looking into if there are any procedures that improve interaction and collaboration between taxpayers and the tax authorities, have the potential to lower noncompliance, and can be categorized as "dispute prevention". In this regard, the existence of practices that foster a cooperative environment where taxpayers can share information, get advice from the administration, and the administration can have comprehensive knowledge of the transactions of the taxpayers, can be interpreted as a sign that measures have been taken to ensure cooperative compliance in Turkey and that the said cooperation can be carried to the international arena.

First of all, the modification made to the revenue administration's organizational structure in 2007 might be seen as the first action that falls under the purview of cooperative compliance. The Revenue Administration's Large Taxpayer Office was founded in 2006 in Istanbul but started operating in early 2007. The following stands out in light of the establishment goals of this unit: Reaching out to significant taxpayers without waiting to get into touch with the administration, better understanding better their needs and preparing for potential issues, doing research to promote voluntary compliance by attentively observing each taxpayer, enlightening them about the laws, rules, and legal interpretations, and identifying taxpayers who fail to comply with the law on time and taking immediate action, creating and tracking various adaption techniques to improve their compliance (Vergi Dairesi Başkanlığı, 2022).

The service idea is described as "to serve with a public administration strategy that prioritizes taxpayer happiness, encourages positive tax compliance, and is trusted by taxpayers" on the website of the aforementioned tax office. It adopts the idea of "service and client" as opposed to "cops and robbers," as may be observed from this. As a result, the establishment of such a tax office can be viewed as an illustration of cooperative compliance, which incorporates the notion of forging closer ties to guarantee the tax compliance of the category known as large taxpayers.

On the other hand, tax authorities provide early disclosure and issue resolution, which increases certainty regarding the tax status, minimise unnecessary audit time, and lowers compliance costs in the cooperative compliance programmes (Goslinga, S., Jonge, J. D., Siglé, M., & Dijk, H. V., 2021: 95). From this point of view, the application of advanced ruling can be regarded as another instance in the field of issue resolution. In this procedure, the tax administration provides the taxpayers with a written opinion known as a tax ruling in response to their request for a written explanation of the matters that are confused and hesitant in terms of their tax status and tax application. In this regard, the government's view on a risk-creating issue is taken before any dispute between the administration and the taxpayers emerges. This procedure can increase tax certainty and decrease compliance costs.

Other techniques applied in Turkey include APA and MAP, which can be assessed as part of cooperative compliance. An agreement is reached between the administration and the taxpayers in these ways, which are implemented based on legislation, and it is evident that there is cooperation in this area. At this point, it's crucial to note the following quotes from the article on APA's justification (Article 13 of Turkish Corporate Income Tax Law no. 5520); "The biggest benefit of this system is that taxpayers know exactly how the system will be used for a specific amount of time, which allows them to plan and look ahead without fear of rebuke or punishment. The subject is chosen with a specific agreement at the outset and commencing with the critique process, which is the administration's major benefit. A technique that can be extended to the judicial process results in time and workload savings " (Legalbank, 2022).

On the other hand, MAP was introduced in Turkish tax system by Tax Procedure Law and purpuant to this law, taxpayers can apply to the Revenue Administration in accordance with the provisions of the MAP, claiming that they are taxed in violation of the provisions of a duly enacted double taxation agreement or that there are strong indications that they will be taxed in this way.

ICAP as a Policy Tool for Turkey

As mentioned above, ICAP has a voluntary basis. Thus, in principle, countries are not necessitated to meet any requirements or commitments in order to participate in this program. Even if this were the case, it is determined that at least some requirements must be completed in order to hold the title of "lead tax administration" in light of the requirements that MNE groups that can apply to the ICAP risk assessment procedure must meet.

Some qualifications for MNE groups applying to the program are introduced in the ICAP Handbook. On the basis of this, it can be recognized that the "lead tax administration" must also adhere to the minimum requirements. The country in question must be a member of the OECD Forum on Tax Administration and also legal regulations regarding CbCR, master file and local file obligations must be in effect in that country.

If the situation in Turkey is examined in light of these two criteria, it is possible to conclude that both criteria have been satisfied. since, as was previously noted, Turkey is one of the participants of the OECD Forum on Tax Administration and has paralleled the CbCR, the master file, and the local file by incorporating the laws into its domestic law.

Actually, Turkey complies with international standards because it has put into practice the transfer pricing documentation requirements of the OECD's BEPS Action 13. There are currently six different categories of documentation requirements regarding transfer pricing. Within this context, in Turkey, master file should be prepared by corporate taxpayers affiliated with the multinational enterprises group, whose asset size in the balance sheet and net sales in the income statement are both 500 million Turkish

Liras and above. In addition, all corporate income taxpayers are required to prepare an Annual Transfer Pricing Report, which means the local file, up until the deadline for filing the annual corporate income tax return. On the other hand, the ultimate parent entity or surrogate parent entity of the multinational enterprise group whose total consolidated group revenue equals or exceeds 750 million Euro prepares CbCR until the end of the twelfth month after the reporting fiscal year and electronically submits it to the Turkish Revenue Administration, according to the consolidated financial statements of the immediately preceding fiscal year prior to the reporting fiscal year. Thus, Turkey's master file, local file, and CbCR application processes run concurrently with OECD requirements.

As far as ICAP is concerned, the circumstances are ideal for Turkey to be a participating country in the programme. Evaluations of the advantages of Turkey's involvement in this programme as a policy objective might be made in this regard. The focus of this assessment will be on the value of cooperation in guaranteeing voluntary tax compliance and the advantages Turkey can get from ICAP, which stands for "dispute prevention" in terms of tax risks like transfer pricing or PE.

First of all, Turkey's participation in this program is eligible for all the advantages offered by ICAP that are described in the heading 3.1.2. However, if this matter is considered in a broader context, it can be claimed that the main purpose of ICAP for Turkey is to offer fiscal certainty as a country that seeks to draw in foreign investment. There is a risk that uncertainty will deter investment. Even though it is scant, the empirical evidence is more conclusive and does point to negative consequences of tax uncertainty on trade and investment. Tax certainty is a key factor in investment decisions for MNEs and can have a significant impact on economic growth. Assessments of investment risk and investment finance are negatively impacted by inconsistent and occasionally coercive execution techniques in tax administrations (Diaz de Sarralde, S., von Haldenwang, C., Hentze, T., & Monkam, N., 2018). In this context, with a more certain and predictable tax environment, cross-border trade and investment are supported, and governments are able to obtain a more steady and predictable income stream (IMF/OECD, 2017: 7).

Moving from this point, the IMF and OECD outlined some policy steps and practical tools to enhance tax certainty in the Tax Certainty Report of 2017. In this report, cooperative compliance programmes and APA's, as well as simultaneous and joint audits are highlighted as instances of tools in tackling tax uncertaincy. We can add ICAP to the list. In other words, ICAP can be seen as a practical to remove tax uncertainty at an early stage. This effectively increases the country's tax system's predictability, which fosters a favorable climate for investments. In this scenario, Turkey's ability to attract investment in the framework of tax certainty is expected to improve in case of its participation in ICAP.

As another point, tax authorities have a variety of methods at their disposal to modify behavior but most of them are costly. Understanding and influencing behavior is a powerful strategy that can be used in conjunction with other approaches (Walsh, K., 2012) and this causes efficient and effective administration. Based on this, ICAP enables tax authorities to carry out a more efficient and multidimensional risk assessment and thus helps administrative costs decrease. In addition, this programme makes better use of the limited resources available. In this regard, Turkey's membership in ICAP will make it feasible to analyze risks effectively, and by focusing resources on noncompliant taxpayers, a decline in the tax base and tax revenues will be avoided. As it is for every country, one of Turkey's most significant policy goals is the protection of tax revenues, so joining ICAP is seen as a step in the right direction.

Additionally, it should be recognized that strengthening taxpayer rights and expanding services to make paying taxes easier might likely increase voluntary compliance. All tax administrations prefer voluntary compliance since it is the most straightforward method of tax collection. According to the fiscal policy aims in Turkey's Eleventh Development Plan for the years 2019–2023, "voluntary compliance, predictability and transparency in taxation will be increased, taxation services will be activated". This illustrates that voluntary compliance is a policy goal for Turkey and an approach is taken to achieve this. In order to attain the development plan's goal of voluntary compliance on a global and MNE group level, participation in ICAP may be a valuable endeavor.

The application of the principle of equal taxation of those who are equal before the law is impacted by maintaining tax compliance as well. The positive result arising from this principle, which is one of the basic taxation principles, also has the effect of ensuring the tax system's consistency and boosting public confidence in it. Turkey's involvement in ICAP, which encourages voluntary compliance, will be advantageous for equality in this situation.

CONCLUSION

Studies into taxpayers responses to taxes go back as far as studies on taxation itself. Taxpayers are looking for ways to reduce the amount of taxes paid, since taxes cause a decrease in the assets of them and create a burden. Tax avoidance in the form of tax planning, as well as tax evasion by going beyond the law, have been the areas where these reactions became concrete. On the other hand, these responses lead to a decline in tax revenues and have negative effects on maintaining the consistency in the tax system. States aim to ensure that taxpayers comply with the law in regard to collecting taxes.

The concept of tax compliance of taxpayers is a subject that has always been discussed. While the perspectives that directly link tax compliance to cost-benefit analysis or the state's punishment authority have long dominated the debate, over time the idea that social and psychological factors also play a role in taxpayers' tax compliance has grown in popularity. As a result of the development of this understanding, it is thought that it is possible to implement tax compliance policies that promote rather than penalize taxpayers. Within this context, voluntary tax compliance has now become a policy target.

It should be noted that one of the concepts that emerged within the framework of the search for voluntary compliance has been "enhanced relationship". This idea focuses on creating a connection between the taxpayer and the administration that is built on collaboration and trust in order to ensure voluntary compliance. The OECD introduced this idea to the literature in the early 2000s, and it primarily addressed three key issues: the need to adapt to societal changes and development, aggressive tax planning, and the outdated of tax collection and compliance.

In order to provide a more definitive statement, the OECD substituted the term "cooperative compliance" for "enhanced relationship". The goal of the cooperative compliance idea is to enhance tax compliance by forging a friendly working relationship between taxpayers and tax authorities. A review of the procedures in 24 countries that may fall under the purview of this idea is also included in the 2013 report of the OECD, which also incorporates this concept. Based on this, the OECD, which drew several conclusions regarding the cooperative compliance, called attention to the need for action to bring voluntary tax compliance to the international level.

After the preparations in 2016 and 2017, the first implementation in the international arena within the scope of cooperative compliance could only be completed in 2018. This programme is called ICAP. In accordance with ICAP, MNE groups assess particular tax risks in conjunction with the tax authorities of the many nations in which they conduct business. This initiative, which is a voluntary tax "compliance" practice because it is applied by the MNE groups themselves, also exemplifies "cooperation" because it fosters communication between tax authorities and taxpayers.

Several MNE firms and the tax administrations of eight countries joined together to undertake the first pilot project, ICAP 1.0, in 2018. Following that, tax administrations and MNE groups provided feedback, and the foundation for the second pilot application was laid. The stages in the programme were condensed into three within this structure, and new countries participate. As a result, ICAP 2.0, the second pilot program, went into effect in 2019.

In general, the implementation occurs as follows in both the ICAP 1.0 and ICAP 2.0 frameworks: After the MNE group applies to the UPE tax administration and this country is approved to serve as the lead tax administration, the application procedure, which consists of four steps in total, begins with the submission of the relevant documentation by the MNE group. In this framework, the participating nations that the MNE group can operate in consent to be included in the program, and all tax administrations analyze the supplied documentation. Each administration decides whether to classify the examined tax hazards as low risk as a result of this examination. The MNE group finds comfort in the assessment of covered tax risks as low risk.

ICAP is comparable to several applications that emerge within the context of the notion of cooperative compliance because it has been implemented in this fashion and has generally gotten positive feedback from tax administrations and MNE groups. These implementations, including tax audits, advanced rulings, APA's, and MAP's, are crucial in practice, particularly when it comes to building relationships between the taxpayer and the administration and, in some cases, as a means of dispute resolution. The key difference between ICAP and these applications is that it has a procedure that involves both many tax administrations and the taxpayer in an international context. It also has a "dispute prevention" feature as another key factor.

With tax certainty and international collaboration, ICAP has main drivers such as ensuring tax compliance, effective dispute resolution and standardized approach in transfer pricing risk assessment. On the other hand, it includes benefits mainly efficient use of resources, multilateral tax certainty, cooperative relationships and fewer disputes. It is anticipated that many nations will engage in this practice in the future as a result of the macroeconomic analysis of these advantages. Turkey, which has not yet joined the program, might therefore benefit from considering a potential participation.

Although there are no explicit requirements listed in the OECD's ICAP Handbook for a country's participation, it is possible to deduce that there are two requirements based on a set of standards for determining if MNE organizations are suitable for the program. These requirements include the country's membership in the OECD's Forum on Tax Administration and the existence of CbCR, master file, and local file, as transfer pricing documentation obligations. Given that Turkey satisfies the aforementioned requirements, there is no obstacle to the membership in ICAP in this context.

Evaluations of the outcomes and advantages of this program can be utilized to support Turkey's participation. Within this context, tax certainty and consistency in the tax system can be evaluated primarily as consequences. The provision of tax certainty and consistency as a result of risk assessment and utilizing the opinions of more than one tax administration is one of the most significant effects of ICAP. These outcomes benefit countries by encouraging foreign investment and promoting voluntary compliance. ICAP, on the other hand, has an impact on ensuring taxpayer equality, which is also beneficial for voluntary compliance. As a country which has an emerging economy, Turkey aims to attract foreign investment. Thus, Turkey will gain from the favorable environment that ICAP has created for further investment. Based on this, it is suggested that Turkey take part in ICAP.

ICAP also makes sure that tax auditing is efficient and effective. Existence of a tax administration with these attributes prevents the erosion in the tax base and loss of tax revenues while also lowering administrative costs. In this regard, participation in ICAP can again lead to positive results for Turkey.

REFERENCES

Aim, J., & Torgler, B. (2011). Do Ethics Matter? Tax Compliance and Morality. *Journal of Business Ethics*, *101*(4), 635–651. doi:10.100710551-011-0761-9

Allingham, M. G., & Sandmo, A. (1972). Income Tax Evasion: A Theoretical Analysis. *Journal of Public Economics*, *1*(3–4), 323–338. doi:10.1016/0047-2727(72)90010-2

Alm, J. (2019). What Motivates Tax Compliance? *Journal of Economic Surveys*, *33*(2), 353–388.

Alm, J., & Torgler, B. (2011). Do Ethics Matter? Tax Compliance and Morality. *Journal of Business Ethics*, *101*(4), 635–651. doi:10.100710551-011-0761-9

Andreoni, J., Erard, B., & Feinstein, J. (1998). Tax Compliance. *Journal of Economic Literature*, *36*(2), 818–860.

Baird, D., Gertner, R., & Picker, R. (1994). *Game Theory and the Law*. Harvard University Press.

Baisalbayeva, K. (2014). Counteracting Tax Evasion and Avoidance (Focus on Non-compliance by High-net-worth Individuals). In K. Spies & R. Petruzzi (Eds.), *Tax Policy Challenges in the 21st Century: Schriftenreihe IStR* (Vol. 86, pp. 365–390). Linde Verlag GmbH.

Becker, G. S. (1968). Crime and Punishment – An Economic Approach. *Journal of Political Economy*, *76*(2), 169–217. doi:10.1086/259394

Bergman, M., & Nevârez, A. (2005). ¿Evadir o Pagar Impuestos? Una Aproximacion a Los Mecanismos Sociales del Cumplimiento. *Política y Gobierno*, *12*(1), 9–40.

Boudon, R. (2009). *La Rationalité*. Presses Universitaires de France. doi:10.3917/puf.boudo.2009.01

Braithwaite, V. (2003). Perceptions of Who's not Paying Their Fair Share. *The Australian Journal of Social Issues*, *38*, 335–362. doi:10.1002/j.1839-4655.2003.tb01149.x

Bronżewska, K. (2016). *Cooperative Compliance: A New Approach to Managing Taxpayer Relations*. IBFD. [e-book]

Büyük Mükellefler Vergi Dairesi Başkanlığı. (n.d.). https://bmvdb.gib.gov.tr/kurumsal/kurulu%C5%9F-nedenleri

de Widt, D., & Oats, L. (2019). *Co-operative Compliance: Views of Large Businesses in the Netherlands and UK*. H2020-EURO-SOCIETY-2014: Deliverable D7.2/Oats.

Diaz de Sarralde, S., von Haldenwang, C., Hentze, T., & Monkam, N. (2018). *Tax Certainty: Proposals for the Short Term and the Long Term* (No. 2018-16). Economics Discussion Papers.

Elster, J. (2007). *Explaining Social Behavior: More Nuts and Bolts for the Social Sciences*. Cambridge University Press. doi:10.1017/CBO9780511806421

Glossary of Tax Terms. (n.d.). https://www.oecd.org/ctp/glossaryoftaxterms.htm

Goslinga, S., Jonge, J. D., Siglé, M., & Dijk, H. V. (2021). Cooperative Compliance Programmes: Who Participates and Why? *Journal of Tax Administration*, *6*(2), 95–121.

Graetz, M. J., Reinganum, J. F., & Wilde, L. L. (1986). The Tax Compliance Game: Toward An Interactive Theory of Law Enforcement. *JL Econ. & Org.*, *2*(1), 1–32.

Hedström, P., & Ibarra, R. (2010). On the Contagiousness of Non-Contagious Behavior: The Case of Tax Avoidance and Tax Evasion. In H. Joas & B. Klein (Eds.), *The Benefit of Broad Horizons: Intellectual and Institutional Preconditions for a Global Social Science*. Brill. doi:10.1163/ej.9789004192843.i-436.103

Huiskers-Stoop, E., & Gribnau, H. (2019). Cooperative Compliance and the Dutch Horizontal Monitoring Model. *Journal of Tax Administration*, *5*(1), 66–110.

IMF/OECD. (2017). *Tax Certainty, Report for the G20 Finance Ministers*. https://www.oecd.org/tax/tax-policy/tax-certainty-report-oecd-imf-report-g20-finance-ministers-march-2017.pdf

Katz, A. W. (1998). *Foundations of the Economic Approach to Law*. Oxford University Press.

Kirchler, E. (2007). *The Economic Psychology of Tax Behaviour*. Cambridge University Press. doi:10.1017/CBO9780511628238

Kirchler, E., Hoelzl, E., & Wahl, I. (2008). Enforced versus Voluntary Tax Compliance: The "Slippery Slope" Framework. *Journal of Economic Psychology*, *29*(2), 210–225. doi:10.1016/j.joep.2007.05.004

Klepper, S., & Nagin, D. (1989). The Role of Tax Preparers in Tax Compliance. *Policy Sciences*, *22*(2), 167–194. doi:10.1007/BF00141383

Legalbank. (n.d.). https://legalbank.net/belge/kurumlar-vergisi-kanunu/1350080/kurumlar+vergisi

Majdańska, A. (2021). *An Analysis of Cooperative Compliance Programmes*. IBFD.

Martini, M. (2020). Introduction: An Investigation of Co-operative Compliance Regimes and the ICAP. In Co-operative Compliance and the OECD's International Compliance Assurance Programme. Kluwer Law International BV.

Mehta, G. B. (1998). Preference and Utility. In S. Barberà, P. J. Hammond, & H. Seidl (Eds.), *Handbook of Utility Theory* (pp. 1–47). Kluwer Academic Publishers.

Murphy, K. (2004). The Role of Trust in Nurturing Compliance: A Study of Accused Tax Avoiders. *Law and Human Behavior*, *28*(2), 187–209. doi:10.1023/B:LAHU.0000022322.94776.ca PMID:15141778

Noguera, J. A., Quesada, F. J. M., Tapia, E., & Llàcer, T. (2014). Tax Compliance, Rational Choice, and Social Influence: An Agent-based Model. *Revue Française de Sociologie*, *55*(4), 765–804. doi:10.3917/rfs.554.0765

OECD. (2013). *Co-operative Compliance: A Framework: From Enhanced Relationship to Co-operative Compliance*. OECD Publishing.

OECD. (2019). *International Compliance Assurance Programme Pilot Handbook 2.0*. OECD.

OECD. (2021). *International Compliance Assurance Programme – Handbook for Tax Administrations and MNE Groups*. OECD.

Posner, R. A. (1998). *Economic Analysis of Law* (5th ed.). Aspen.

Pross, A., & Johnson, M. (2020). International Compliance Assurance Programme. In Co-operative Compliance and the OECD's International Compliance Assurance Programme. Kluwer Law International BV.

Pross, A., Knaepen, S., & Johnson, M. (2017). Embracing Tax Certainty Through Improved Dispute Resolution. *Int'l Tax Rev.*, *28*, 16–22.

Rijsbergen, H. (2020). The ICAP Experience: From A Tax Authority Perspective. In Co-operative Compliance and the OECD's International Compliance Assurance Programme. Kluwer Law International BV.

Srinivasan, T. N. (1973). Tax Evasion: A Model. *Journal of Public Economics*, *2*(4), 339–346.

Torgler, B. (2007). *Tax Compliance and Tax Morale: A Theoretical and Empirical Analysis*. Edward Elgar. doi:10.4337/9781847207203

Turkey Transfer Pricing Country Profile. (n.d.). https://www.oecd.org/tax/transfer-pricing/transfer-pricing-country-profile-turkey.pdf

Türkiye Cumhuriyeti Cumhurbaşkanlığı Strateji ve Bütçe Başkanlığı, On Birinci Kalkınma Planı (2019-2023). (n.d.). https://www.sbb.gov.tr/wp-content/uploads/2022/07/On_Birinci_Kalkinma_Plani-2019-2023.pdf

Walsh, K. (2012). Understanding Taxpayer Behaviour–New Opportunities for Tax Administration. *The Economic and Social Review*, *43*(3), 451–475.

Wenzel, M. (2002). The Impact of Outcome Orientation and Justice Concerns on Tax Compliance: The Role of Taxpayers' Identity. *The Journal of Applied Psychology*, *87*(4), 629–645. doi:10.1037/0021-9010.87.4.629 PMID:12184568

Wenzel, M. (2003). Tax Compliance and the Psychology of Justice: Mapping the Field. In Taxing Democracy: Understanding Tax Avoidance and Tax Evasion. Routledge.

ADDITIONAL READING

Andiko, T. (2018). Cooperative Compliance: Improving Relationship between Taxpayers and Tax Authorities. *International Journal of English Literature and Social Sciences*, *3*(3), 368–373. doi:10.22161/ijels.3.3.14

Broek, H. V. D. (2019). Tax Risk Management and the Paradox of Cooperative Compliance. In P. De Vincentiis, F. Culasso, & S. A. Cerrato (Eds.), *The Future of Risk Management* (Vol. I, pp. 305–327). Palgrave Macmillan.

De Widt, D., Oats, L., & Mulligan, E. (2019). The US compliance assurance process: A relational signalling perspective. *Journal of Tax Administration*, *5*(1), 145–161.

Eberhartinger, E., & Zieser, M. (2021). The Effects of Cooperative Compliance on Firms' Tax Risk, Tax Risk Management and Compliance Costs. *Schmalenbach Journal of Business Research*, *73*(1), 125–178. doi:10.100741471-021-00108-6 PMID:34803211

Hani, G. A., & Brown, C. C. (2019). The Revitalized Compliance Assurance Process. *J. Tax Prac. &. Proc.*, *21*, 5–9.

Kennan, J., Hoshall, D., & Harkins, C. (2018). More than Tax Reform: The Issues LB&I Transfer Pricing Directives, the Organisation for Economic Co-Operation and Development Launches the International Compliance Assurance Programme, and the National Taxpayer Advocate Released 2017 Annual Report to Congress. *J. Tax Prac. & Proc.*, *20*, 13.

Kirchler, E., Kogler, C., & Muehlbacher, S. (2014). Cooperative tax compliance: From deterrence to deference. *Current Directions in Psychological Science*, *23*(2), 87–92. doi:10.1177/0963721413516975

Lambooij, M. (2013). Multilateral Cooperative Compliance: The Next Step. *Int'l Tax Rev.*, *24*, 46–57.

Owens, J., & Pemberton, J. L. (2021). Cooperative Compliance: A Multi-stakeholder and Sustainable Approach to Taxation. Kluwer Law International BV.

van der Hel-van Dijk, E. C. J. M., & Siglé, M. (2016). Cooperative Compliance: Tax Risk Management and Monitoring. *Intertax*, *44*(8/9), 642–650.

KEY TERMS AND DEFINITIONS

Advanced Ruling: It is a written document that tax authorities send to a taxpayer in which they interpret and apply the tax law to a particular set of facts.

Avoidance: It is typically used to describe how a taxpayer arranges his affairs in order to lower his tax bill. Although the arrangement may be strictly lawful, it is usually at conflict with the spirit of the law it claims to serve.

Cooperative Compliance: It is the development of a trust-based relationship between tax administrations and taxpayers that is based on voluntary tax compliance and results in the timely payment of the correct amount.

Evasion: Typically used to refer to illegal arrangements where tax liability is concealed or neglected, i.e., the taxpayer pays less tax than he is legally required to pay by concealing income or information from the tax authorities.

Multinational Enterprises (MNE's): MNE's are enterprises or groups of enterprises with locations in two or more different countries.

OECD: The Organization for Economic Co-operation and Development (OECD) is a multilateral organization consisting of 30 jurisdictions, the majority of which are Western European industrialized countries such as the US and Japan. The OECD, which was established in 1961, offers a platform for national leaders to talk about and make an effort to coordinate economic and social policies. In problems of international taxation, it plays a particularly important role.

Parent Company: The parent company is the enterprise that holds a sizable portion of the capital stock of the subsidiary.

Permanent Establishment: Term used in double taxation agreements or tax legislation to describe a circumstance where a non-resident entrepreneur is taxable in one nation but is exempt from paying income taxes in the other country unless they have a "permanent establishment" through which they conduct business there. Even if it has a PE, only the portion of the income that is "attributable" to the PE will be subject to taxation.

Tax Authority: The organization in charge of enforcing a jurisdiction's, region's, or municipal government's tax rules.

Tax Compliance: The extent to which a taxpayer complies (or does not comply) with the tax laws of his country, such as by disclosing income, filing a return, and timely paying the taxes required.

Transfer Pricing: A transfer price is the amount an enterprise charges a subsidiary or other affiliated entity for goods, services, or intangible property. When revenue and expenses are unfairly allocated to lower taxable income, this is considered abusive transfer pricing.

Compilation of References

Abbott, C. (2003). *ICT: Changing Education*. Taylor & Francis. doi:10.4324/9780203400197

Abecker, A., Papavassiliou, G., Mentzas, G., & Müller, S. (2014). *Methods and tools for business-process oriented knowledge management: Experiences from case studies*. Available online at: https://www.researchgate.net/publication/228769241_Methods_and_Tools_for_Business-Process_Oriented_Knowledge_Management_Experiences_from_Case_Studies

Acilar, A., & Sæbø, Ø. (2021). *Towards understanding the gender digital divide: A systematic literature review*. Global Knowledge, Memory and Communication.

Adeleye, N., & Eboagu, C. (2019). Evaluation of ICT development and economic growth in Africa. *NETNOMICS: Economic Research and Electronic Networking*, *20*(1), 31–53. doi:10.100711066-019-09131-6

Adshade, M., & Keay, I. (2010). Technological and Organisational Change and the Employment of Women: Early Twentieth-Century Evidence from the Ohio Manufacturing Sector. *Feminist Economics*, *16*(1), 129–157. doi:10.1080/13545700903382711

Afawubo, K., & Noglo, Y. A. (2022). ICT and entrepreneurship: A comparative analysis of developing, emerging and developed countries. *Technological Forecasting and Social Change*, *175*, 121312. doi:10.1016/j.techfore.2021.121312

Aghaei, M., & Rezagholizadeh, M. (2017). The impact of information and communication technology (ICT) on economic growth in the OIC countries. *Economic and Environmental Studies (E&ES)*, *17*(2), 257-278.

Aguiar, M., & Hurst, E. (2013). Deconstructing Life Cycle Expenditure. *Journal of Political Economy*, *121*(3), 437–492. doi:10.1086/670740

Ahadi, S., & Kasraie, S. (2020). Contextual factors of entrepreneurship intention in manufacturing SMEs: The case study of Iran. *Journal of Small Business and Enterprise Development*, *27*(4), 633–657. doi:10.1108/JSBED-02-2019-0074

Aim, J., & Torgler, B. (2011). Do Ethics Matter? Tax Compliance and Morality. *Journal of Business Ethics*, *101*(4), 635–651. doi:10.100710551-011-0761-9

Ainissyifa, H. (2012). The influence of knowledge management implementation toward the quality of high schools. *International Journal of Development and Sustainability*, *1*(2), 79–84.

Ajide, F. M. (2021). Financial Inclusion and Labour Market Participation of Women in Selected Countries in Africa. *Economics and Culture*, *18*(1), 15–31. doi:10.2478/jec-2021-0002

Akgul, A., Akbas, H. E., & Taskin Gumus, A. (2018). A Survey of Students' Perceptions on Industry 4.0 in a Large Public University in Turkey. In F. B. Candan & H. Kapucu (Eds.), *Current Debates In Business Studies* (pp. 237–247). IJOPEC Publication Limited.

Akhavan, P., Hosseini, M., Abbasi, M., & Manteghi, M. (2015). Knowledge sharing determinants, behaviours and innovative work behaviours: An integrated theoretical view and empirical examination. *Aslib Journal of Information Management*, *67*(5), 562–591. doi:10.1108/AJIM-02-2015-0018

Akkoç, I. (2012). Gelişim Kültürü ve Etik İklimin Yenilikçiliğe Etkisinde Dağıtım Adaletinin Rolü. *Uluslararası Alanya İşletme Fakültesi Dergisi*, *4*(3), 45–60.

Akram, T., Lei, S., Haider, M. J., & Hussain, S. T. (2018). Exploring the impact of knowledge sharing on the innovative work behavior of employees: A study in China. *International Business Research*, *11*(3), 186–194. doi:10.5539/ibr.v11n3p186

Aktharsha, S. U., & Anisa, H. (2011). Knowledge Management System and Learning Organization: An Empirical Study in an Engineering Organization. *The IUP Journal of Knowledge Management*, *6*(2), 33–51.

Al Ahmar, G. O., Rofiqb, A., & Hadiwidjojoc, D. (2014). The Impact of Knowledge Management, Learning Organization, and Educations Organization on Organization Performance: A Case in Brawijaya University. *Asia-Pacific Management and Business Application*, *3*(1), 28–47. doi:10.21776/ub.apmba.2014.003.01.3

Al-Amin, B. M., & Benlahabib, T. (2018). *The impact of information and communication technology on economic growth (case study of Arab countries 2018). In Modern Technology and Quality of Life*. Hamma Lakhdar Eloued University.

Alavi, M., Kayworth, T. R., & Leidner, E. L. (2006). An empirical examination of the influence of organizational culture on knowledge management practices. *Journal of Management Information Systems*, *22*(3), 191–224. doi:10.2753/MIS0742-1222220307

Albulescu, C. T., & Draghici, A. (2016). Entrepreneurial activity and national innovative capacity in selected European countries. *International Journal of Entrepreneurship and Innovation Management*, *17*(3), 155–172. doi:10.1177/1465750316655902

Aldan, A. (2021). Rising Female Labor Force Participation and Gender Wage Gap: Evidence From Turkey. *Social Indicators Research*, *155*(3), 865–884. doi:10.100711205-021-02631-9

Alderete, M. V. (2014). ICT incidence on the entrepreneurial activity at country level. *International Journal of Entrepreneurship and Small Business*, *21*(2), 183–201. doi:10.1504/IJESB.2014.059472

Alderete, M. V. (2017). Mobile broadband: A key enabling technology for entrepreneurship? *Journal of Small Business Management*, *55*(2), 254–269. doi:10.1111/jsbm.12314

Allingham, M. G., & Sandmo, A. (1972). Income Tax Evasion: A Theoretical Analysis. *Journal of Public Economics*, *1*(3–4), 323–338. doi:10.1016/0047-2727(72)90010-2

Alm, J. (2019). What Motivates Tax Compliance? *Journal of Economic Surveys*, *33*(2), 353–388.

Alozie, N. O., & Akpan-Obong, P. (2017). The digital gender divide: Confronting obstacles to women's development in Africa. *Development Policy Review*, *35*(2), 137–160. doi:10.1111/dpr.12204

Alrefaai, M. R., & Khalil, A. H. A. L. (2019). The Impact of Knowledge Management on the Learning Organization: An Empirical Study on the Telecommunications Sector in Damascus City. *International Journal of Academic Research in Business and Social Sciences*, *9*(12), 774–783.

Alsabbagh, M., & Al Khalil, A. H. (2017). The Impact of Knowledge Management on Organizational Learning (An Empirical Study on the Education Sector in Damascus City). *International Journal of Academic Research in Business and Social Sciences*, *7*(4), 560–578. doi:10.6007/IJARBSS/v7-i4/2833

Al-Shorbaji, N. (2021). *Improving Healthcare Access through Digital Health: The Use of Information and Communication Technologies. Healthcare Access.* doi:10.5772/intechopen.94675

Altındal, M. (2013). Dijital Pazarlamada Marka Yönetimi ve Sosyal Medyanın Etkileri. Akademik Bilişim 2013- XV. Akademik Bilişim konferans Bildirileri 23-25 Ocak. Antalya.

Alves, E. E. C., & Steiner, A. Q. (2017). Globalisation, Technology and Female Empowerment: Breaking Rights or Connecting Opportunities? *Social Indicators Research, 133*(3), 859–877. doi:10.100711205-016-1395-1

Amorós, J. E., Fernández, C., & Tapia, J. (2012). Quantifying the relationship between entrepreneurship and competitiveness development stages in Latin America. *The International Entrepreneurship and Management Journal, 8*(3), 249–270. doi:10.100711365-010-0165-9

Andreoni, J., Erard, B., & Feinstein, J. (1998). Tax Compliance. *Journal of Economic Literature, 36*(2), 818–860.

Anggraeni, C. W. (2018). Promoting Education 4.0 in English for survival class: What are the challenges? Metathesis. *Journal of English Language, Literature, and Teaching, 2*(1), 12–24. doi:10.31002/metathesis.v2i1.676

Anguera, J. A., Jordan, J. T., Castaneda, D., Gazzaley, A., & Arean, P. A. (2016). Conducting a Fully Mobile Randomized Clinical Trial for Depression: Access, Engagement, and Expense. *British Medical Journal Innovations, 2,* 14-21. Retrieved from https://innovations.bmj.com/content/bmjinnov/2/1/14.full.pdf

Anser, M. K., Yousaf, Z., Khan, A., & Usman, M. (2020). Towards innovative work behavior through knowledge management infrastructure capabilities: Mediating role of functional flexibility and knowledge sharing. *European Journal of Innovation Management, 24*(5), 1832–1859.

Antonio, J. M., Gomes, M. C. A., & Marcelino, M. J. (2015). *ICT in Education: Multiple and Inclusive Perspectives.* Springer Publication.

Appel, J., Race, J., Yang, E., & Jumreornvong, O. (2020). Telemedicine and Medical Education in the Age of COVID-19. *Academic Medicine, 95*(12), 1838–1843. doi:10.1097/ACM.0000000000003711 PMID:32889946

Appiah-Otoo, I., & Song, N. (2021). The impact of ICT on economic growth-Comparing rich and poor countries. *Telecommunications Policy, 45*(2), 102082. doi:10.1016/j.telpol.2020.102082

Ardhi, M.K., Mulyo, J.H., & Irham. (2021). How does entrepreneurial orientation affect the business performance of coffee shop MSMEs in Indonesia? *E3S Web of Conferences, 306*, 03011. doi:10.1051/e3sconf/202130603011

Arellano, M., & Bond, S. (1991). Some Tests of Specification for Panel Data - Monte-Carlo Evidence and an Application to Employment Equations. *The Review of Economic Studies, 58*(2), 277–297. doi:10.2307/2297968

Argote, L. (2013). *Organizational learning: Creating, retaining and transferring knowledge.* Springer. doi:10.1007/978-1-4614-5251-5

Asongu, S. A., & Odhiambo, N. M. (2022). *Enhancing ICT for Female Economic Participation in Sub-Saharan Africa.* AGDI Working Paper. African Governance and Development Institute Research Department.

Asongu, S. A., Amankwah-Amoah, J., Nting, R. T., & Afrifa, G. A. (2021). Information Technology and Gender Economic Inclusion in Sub-Saharan Africa. *Journal of Global Information Technology Management, 24*(2), 120–133. doi: 10.1080/1097198X.2021.1914497

Asongu, S. A., & Nwachukwu, J. C. (2018). Openness, ICT and entrepreneurship in sub-Saharan Africa. *Information Technology & People, 31*(1), 278–303. doi:10.1108/ITP-02-2017-0033

Asongu, S. A., & Odhiambo, N. M. (2020). Inequality and gender inclusion: Minimum ICT policy thresholds for promoting female employment in Sub-Saharan Africa. *Telecommunications Policy*, *44*(4), 101900. doi:10.1016/j.telpol.2019.101900

Atasoy, H., Banker, R. D., & Pavlou, P. A. (2021). Information Technology Skills and Labor Market Outcomes for Workers. *Information Systems Research*, *32*(2), 437–461. doi:10.1287/isre.2020.0975

Aydemir, A. (2020). *Bitcoin Yeşil Bir Dünyanın Neresinde?* 12 Haziran 2022. https://www.paribu.com/blog/ perspektif/bitcoin-yesil-bir-dunyanin-neresinde-akar-aydemir/

Azadnia, M., Zahedi, S., Majjedin, A., & Reza, M. (2017). Analysis of the impact of ICT on sustainable development using sustainability indicators. *International Journal of Computers and Applications*, *169*(6), 13–24. doi:10.5120/ijca2017914766

Babur, Y. (2022). Metaverse Ağında Turizm Endüstrisinin Rolü. *Turizm Ekonomi ve İşletme Araştırmaları Dergisi*, *4*(1), 91–101.

Bahrini, R., & Qaffas, A. A. (2019). Impact of Information and Communication Technology on Economic Growth: Evidence from Developing Countries. *Economies*, *7*(21), 1–13. doi:10.3390/economies7010021

Baird, D., Gertner, R., & Picker, R. (1994). *Game Theory and the Law*. Harvard University Press.

Baisalbayeva, K. (2014). Counteracting Tax Evasion and Avoidance (Focus on Non-compliance by High-net-worth Individuals). In K. Spies & R. Petruzzi (Eds.), *Tax Policy Challenges in the 21st Century: Schriftenreihe IStR* (Vol. 86, pp. 365–390). Linde Verlag GmbH.

Bakhareva, A. A. (2017). Prospects for the Development of the Banking Sector in the Context of the Introduction of Modern Financial Technologies. *Symbol of Science*, *1*, 12–14.

Baliamoune, M. (2021). *Trade, Infrastructure, and Female Participation in Labor Markets*. Research Paper. Policy Center for the New South.

Bánhidi, Z., Dobos, I., & Nemeslaki, A. (2020). What the overall Digital Economy and Society Index reveals: A statistical analysis of the DESI EU28 dimensions. *Regional Statistics*, *10*(2), 42–62. doi:10.15196/RS100209

Bankole, F. O., Osei-Bryson, K.-M., & Brown, I. (2013). The Impact of ICT Investments on Human Development: A Regression Splines Analysis. *Journal of Global Information Technology Management*, *16*(2), 59–85. doi:10.1080/1097198X.2013.10845636

Bankole, F. O., Shirazi, F., & Brown, I. (2011). Investigating the Impact of ICT Investments on Human Development. *The Electronic Journal on Information Systems in Developing Countries*, *48*(1), 1–19. doi:10.1002/j.1681-4835.2011.tb00344.x

Barnett, W. A., Hu, M., & Wang, X. (2018). *Does the utilization of information communication technology promote entrepreneurship: Evidence from rural China*. http://www2.ku.edu/~kuwpaper/2018Papers/201802.pdf

Batinić, I. (2015). The Role and Importance of Internet Marketıng in Modern Hotel Industry. *Journal of Process Management- New Tecnologies International*, *3*(3), 34-38.

Bauer, W., Hammerle, M., Schlund, S., & Vocke, C. (2015). Transforming to a hyperconnected society and economy-towards an "Industry 4.0". *Procedia Manufacturing*, *3*, 417–424. doi:10.1016/j.promfg.2015.07.200

Beach, S. N., & Hall, L. R. (2011). *The Connected Educator Learning and Leading in a Digital Age*. Solution Tree Press.

Bec, A., Moyle, B., Timms, K., Schaffer, V., Skavronskaya, L., & Little, C. (2019). Management of Immersive Heritage Tourism Experiences: A Conceptual Model. *Tourism Management, 72*, 117–120. doi:10.1016/j.tourman.2018.10.033

Becker, G. S. (1965). A Theory of the Allocation of Time. *Economic Journal (London), 75*(299), 493–517. doi:10.2307/2228949

Becker, G. S. (1968). Crime and Punishment – An Economic Approach. *Journal of Political Economy, 76*(2), 169–217. doi:10.1086/259394

Becker, G. S. (1973). A Theory of Marriage: Part I. *Journal of Political Economy, 81*(4), 813–846. doi:10.1086/260084

Becker, G. S. (1974). A Theory of Marriage: Part II. *Journal of Political Economy, 82*(2, Part 2), 511–526. doi:10.1086/260287

Belianca, J. A. (2010). *21st Century Skills: Rethinking How Students Learn.* Solution Tree Press.

Belloumi, M., & Touati, K. (2022). Do FDI Inflows and ICT Affect Economic Growth? An Evidence from Arab Countries. *Sustainability, 14*(6293), 1–21. doi:10.3390u14106293

Ben Yaba, M., Bouaziz, A., & Lnassari, A. (2021). A Standard Study of The Impact of Information and Communication Technology on Economic Growth in the Arab Countries During The Period (2004-2018). *Journal of International Economy and Globalization, 4*(1), 39–50.

Ben Youssef, A., Boubaker, S., & Omri, A. (2018). Entrepreneurship and sustainability: The need for innovative and institutional solutions. *Technological Forecasting and Social Change, 129*, 232–241. doi:10.1016/j.techfore.2017.11.003

Benesova, A., & Tupa, J. (2017). Requirements for education and qualification of people in Industry 4.0. *Procedia Manufacturing, 11*, 2195–2202. doi:10.1016/j.promfg.2017.07.366

Benlahbib, T. (2018). The impact of ICTs on economic growth in developing countries econometric study during 2005-2015. *Journal of Economic and Financial Research, 5*(1), 559–580.

Berardelli, J. (2022). *Bitcoin Emissions Could Propel Planet To Critical Climate Threshold.* https://www.cbsnews.com/news/bitcoin-emissions-could-propel-planet-to-criticalclimate-threshold-study/

Bergman, M., & Nevârez, A. (2005). ¿Evadir o Pagar Impuestos? Una Aproximacion a Los Mecanismos Sociales del Cumplimiento. *Política y Gobierno, 12*(1), 9–40.

Berman, N. B., Durning, S. J., Fischer, M. R., Huwendiek, S., & Triola, M. M. (2016). The Role of Virtual Patients in the Future of Medical Education. *Academic Medicine, 91*(9), 1217–1222. doi:10.1097/ACM.0000000000001146 PMID:26959224

Berman, S. J. (2012). Digital transformation: Opportunities to create new business models. *Strategy and Leadership, 40*(2), 16–24. doi:10.1108/10878571211209314

Biagi, F. (2013). *ICT and productivity: A review of the literature (JRC-IPTS working papers on digital economy No. 2013–09).* Institute of Prospective Technological Studies, Joint Research Centre.

Bican, P. M., & Brem, A. (2020). Digital Business Model, Digital Transformation, and Digital Entrepreneurship: Is There a Sustainable "Digital". *Sustainability, 12*(1), 5239–5240. doi:10.3390u12135239

Bihari, S. C. (2011). Internet Technology and Customer Profitability in Banks. *Journal of Contemporary Research in Management, 6*, 33-48.

Bitcoin Block Reward Halving Countdown. (2022). https://www.bitcoinblockhalf.com/

Bitcoin Miners vs Bitcoin Nodes. (2022). https://blog.unocoin.com/bitcoin-miners-vs-bitcoin-nodes-6a4d35be9712

Björkdahl, J. (2020). Strategies for Digitalization in Manufacturing Firms. *California Management Review*, *62*(4), 17–36. doi:10.1177/0008125620920349

Blandin, A., Pieters, G., Wu, Y., Eisermann, T., Dek, A., Taylor, S., & Njoki, D. (2020). *3rd Global Cryptoasset Benchmarking Study*. University of Cambridge, Judge Business School. doi:10.2139srn.3700822

Bloomberght. (2022). https://www.bloomberght.com/bitcoin-cin-den-gelen-aciklamalarla-sert-dustu-2280652

Blundell, R., & Bond, S. (1998). Initial conditions and moment restrictions in dynamic panel data models. *Journal of Econometrics*, *87*(1), 115–143. doi:10.1016/S0304-4076(98)00009-8

Bouallaga, L., & Kebir, M. (2021). Measuring the impact of information and communication technology on economic growth in Algeria and some Arab countries Standard study accrediting Analysis - Dynamic Panel for the period: 2000-2018. *The Journal of Business Administration and Economic Studies*, *7*(1), 205–526.

Boudon, R. (2009). *La Rationalité*. Presses Universitaires de France. doi:10.3917/puf.boudo.2009.01

Boyd, D., & Ellison, N. (2008). Social network sites: Definition, history, and scholarship. *Journal of Computer-Mediated Communication*, *13*(1), 210–230. doi:10.1111/j.1083-6101.2007.00393.x

Braithwaite, V. (2003). Perceptions of Who's not Paying Their Fair Share. *The Australian Journal of Social Issues*, *38*, 335–362. doi:10.1002/j.1839-4655.2003.tb01149.x

Breusch, T. S., & Pagan, A. R. (1980). The Lagrange Multiplier Test and its Applications to Model Specification in Econometrics. *The Review of Economic Studies*, *47*(1), 239–253. doi:10.2307/2297111

Bronżewska, K. (2016). *Cooperative Compliance: A New Approach to Managing Taxpayer Relations*. IBFD. [e-book]

Bruno, G. S. F. (2005). Estimation and inference in dynamic unbalanced panel-data models with a small number of individuals. *The Stata Journal*, *5*(4), 473–500. doi:10.1177/1536867X0500500401

Brunswicker, S., Bertino, E., & Matei, S. (2015). Big data for open digital innovation-a research roadmap. *Big Data Research*, *2*(2), 53–58. doi:10.1016/j.bdr.2015.01.008

Brussevich, M., Dabla-Norris, E., Kamunge, C., Karnane, P., Khalid, S., & Kochhar, K. (2018). *Gender, Technology, and the Future of Work*. Retrieved August 21, 2022, from https://www.imf.org/-/media/Files/Publications/SDN/2018/SDN1807.ashx

Büberkökü, Ö. (2022). Financial Analysis Based on Metaverse Coins. *International Izmir Economics Congress*.

Bun, M. J. G., & Kiviet, J. F. (2003). On the diminishing returns of higher order terms in asymptotic expansions of bias. *Economics Letters*, *79*(2), 145–152. doi:10.1016/S0165-1765(02)00299-9

Büyük Mükellefler Vergi Dairesi Başkanlığı. (n.d.). https://bmvdb.gib.gov.tr/kurumsal/kurulu%C5%9F-nedenleri

Byrne, D. M., & Corrado, C. A. (2017). *ICT Services and their Prices: What do they tell us about Productivity and Technology?* Available at: https://ssrn.com/abstract=2920451

Calışır, E. Ç. (2018, January 9). *WEB 1.0-WEB 2.0-WEB 3.0-WEB 4.0 "NEDİR BU WEB?* Retrieved September 2022, from Esma Ç. Çalışır: https://esmacalisir.com/2018/01/09/web-1-0-web-2-0-web-3-0-web-4-0-nedir-bu-web/

Çalışkan, Ş., Karabacak, M., & Meçik, O. (2017). Türkiye Ekonomisinde Eğitim Harcamaları ve Ekonomik Büyüme İlişkisi: Bootstrap Toda-Yamamoto Nedensellik Testi Yaklaşımı. *Kocaeli Üniversitesi Sosyal Bilimler Dergisi*, (33), 45-56.

Calvino, J. (2021, May). *The Guide to Remarketing for Tour & Activities -Part 1*. October 2022 tarihinde Blend: https://blend.travel/blog/the-guide-to-remarketing-for-tour-activities-part-1/adresinden alındı

Cambridge Center For Alternative Finance. (2022). https://ccaf.io/cbeci/mining_map

Cancer. (n.d.). In *Merriam-Webster Dictionary*. Retrieved July 24, 2022, from https://www.merriam-webster.com/dictionary/cancer

Carre, A., Dubois, A., Partarakis, N., Zabulis, X., Patsiouras, N., Mantinaki, E., Zidianakis, E., Cadi, N., Baka, E., Thalmann, N. M., Makrygiannis, D., Glushkova, A., & Manitsaris, S. (2022). Mixed-Reality Demonstration and Training of Glassblowing. *Heritage*, *5*(1), 103–128. doi:10.3390/heritage5010006

Carr, N. G. (2003). IT doesn't matter. *EDUCAUSE Review*, *38*, 24–38. PMID:12747161

Casimir, G., Lee, K., & Loon, M. (2012). Knowledge sharing: Influences of trust, commitment and cost. *Journal of Knowledge Management*, *16*(5), 740–753. doi:10.1108/13673271211262781

Castaño, C., Martín, J., & Luis Martínez, J. (2011). La brecha digital de género en España y Europa: Medición con indicadores compuestos. *Revista Espanola de Investigaciones Sociologicas*, *136*(1), 127–140. doi:10.5477/cis/reis.136.127

Celikkol, Ş. (2022). Metaverse Dünyasının Tüketici Satın Alma Davranışları Açısından Değerlendirilmesi. *İstanbul Kent Üniversitesi İnsan ve Toplum Bilimleri Dergisi*, *3*(1), 64-75.

Cepel, M., Belas, J., & Dvorský, J. (2019). The impact of significant political factors on starting a new business. *14th European Conference on Innovation and Entrepreneurship ECIE 2019*, 220-228. 10.34190/ECIE.19.117

Chaffey, D., & Smith, P. (2008). eMarketing eXcellence: Planning and Optimizing Your Digital Marketing. Butterworth-Heinemann.

Chandrashekar, D., & Bala Subrahmanya, M. H. (2017). Absorptive capacity as a determinant of innovation in SMEs: A study of Bengaluru high-tech manufacturing cluster. *Small Enterprise Research*, *24*(3), 290–315. doi:10.1080/13215906.2017.1396491

Chang, P. (2017). The Importance performance Analysis of Taiwan Tourism Mobile Marketing. *Journal of Tourism Management Research*, *4*(1), 12–16. doi:10.18488/journal.31.2017.41.12.16

Chatterjee, A. (2020). Financial inclusion, information and communication technology diffusion, and economic growth: A panel data analysis. *Information Technology for Development*, *26*(3), 607–635. doi:10.1080/02681102.2020.1734770

Chen, C.-J., Huang, J.-W., & Hsiao, Y.-C. (2010). Knowledge management and innovativeness. The role of organizational climate and structure. *International Journal of Manpower*, *31*(8), 848–870. doi:10.1108/01437721011088548

Cheng, C. Y., Chien, M. S., & Lee, C. C. (2021). ICT diffusion, financial development, and economic growth: An international cross-country analysis. *Economic Modelling*, *94*, 662–671. doi:10.1016/j.econmod.2020.02.008

Cheng, E. C. K., Wu, S. W., & Hu, J. (2017). Knowledge management implementation in the school context: Case studies on knowledge leadership, storytelling, and taxonomy. *Educational Research for Policy and Practice*, *16*(2), 177–188. doi:10.100710671-016-9200-0

Chen, M. X., Qin, L. Y., & Zhao, C. Z. (2021). What Are Reasons of the Significant Change in the Chinese Female Labour Force Participation Rate? Analysis on Influencing Factors Based on the Macroscopic Trend. *Transformations in Business & Economics*, *20*(3c), 405–428.

Chen, Y., Wang, Y., Nevo, S., Benitez-Amado, J., & Kou, G. (2015). IT capabilities and product innovation performance: The roles of corporate entrepreneurship and competitive intensity. *Information & Management*, *52*(6), 643–657. doi:10.1016/j.im.2015.05.003

Chesbrough, H., Vanhaverbeke, W., & West, J. (2006). *Open Innovation: Researching a New Paradigm*. Oxford University Press.

Chinn, M. D., & Fairlie, R. W. (2007). The determinants of the global digital divide: A cross-country analysis of computer and internet penetration. *Oxford Economic Papers*, *59*(1), 16–44. doi:10.1093/oep/gpl024

Choi, S. Y., Lee, H., & Yoo, Y. (2010). The impact of information technology and transactive memory systems on knowledge sharing, application, and team performance: A field study. *Management Information Systems Quarterly*, *34*(1), 855–870. doi:10.2307/25750708

Choudhry, M. T., & Elhorst, P. (2018). Female labour force participation and economic development. *International Journal of Manpower*, *39*(7), 896–912. doi:10.1108/IJM-03-2017-0045

Chou, S. Y., & Ramser, C. (2019). A multilevel model of organizational learning Incorporating employee spontaneous workplace behaviors, leadership capital and knowledge management. *The Learning Organization*, *26*(2), 132–145. doi:10.1108/TLO-10-2018-0168

Cilan, Ç. A., Bolat, B. A., & Coşkun, E. (2009). Analyzing digital divide within and between member and candidate countries of European Union. *Government Information Quarterly*, *26*(1), 98–105. doi:10.1016/j.giq.2007.11.002

Cilesiz, E., & Aydın, N. (2022). Metaverse and Tourism: A Conceptual Approach. *Journal of Academic Studies*, *3*(1), 32–44.

Ciolacu, M., Tehrani, A. F., Beer, R., & Popp, H. (2017). Education 4.0—Fostering student's performance with machine learning methods. In *2017 IEEE 23rd International Symposium for Design and Technology in Electronic Packaging (SIITME)* (pp. 438-443). IEEE.

Clark, C. E., & Greenley, H. L. (2019). *Bitcoin, blockchain, and the energy sector*. Congressional Research Service.

Clinical Trial. (n.d.). In *Merriam-Webster Dictionary*. Retrieved July 24, 2022, from https://www.merriam-webster.com/dictionary/clinical%20trial

Coal Is Fueling Bitcoin's Meteoric Rise. (2022). https://www.bloomberg.com/news/articles/2017-12-15/turning-coal-into-bitcoindirty-secret-of-2017-s-hottest-market

Coban, Ö. (2021). Turizmde Arttırılmış Gerçeklik. In Ş. Aydın & M. Boz (Eds.), *Turizmde Güncel Konu ve Eğilimler* (pp. 44–55). Detay.

Cobb, C. W., & Douglas, P. H. (1928). A Theory of Production. *The American Economic Review*, *18*(1), 139–165. https://www.jstor.org/stable/1811556

CoinMarketCap. (2022). www.coinmarketcap.com

Collins, A., & Halverson, R. (2010). The second educational revolution: Rethinking education in the age of technology. *Journal of Computer Assisted Learning*, *26*(1), 18–27. doi:10.1111/j.1365-2729.2009.00339.x

Conole, G., & Dyke, M. (2004). What are the affordances of information and communication technologies? *ALT-J Research in Learning Technology*, *12*(2), 113–124. doi:10.3402/rlt.v12i2.11246

Cooper, J. (2006). The digital divide: The special case of gender. *Journal of Computer Assisted Learning*, *22*(5), 320–334. doi:10.1111/j.1365-2729.2006.00185.x

Corbet, S., Lucey, B., & Yarovaya, L. (2021). Bitcoin energy markets interrelationships-New evidence. *Resour. Pol.*, *70*(1–12), 101916.

Cortet, M., Rijks, T., & Nijland, S. (2016). Psd2: The digital transformation accelerator for banks. *Journal of Payments Strategy & Systems*, *10*, 13–27.

Cronin, F. J., Parker, E. B., Colleran, E. K., & Gold, M. A. (1991). Telecommunications infrastructure and economic growth: An analysis of causality. *Telecommunications Policy*, *15*(6), 529–535. doi:10.1016/0308-5961(91)90007-X

Cruz-Jesus, F., Oliveira, T., & Bacao, F. (2012). Digital divide across the European Union. *Information & Management*, *49*(6), 278–291. doi:10.1016/j.im.2012.09.003

Cuervo, M. R. V., & Menéndez, A. J. L. (2006). A multivariate framework for the analysis of the digital divide: Evidence for the European Union-15. *Information & Management*, *43*(6), 756–766. doi:10.1016/j.im.2006.05.001

Cuevas-Vargas, H., Estrada, S., & Larios-Gómez, E. (2016). The effects of ICTs as innovation facilitators for a greater business performance. Evidence from Mexico. *Procedia Computer Science*, *91*, 47–56. doi:10.1016/j.procs.2016.07.040

Cummings, J. (2004). Work Groups, Structural Diversity and Knowledge Sharing in a Global Organisation. *Management Science*, *50*(3), 352–364. doi:10.1287/mnsc.1030.0134

Curado, C., & Vieira, S. (2019). Trust, knowledge sharing and organizational commitment in SMEs. *Personnel Review*, *48*(6), 1449–1468. doi:10.1108/PR-03-2018-0094

Czernich, N., Falck, O., Kretschmer, T., & Woessmann, L. (2011). Broadband infrastructure and economic growth. *Economic Journal (London)*, *121*(552), 505–532. doi:10.1111/j.1468-0297.2011.02420.x

Daellenbach, H., McNickle, D., & Dye, Sh. (2012). *Management Science. Decision-making through systems thinking* (2nd ed.). Palgrave Macmillan.

Daher, J., Vijh, R., Linthwaite, B., Dave, S., Kim, J., Dheda, K., Peter, T., & Pai, N. P. (2017). Do digital innovations for HIV and sexually transmitted infections work? Results from a systematic review (1996-2017). *BMJ Open*, *7*(11), e017604. doi:10.1136/bmjopen-2017-017604 PMID:29101138

Dahou, K., Hacini, I., & Burgoyne, J. (2019). Knowledge management as a critical success factor in developing international companies' organizational learning capability. *Journal of Workplace Learning*, *31*(1), 2–16. doi:10.1108/JWL-12-2017-0118

Danish, K. (2012). Impact of Knowledge Management Practices on Organizational Performance, evidence from Pakistan. *International Journal of Scientific and Engineering Research*, *3*(8), 1–7.

Davenport, T. H. (1997). Ten Principles of Knowledge Management and Four Case Studies. *Knowledge and Process Management*, *4*(3), 187–208. doi:10.1002/(SICI)1099-1441(199709)4:3<187::AID-KPM99>3.0.CO;2-A

Davenport, T. H., & Völpel, S. C. (2001). The rise of knowledge towards attention management. *Journal of Knowledge Management*, *5*(3), 212–221. doi:10.1108/13673270110400816

De Jong, J. P. J., & Den Hartog, D. N. (2008). *Innovative Work Behavior: Measurement and Validation.* A working paper. EIM Business and Policy Research. Working Paper, 1-27.

De la Hoz-Rosales, B., Camacho-Ballesta, J. A., Tamayo-Torres, I., & Buelvas-Ferreira, K. (2019). Effects of information and communication technology usage by individuals, businesses, and government on human development: An international analysis'. *IEEE Access: Practical Innovations, Open Solutions*, *7*, 129225–129243. doi:10.1109/ACCESS.2019.2939404

De Vries, A. (2018). Bitcoin's growing energy problem. *Joule*, *2*(5), 801–880. doi:10.1016/j.joule.2018.04.016

de Widt, D., & Oats, L. (2019). *Co-operative Compliance: Views of Large Businesses in the Netherlands and UK*. H2020-EURO-SOCIETY-2014: Deliverable D7.2/Oats.

Del Giudice, M., & Della Peruta, M. R. (2016). The impact of IT-based knowledge management systems on internal venturing and innovation: A structural equation modeling approach to corporate performance. *Journal of Knowledge Management*, *20*(3), 484–498. doi:10.1108/JKM-07-2015-0257

Delfmann, H., & Koster, S. (2016). The effect of new business creation on employment growth in regions facing population decline. *The Annals of Regional Science*, *56*(1), 33–54. doi:10.100700168-015-0738-1

Demir, Y. A. (2022, June 10). Retrieved August 2022, from Internetin Geleceği: Web 3.0 Nedir?: https://shiftdelete.net/internetin-gelecegi-web-3-0-nedir

Denner, M.-S., Püschel, L. C., & Maximilian, R. (2018). How to Exploit the Digitalization Potential of Business Processes. *Business & Information Systems Engineering*, *60*(4), 331–349. doi:10.100712599-017-0509-x

DeSerpa, A. C. (1971). A Theory of the Economics of Time. *Economic Journal (London)*, *81*(324), 828–846. doi:10.2307/2230320

Dettling, L. J. (2017). Broadband in the Labor Market: The Impact of Residential High-Speed Internet on Married Women's Labor Force Participation. *Industrial & Labor Relations Review*, *70*(2), 451–482. doi:10.1177/0019793916644721

Dewan, S., Ganley, D., & Kraemer, K. L. (2010). Complementarities in the diffusion of personal computers and the Internet: Implications for the global digital divide. *Information Systems Research*, *21*(4), 925–940. doi:10.1287/isre.1080.0219

Dhahri, S., & Omri, A. (2018). Entrepreneurship contribution to the three pillars of sustainable development: What does the evidence really say? *World Development*, *106*, 64–77. doi:10.1016/j.worlddev.2018.01.008

Diaz de Sarralde, S., von Haldenwang, C., Hentze, T., & Monkam, N. (2018). *Tax Certainty: Proposals for the Short Term and the Long Term* (No. 2018-16). Economics Discussion Papers.

Díaz-Chao, Á., Sainz-González, J., & Torrent-Sellens, J. (2015). ICT, Innovation, and Firm Productivity: New Evidence from Small Local Firms. *Journal of Business Research*, *68*(7), 1439–1444. doi:10.1016/j.jbusres.2015.01.030

Digiconomist. (2022). *Cambridge Bitcoin Energy Consumption Index.* 10 Haziran 2022, www.digiconomist.net

Digital Technology. (n.d.). In *Dictionary.com*. Retrieved July 24, 2022, from https://www.dictionary.com/browse/digital-technology

Distance Learning. (n.d.). In *Merriam-Webster Dictionary*. Retrieved July 24, 2022, from https://www.merriam-webster.com/dictionary/distance%20learning

Dodd, L., & Johnson, K. (2009). Virtual Tourism: Complementary or Contradictory to Traditional Tourism. *27th EuroCHRIE Annual Conference*, 672-679.

Dogan, E., Majeed, M. T., & Luni, T. (2022). Are clean energy and carbon emission allowances caused by bitcoin? A novel time-varying method. *Journal of Cleaner Production*, *347*, 347. doi:10.1016/j.jclepro.2022.131089

Dogan, O., & Baloglu, N. (2020). Industrial 4.0 Conceptual Awareness Levels of University Students. *Turkish Science Research Foundation, 13*(1), 126–142.

Donate, M. J., & Pablo, J. D. S. (2015). The role of knowledge-oriented leadership in knowledge management practices and innovation. *Journal of Business Research, 68*(2), 360–370. doi:10.1016/j.jbusres.2014.06.022

Dragulanescu, N. G. (2002). Social impact of the "Digital Divide" in a Central–Eastern European country. *The International Information & Library Review, 34*(2), 139–151. doi:10.1080/10572317.2002.10762570

Dvouletý, O. (2017). Can policy makers count with positive impact of entrepreneurship on economic development of the Czech regions? *Journal of Entrepreneurship in Emerging Economies, 9*(3), 286–299. doi:10.1108/JEEE-11-2016-0052

Dzingirai, M. (2021). The role of entrepreneurship in reducing poverty in agricultural communities. *Journal of Enterprising Communities: People and Places in the Global Economy, 15*(5), 665–683. doi:10.1108/JEC-01-2021-0016

Easterbrook, S., Singer, J., Storey, M., & Damian, D. (2008). *Guide to Advanced Empirical Software Engineering-Selecting Empirical Methods for Software Engineering Research.* Springer.

Efobi, U. R., Tanankem, B. V., & Asongu, S. A. (2018). Female Economic Participation with Information and Communication Technology Advancement: Evidence from Sub-Saharan Africa. *The South African Journal of Economics, 86*(2), 231–246. doi:10.1111aje.12194

Ejemeyovwi, J. O., & Osabuohien, E. S. (2018). Investigating the relevance of mobile technology adoption on inclusive growth in West Africa. *Contemporary Social Science.*

Ejemeyovwi, J. O., Osabuohien, E. S., Bowale, E. K., Abuh, O. O., Adedoyin, J. P., & Ayanda, B. (2019, December 1). Information and communication technology adoption and innovation for sustainable entrepreneurship. *Journal of Physics: Conference Series, 1378*(2), 022085. doi:10.1088/1742-6596/1378/2/022085

Elena-Bucea, A., Cruz-Jesus, F., Oliveira, T., & Coelho, P. S. (2021). Assessing the Role of Age, Education, Gender and Income on the Digital Divide: Evidence for the European Union. *Information Systems Frontiers, 23*(4), 1007–1021. doi:10.100710796-020-10012-9

Elster, J. (2007). *Explaining Social Behavior: More Nuts and Bolts for the Social Sciences.* Cambridge University Press. doi:10.1017/CBO9780511806421

Emirmahmutoglu, F., & Kose, N. (2011). Testing for granger causality in heterogeneous mixed panels. *Economic Modelling, 28*(3), 870–876. doi:10.1016/j.econmod.2010.10.018

Erkişi, K., & Boğa, S. (2019). High-technology products export and economic growth: A panel data analysis for EU-15 countries. *Bingöl Üniversitesi Sosyal Bilimler Enstitüsü Dergisi, 9*(18), 669–683.

European Commission. (2020). *Digital Economy and Society Index 2021 Methodological Note.* European Commission.

EUROSTAT. (2022). *Eurostat data.* Available at: https://ec.europa.eu/eurostat/metadata/

Evans, A. W. (1972). On the Theory of the Valuation and Allocation of Time. *Scottish Journal of Political Economy, 19*(1), 1–17. doi:10.1111/j.1467-9485.1972.tb00504.x

Eze, S. C., Chinedu-Eze, V. C., & Bello, A. O. (2018). Actors and emerging information, communications and technology (EICT) adoption: A study of UK small and medium services enterprises. *Cogent Business & Management, 5*(1), 1–19. doi:10.1080/23311975.2018.1480188

Farhadi, M., Ismail, R., & Fooladi, M. (2012). Information and Communication Technology Use and Economic Growth. *PLoS One, 7*(11), e48903. doi:10.1371/journal.pone.0048903 PMID:23152817

Fichman, R. G., Dos Santos, B. L., & Zheng, Z. E. (2014). Digital innovation as a fundamental and powerful concept in the information systems curriculum. *Management Information Systems Quarterly*, *38*(2), 329–353. doi:10.25300/MISQ/2014/38.2.01

Flavián, C., Ibáñez-Sánchez, S., & Orús, C. (2019). The Impact of Virtual, Augmented and Mixed Reality Technologies on the Customer Experience. *Journal of Business Research*, *100*, 547–560. doi:10.1016/j.jbusres.2018.10.050

Foteinis, S. (2018). Bitcoin's alarming carbon footprint. *Nature*, *554*(7690), 169–170. doi:10.1038/d41586-018-01625-x

Fountain, J. E. (2000). Constructing the information society: Women, information technology, and design. *Technology in Society*, *22*(1), 45–62. doi:10.1016/S0160-791X(99)00036-6

Fowler, M. (2003). *Catalog of Patterns of Enterprise Application Architecture*. Fowler. https://martinfowler.com/eaaCatalog/

Franco, M. (2020). Entrepreneurship, competitiveness and innovation. *Sustainability*, *12*(16), 6576. doi:10.3390u12166576

Frankenfield, J., Brown, J. R., & Kazel, M. (2022). *Bitcoin Mining*. 11 Mayıs 2022, https://www.investopedia.com/terms/b/bitcoin-mining.asp

Fu, J. S. (2013). Article. *International Journal of Education and Development Using Information and Communication Technology*, *9*(1), 112–125.

G20 Argentina Presidency. (2018). *Toolkit for Measuring the Digital Economy*. https://www.oecd.org/g20/summits/buenos-aires/g20-detf-toolkit.pdf

Gaberli, U. (2019). Tourism in Digital Age: An Explanation for the Impact of Virtual, Augmented and Mixed Reality Tecnologies on Tourist Experiences. *Journal of Tourism Intelligence and Smartness*, *2*(2), 61–69.

Gaddis, I., & Klasen, S. (2014). Economic development, structural change, and women's labor force participation: A reexamination of the feminisation U hypothesis. *Journal of Population Economics*, *27*(3), 639–681. doi:10.100700148-013-0488-2

Galindo-Martín, M. Á., Castaño-Martínez, M. S., & Méndez-Picazo, M. T. (2019). Digital transformation, digital dividends and entrepreneurship: A quantitative analysis. *Journal of Business Research*, *101*, 522–527. doi:10.1016/j.jbusres.2018.12.014

Garvin, D. A. (1993). Building a Learning Organization. *Harvard Business Review*, *71*, 78–91. PMID:10127041

Genre, V., Salvador, R. G., & Lamo, A. (2010). European women: Why do(n't) they work? *Applied Economics*, *42*(12), 1499–1514. doi:10.1080/00036840701721547

Gerguri-Rashiti, S., Ramadani, V., Abazi-Alili, H., Dana, L., & Ratten, V. (2017). ICT, innovation and firm performance: The transition economies context. *Thunderbird International Business Review*, *59*(1), 93–102. doi:10.1002/tie.21772

Ghalandari, K. (2013). The Effect of Information and Communication Technology on Export Performance of Iranian Firms. *Research Journal of Applied Sciences, Engineering and Technology*, *5*(17), 4367–4372. doi:10.19026/rjaset.5.4431

Gholami, R., Higón, D. A., Hanafizadeh, P., & Emrouznejad, A. (2010). Is ICT The Key to Development? *Journal of Global Information Management*, *18*(1), 66–83. doi:10.4018/jgim.2010091104

Giesen, E., Riddleberger, E., Christner, R., & Bell, R. (2010). When and how to innovate your business model. *Strategy and Leadership*, *38*(4), 17–26. doi:10.1108/10878571011059700

Giniuniene, J., & Jurksiene, L. (2015). Dynamic capabilities, innovation and organizational learning: Interrelations and impact on firm performance. *Procedia: Social and Behavioral Sciences*, *213*, 985–991. doi:10.1016/j.sbspro.2015.11.515

Giudice, M., & Straub, D. (2011). Editor's comments: IT and entrepreneurism: An on-again, off-again love affair or a marriage? *Management Information Systems Quarterly*, *35*(4), 3–8. doi:10.2307/41409961

Global Entrepreneurship Monitor. (2022). *Entrepreneurial Behaviour and Attitudes*. https://www.gemconsortium.org/data

Glossary of Tax Terms. (n.d.). https://www.oecd.org/ctp/glossaryoftaxterms.htm

Gold, A. H., Malhotra, A., & Segars, A. H. (2001). Knowledge management: An organizational capabilities perspective. *Journal of Management Information Systems*, *18*(1), 185–214. doi:10.1080/07421222.2001.11045669

Gomber, P., Koch, J. A., & Siering, M. (2017). Digital Finance and FinTech: Current research and future research directions. *Journal of Business Economics*, *87*(5), 537–580. doi:10.100711573-017-0852-x

Gomes, S., & Lopes, J. M. (2022). ICT access and entrepreneurship in the open innovation dynamic context: Evidence from OECD countries. *Journal of Open Innovation*, *8*(2), 102. doi:10.3390/joitmc8020102

Goodarzi, M., Ebrahimzadeh, I., Rabi, A., Saedipoor, B., & Jafarabadi, M. A. (2012). Impact of Distance Education via Mobile Phone Text Messaging on Knowledge, Attitude, Practice and Self-efficacy of Patients with Type 2 Diabetes Mellitus in Iran. *Journal of Diabetes and Metabolic Disorders*, *11*(10), 10. doi:10.1186/2251-6581-11-10 PMID:23497632

Gorkey-Aydinoglu, S., & Yalki-Berker, Ġ. (2016). The Comparative Development of ICT in BRICS: A cluster analysis. *The Journal of Knowledge Economy and Knowledge Management*, *11*(1), 17–28.

Goslinga, S., Jonge, J. D., Siglé, M., & Dijk, H. V. (2021). Cooperative Compliance Programmes: Who Participates and Why? *Journal of Tax Administration*, *6*(2), 95–121.

Goyal, A. (2011). Developing women: Why technology can help. *Information Technology for Development*, *17*(2), 112–132. doi:10.1080/02681102.2010.537252

Graetz, M. J., Reinganum, J. F., & Wilde, L. L. (1986). The Tax Compliance Game: Toward An Interactive Theory of Law Enforcement. *JL Econ. & Org.*, *2*(1), 1–32.

Gulicheva, E., Lisin, E., Osipova, M., & Khabdullin, A. (2017). Leading factors in the formation of innovative education environment. *Journal of International Studies*, *10*(2), 129–137. doi:10.14254/2071-8330.2017/10-2/9

Gulli, A. (2020). (un)sustainability of bitcoin mining. *Rutgers Computer & Technology Law Journal*, *46*(1), 95–123.

Guze, P. A. (2015). Using Technology to Meet the Challenges of Medical Education. *Transactions of the American Clinical and Climatological Association*, *126*, 260–270. https://www.ncbi.nlm.nih.gov/pmc/articles/PMC4530721/ PMID:26330687

Haber, J. (2014). MOOCs. MIT Press.

Hagberg, J., Sundstrom, M., & Egels-Zandén, N. (2016). The digitalization of retailing: An exploratory framework. *International Journal of Retail & Distribution Management*, *44*(7), 694–712. doi:10.1108/IJRDM-09-2015-0140

Halili, S, H. (2019). Technological advancements in education 4.0. *The Online Journal of Distance Education and e-Learning*, *7*(1), 63-69.

Haller, S. A., & Lyons, S. (2015). Broadband adoption and firm productivity: Evidence from Irish manufacturing firms. *Telecommunications Policy*, *39*(1), 1–13. doi:10.1016/j.telpol.2014.10.003

Hamburg, I., & Lütgen, G. (2019). Digital divide, digital inclusion and inclusive education. *Adv. Soc. Sci. Res. J., 6*, 193–206. doi:10.14738/assrj.64.6457

Hanif, A., & Bukhari, I. (2015). Relationship between innovative work behavior and job involvement among the employees of telecom sector. *Pakistan Journal of Social and Clinical Psychology, 13*(2), 23–29.

Hannemann-Weber, H., Kessel, M., Budych, K., & Schultz, C. (2011). Shared communication processes within healthcare teams for rare diseases and their influence on healthcare professionals' innovative behavior and patient satisfaction. *Implementation Science; IS, 6*(40), 1–7. doi:10.1186/1748-5908-6-40 PMID:21510848

Hansen, R., & Sia, S. K. (2015). Hummel's digital transformation toward omnichannel retailing: Key lessons learned. *MIS Quarterly Executive, 14*(2), 51–66.

Hausman, J. A. (1978). Specification tests in econometrics. *Econometrica, 46*(6), 1251–1271. doi:10.2307/1913827

Hau, Y. S., Kim, B., Lee, H., & Kim, Y. G. (2013). The effects of individual motivations and social capital on employees' tacit and explicit knowledge sharing intentions. *International Journal of Information Management, 33*(2), 256–266. doi:10.1016/j.ijinfomgt.2012.10.009

Hedström, P., & Ibarra, R. (2010). On the Contagiousness of Non-Contagious Behavior: The Case of Tax Avoidance and Tax Evasion. In H. Joas & B. Klein (Eds.), *The Benefit of Broad Horizons: Intellectual and Institutional Preconditions for a Global Social Science*. Brill. doi:10.1163/ej.9789004192843.i-436.103

Hemmati, M. (2022). The Metaverse: An Urban Revolution Effect of the Metaverse on the Perceptions of Urban Audience. *Tourism of Culture, 2*(7), 53–60.

Henderson, R., & Clark, K. (1990). Architectural Innovation: The Reconfiguration of Existing Product Technologies and the Failure of Established Firms. *Administrative Science Quarterly, 35*(1), 9–30. doi:10.2307/2393549

Hern, A. (2017). *Bitcoin Mining Consumes More Electricity A Year Than Ireland*. 12 Mayıs 2022, https://www.theguardian.com/technology/2017/nov/27/bitcoin-mining-consumeselectricity-ireland

Hess, T., Matt, C., Benlian, A., & Wiesböck, F. (2016). Options for Formulating a Digital Transformation Strategy. *MIS Quarterly Executive, 15*(1), 137–139.

Higher Education Council (HEC). (2021). Retrieved from: https://istatistik.yok.gov.tr/

Higher Education Council (HEC). (2021). Retrieved from: https://istatistik.yok.gov.tr/, (12.09.2022).

Hilbert, M. (2011). Digital gender divide or technologically empowered women in developing countries? A typical case of lies, damned lies, and statistics. *Women's Studies International Forum, 34*(6), 479–489. doi:10.1016/j.wsif.2011.07.001

Hina, S. M., Hassan, G., Parveen, M., & Arooj, S. (2021). Impact of entrepreneurial orientation on firm performance through organizational learning: The moderating role of environmental turbulence. *Performance Improvement Quarterly, 34*(1), 77–104. doi:10.1002/piq.21343

Hinings, B., Gegenhuber, T., & Greenwood, R. (2018). Digital Innovation and Transformation: An institutional perspective. *Information and Organization, 28*(1), 52–61. doi:10.1016/j.infoandorg.2018.02.004

Hirschi, A. (2018). The fourth industrial revolution: Issues and implications for career research and practice. *The Career Development Quarterly, 66*(3), 192–204. doi:10.1002/cdq.12142

Hodrab, R., Maitah, M., & Luboš, S. (2016). The Effect of Information and Communication Technology on Economic Growth: Arab World Case. *International Journal of Economics and Financial Issues, 6*(2), 765–775.

Hong, J. P. (2017). Causal relationship between ICT R&D investment and economic growth in Korea. *Technological Forecasting and Social Change*, *116*, 70–75. doi:10.1016/j.techfore.2016.11.005

Hosiaisluoma. (2022). *Holistic Enterprise Development*. https://www.hosiaisluoma.fi/blog/archimate-examples

Howson, P., & Vries, A. (2022). Preying on the poor? Opportunities and challenges for tackling the social and environmental threats of cryptocurrencies for vulnerable and low-income communities. *Energy Research & Social Science*, *84*, 102394. doi:10.1016/j.erss.2021.102394

Hsu, T. (2014). Effects of organization culture, organizational Learning and IT Strategy on Knowledge Management and Performance. *Journal of International Management Studies*, *9*(1), 50–57.

Huang, C.-E., & Liu, C.-H. S. (2015). Employees and creativity: Social ties and access to heterogeneous knowledge. *Creativity Research Journal*, *27*(2), 206–213. doi:10.1080/10400419.2015.1030309

Huang, K. F., Lin, K. H., Wu, L. Y., & Yu, P. H. (2015). Absorptive capacity and autonomous R&D climate roles in firm innovation. *Journal of Business Research*, *68*(1), 87–94. doi:10.1016/j.jbusres.2014.05.002

Hughes, M., Rigtering, J. P. C., Covin, J. G., Bouncken, R. B., & Kraus, S. (2018). Innovative behaviour, trust and perceived workplace performance. *British Journal of Management*, *29*(4), 750–768. doi:10.1111/1467-8551.12305

Huiskers-Stoop, E., & Gribnau, H. (2019). Cooperative Compliance and the Dutch Horizontal Monitoring Model. *Journal of Tax Administration*, *5*(1), 66–110.

Hussein, E. F. (2020). The impact of information and communication technology on inclusive growth: An empirical study on developing and Arab countries. *Journal of the Faculty of Economics and Political Science*, *21*(3), 47–77.

Hussin, A. A. (2018). Education 4.0 made simple: Ideas for teaching. *International Journal of Education and Literacy Studies*, *6*(3), 92–98. doi:10.7575/aiac.ijels.v.6n.3p.92

Ignatov, A. (2020). The digital economy of BRICS: Prospects for multilateral cooperation. *International Organisations Research Journal*, *15*(1), 31–62. doi:10.17323/1996-7845-2020-01-02

IMF. (2022). *GDP per capita, current prices*. Retrieved from IMF DATAMAPPER: https://www.imf.org/external/datamapper/NGDPDPC@WEO/MEQ

IMF. (2022a). *IMF DATAMAPPER*. Retrieved from Real GDP growth: https://www.imf.org/external/datamapper/NGDP_RPCH@WEO/OEMDC/ADVEC/WEOWORLD

IMF/OECD. (2017). *Tax Certainty, Report for the G20 Finance Ministers*. https://www.oecd.org/tax/tax-policy/tax-certainty-report-oecd-imf-report-g20-finance-ministers-march-2017.pdf

Im, K. S., Pesaran, M. H., & Shin, Y. (2003). Testing for unit roots in heterogeneous panels. *Journal of Econometrics*, *115*(1), 53–74. doi:10.1016/S0304-4076(03)00092-7

Inan, Ü. S. E. (2022). Dijital Pazarlama Karması Bileşenleri ve Dijital Pazarlama Kanalları. In Ü. S. E. İnan (Ed.), *İşletme Bilimlerinde Dijital Dönüşüm* (pp. 1–18). Nobel Yayıncılık.

Inan, Ü. S. E. (2022). Use of Augmented Reality Technology in Marketing. In Y. Bayar (Ed.), *TechnologicalDevelopment and Impact onEconomic and Environmental Sustainability* (pp. 168–182). IGI Global. doi:10.4018/978-1-7998-9648-7.ch013

Iqbal, R., Doctor, F., More, B., Mahmud, S., & Yousuf, U. (2018). Big data analytics: Computational intelligence techniques and application areas. *Technological Forecasting and Social Change*, *153*(1), 1–16.

Ishida, H. (2015). The effect of ICT development on economic growth and energy consumption in Japan. *Telematics and Informatics*, *32*(1), 79–88. doi:10.1016/j.tele.2014.04.003

ITU. (2018). *Measuring the Information Society Report: Executive summary.* ITU Publications.

ITU. (2019). *ITU ICT Facts and Figures 2017.* https://www.itu.int/en/ITUD/Statistics/Documents/facts/ICTFactsFigures2017.pdf

ITU. (2022). *The ICT Development Index.* Retrieved from Committed to connecting the world: https://www.itu.int/en/ITU-D/Statistics/Pages/IDI/default.aspx

ITU. (2022a). *The ICT Development Index (IDI): Conceptual framework and methodology.* Retrieved from Committed to connecting the world: https://www.itu.int/en/ITU-D/Statistics/Pages/publications/mis2017/methodology.aspx

ITU. (2022b). *Select an option IDI 2017 Rank.* Retrieved from ICT development Index ITUdata: https://www.itu.int/net4/ITU-D/idi/2017/index.html

Jamalzadeh, M. (2012). The relationship between knowledge management and learning organization of faculty members at Islamic Azad University, Shiraz Branch in Academic year (2010-2011). *Social and Behavioral Sciences*, *62*, 1164–1168. doi:10.1016/j.sbspro.2012.09.199

Jayachandran, S. (2021). Social Norms as a Barrier to Women's Employment in Developing Countries. *IMF Economic Review*, *69*(3), 576–595. doi:10.105741308-021-00140-w

Jerez-Gómez, P., Cespedes-Lorente, J., & Valle, R. (2005). Organizational Learning Capability: A Proposal of Measurement. *Journal of Business Research*, *58*(6), 715–725. doi:10.1016/j.jbusres.2003.11.002

Jiang, S., Li, Y., Lu, Q., Hong, Y., Guan, D., Yiong, Y., & Wang, S. (2021). Policy assessments for the carbon emission flows and sustainability of Bitcoin blockchain operation in China. *Nature Communications*, *12*(1), 1–10. doi:10.103841467-021-22256-3 PMID:33824331

Jin, S., & Cho, C. M. (2015). Is ICT a new essential for national economic growth in an information society? *Government Information Quarterly*, *32*(3), 253–260. doi:10.1016/j.giq.2015.04.007

Jung, M. (2022, October 21). *How to Use Email Marketing to Promote Your Travel & Tourism Business.* Retrieved October 2022, from Constant Contact: https://www.constantcontact.com/blog/email-marketing-travel-industry/#:~:text=Travel%20email%20marketing%20can%20be,business%20stand%20out%20from%20others

Jung, T., Dieck, M. C., Lee, H., & Chung, N. (2016). Effects of Virtual Reality and Augmented Reality on Visitor Experiences in Museum. In A. Inversini & R. Schegg (Eds.), *Information and Communication Technologies in Tourism.* Springer International Publishing.

Kacham, I., & Kebir, M. (2020). Information and Communication Technology and its Impact on Economic Growth in the Organization of Arab Petroleum Exporting Countries (OAPEC) A Standard Study for the period 2000-2018. *The Economic Science Management and Business Science*, *13*(1), 460–479.

Kaiser, U. (2003). *Strategic complementarities between different types of ICT-expenditures* (No. 03-46). ZEW Discussion Papers.

Kanematsu, H., & Barry, D. M. (2015). *STEM and ICT Education in Intelligent Environments.* Springer International Publishing.

Kanter, R. M. (1988). When a Thousand Flowers Bloom: Structural, Collective and Social Conditions for Innovation in Organization. *Research in Organizational Behavior*, *10*, 169–211.

Karaman Aksentijević, N., Ježić, Z., & Zaninović, P. A. (2021). The Effects of Information and Communication Technology (ICT) Use on Human Development—A Macroeconomic Approach. *Economies*, *9*(3), 128. doi:10.3390/economies9030128

Karasneh, A. A.-F. (2019). Reinforcing Innovation through Knowledge Management: Mediating Role of Organizational Learning. *Interdisciplinary Journal of Information, Knowledge, and Management*, *14*(1), 235–252. doi:10.28945/4427

Kastrati, V. (2015). Technological innovation of small and medium enterprises in Kosovo: Challenges and barriers. *European Journal of Sustainable Development*, *4*(3), 145–150.

Katz, A. W. (1998). *Foundations of the Economic Approach to Law*. Oxford University Press.

Kaygısız, E., & Sipahi, H. (2018). Y kuşağı üniversite öğrencilerinin bireysel yenilik ve Endüstri 4.0 bilgi düzeyleri arasındaki ilişkinin incelenmesi. *Gaziantep University Journal of Social Sciences*, *18*(2), 922–936.

Kaytmaz, Y. C. (2022, January 13). *Metaverse'te Sultanahmet Camisini 0.1 Ethereum'a, Ayasofya Camisini 100 Ethereum'a Satıyorlar*. Retrieved August 2022, from Sözcü: https://www.sozcu.com.tr/2022/teknoloji/metaversete-sultanahmet-camisini-0-1-ethereuma-ayasofya-camisini-100-ethereuma-satiyorlar-6885003/

Kendall, D. (2018, May). Four Ways Technology is Helping Mental Health. *Digital Health Today*. Retrieved from https://digitalhealthtoday.com/blog/technology-and-mental-health/

Kerras, H., Sánchez-Navarro, J. L., López-Becerra, E. I., & de-Miguel Gomez, M. D. (2020). The impact of the gender digital divide on sustainable development: Comparative analysis between the European Union and the Maghreb. *Sustainability*, *12*(8), 3347. doi:10.3390u12083347

Ke, W. L., & Wei, K. K. (2006). Organizational Learning Process: Its Antecedents and Consequences in Enterprise System Implementation. *Journal of Global Information Management*, *14*(1), 1–22. doi:10.4018/jgim.2006010101

Khajawi, A. (2017, December 24). *How Travel Brands Use SEM to Attract Potential Tourists?* Retrieved October 2022, from Brighton Gallery Blog: https://blogs.brighton.ac.uk/brightongallery/2017/12/24/how-travel-brands-use-search-engine-marketing-sem-to-attract-potential-tourists/

Khalili, F., Cheong, K. C., & Lau, W. Y. (2014). ICT as a source of economic growth in the information age: Empirical evidence from ICT leading countries. *Research Journal of Economics, Business and ICT*, *9*(1).

Khireddine, B., Noureddine, B., & Bouzelifa, S. (2018). The impact of information and communication technology on economic activity. In *The Third National Forum on Consumers and the Digital Economy: The necessity of transition and protection challenges* (pp. 1-18). Mila: Abdelhafid Boussouf University.

Khmiadashvili, L. (2019). Digital Marketing Strategy Based on Hotel Industry Study in Tbilisi. In *Proceedings of SOCIOINT 24-26 June 2019* (pp. 1032-1037). 6th International Conference on Education, Social Sciences and Humanities. 10.18769/ijasos.592628

Khorakian, A., Mohammadi Shahroodi, H., Jahangir, M., & Nikkhah Farkhani, Z. (2019). Innovative work behavior in public organizations: The roles of ethical and knowledge sharing behaviors. *Creativity Research Journal*, *31*(2), 164–173. doi:10.1080/10400419.2019.1607444

Kidwell, J. J., Linde, K. M. V., & Johnson, S. L. (2000). Applying corporate knowledge management practices in higher education. *EDUCAUSE Quarterly*, *4*, 28–33.

Kiliç, Ç. (2011). Küreselleşen Dünyada Dijital Bölünme Sorunu. *Erzincan Üniversitesi Eğitim Fakültesi Dergisi*, *13*(1), 81–91.

Kim, C. S., Park, S. Y., Kang, J. G., Lee, S. J., Ihm, S. H., Moon, G. C., & Yoo, H. J. (2010). Insulin Dose Titration System in Diabetes Patients Using a Short Messaging Service Automatically Produced by a Knowledge Matrix. *Diabetes Technology & Therapeutics*, *12*(8), 663–669. doi:10.1089/dia.2010.0031 PMID:20615108

Kim, T. T., & Lee, G. (2013). Hospitality Employee Knowledge-Sharing Behaviors in the Relationship between Goal Orientations and Service Innovative Behavior. *International Journal of Hospitality Management*, *34*, 324–337. doi:10.1016/j.ijhm.2013.04.009

Kim, W., & Park, J. (2017). Examining structural relationships between work engagement, organizational procedural justice, knowledge sharing, and innovative work behavior for sustainable organizations. *Sustainability*, *9*(2), 205–220. doi:10.3390u9020205

Kim, Y. J., Kang, H., Sanders, G. L., & Lee, S. T. (2008). Differential Effects of IT Investments: Complementarity and the Effect of GDP Level. *International Journal of Information Management*, *28*(8), 508–516. doi:10.1016/j.ijinfomgt.2008.01.003

Kingsley, A. (2021). *My Secret #EdTech Diary Looking at Educational Technology through a wider lens*. John Catt Educational Limited.

Kingsnorth, S. (2017). Dijital Pazarlama Stratejisi. Nobel Yayınları.

Kirchler, E. (2007). *The Economic Psychology of Tax Behaviour*. Cambridge University Press. doi:10.1017/CBO9780511628238

Kirchler, E., Hoelzl, E., & Wahl, I. (2008). Enforced versus Voluntary Tax Compliance: The "Slippery Slope" Framework. *Journal of Economic Psychology*, *29*(2), 210–225. doi:10.1016/j.joep.2007.05.004

Kırçova, İ. (2012). *İnternette Pazarlama*. Beta Yayınları.

Kitsios, F., Giatsidis, I., & Kamariotu, M. (2021). Digital transformation and strategy in the banking sector: Evaluating the acceptance rate of e-services. *Journal of Open Innovation*, *7*(1), 1–14. doi:10.3390/joitmc7030204

Kiviet, J. F. (1995). On Bias, Inconsistency and Efficiency of Various Estimators in Dynamic Panel Data Models. *Journal of Econometrics*, *68*(1), 53–78. doi:10.1016/0304-4076(94)01643-E

Kiviet, J. F. (1999). Expectation of expansions for estimators in a dynamic panel data model; some results for weakly exogenous regressors. In C. Hsiao, K. Lahiri, L.-F. Lee, & M. H. Pesaran (Eds.), *Analysis of Panels and Limited Dependent Variable Models* (pp. 199–225). Cambridge University Press. doi:10.1017/CBO9780511493140.010

Klasen, S. (2019). What Explains Uneven Female Labor Force Participation Levels and Trends in Developing Countries? *The World Bank Research Observer*, *34*(2), 161–197. doi:10.1093/wbro/lkz005

Klebnikov, S. (2022). *Here's What The Crypto 'Bloodbath' Means For The Stock Market (It's Not Good)*. https://www.forbes.com/sites/sergeiklebnikov/2022/06/13/heres-what-the-crypto-bloodbath-means-for-the-stock-market-its-not-good/?sh=235b51af5fef

Klepper, S., & Nagin, D. (1989). The Role of Tax Preparers in Tax Compliance. *Policy Sciences, 22*(2), 167–194. doi:10.1007/BF00141383

Kmieciak, R., & Michna, A. (2018). Knowledge management orientation, innovativeness, and competitive intensity: Evidence from Polish SMEs. *Knowledge Management Research and Practice, 16*(4), 559–572. doi:10.1080/14778238.2018.1514997

Koenig, J., Rustan, K., & Leino, M. (2016). *Programming Language Features for Refinement.* Stanford University. doi:10.4204/EPTCS.209.7

Koenker, R. (2005). *Quantile Regression, Econometric Society Monographs.* Cambridge University Press. doi:10.1017/CBO9780511754098

Koenker, R., & Bassett, G. Jr. (1978). Regression quantiles. *Econometrica, 46*(1), 33–50. doi:10.2307/1913643

KOF Swiss Economic Institute. (2022). *KOF Globalisation Index.* https://kof.ethz.ch/en/forecasts-and-indicators/indicators/kof-globalisation-index.html

Kolberg, D., & Zuhlke, D. (2015). Lean automation enabled by industry 4.0 technologies. *IFAC-PapersOnLine, 48*(3), 1870–1875. doi:10.1016/j.ifacol.2015.06.359

Kontolaimou, A., & Skintzi, G. (2018). 4.2. digitisation patterns of the Greek economy and society. *Greek Economic Outlook, 2018*(37), 41–48.

Kordab, M., Raudeliuniene, J., & Meidute-Kavaliauskien, I. (2020). Mediating Role of Knowledge Management in the Relationship between Organizational Learning and Sustainable Organizational Performance. *Sustainability, 12*(23), 1–20. doi:10.3390u122310061

Kotler, P., Kartajaya, H., & Setiawan, I. (2017). Pazarlama 4.0 Gelenekselden Dijitale Geçiş. Optimist Yayın Grubu.

Kotler, P., & Keller, K. L. (2006). *Marketing Management* (12th ed.). Prentice Hall.

Kovács, T. Z., Bittner, B., Huzsvai, L., & Nábrádi, A. (2022). Convergence and the Matthew Effect in the European Union Based on the DESI Index. *Mathematics, 10*(4), 613. doi:10.3390/math10040613

Kozak, M. A., Evren, S., & Çakır, O. (2013). Tarihsel Süreç İçinde Turizm Paradigması. *Anatolia:Turizm Araştırmaları Dergisi, 24*(1), 7–22.

Kozinets, R. (2009). *Netnography: Doing Ethnographic Research Online.* Sage.

Krause, M. J., & Tolaymat, T. (2018). Quantification of energy and carbon costs for mining cryptocurrencies. *Nat. Sustain., 1*(11), 711–718. doi:10.103841893-018-0152-7

Kretschmer, T., & Khashabi, P. (2020). Digital Transformation and Organization Design:An Integrated Approach. *California Management Review, 62*(4), 86–104. doi:10.1177/0008125620940296

Krieger-Boden, C., & Sorgner, A. (2018). Labor market opportunities for women in the digital age. *Economics-the Open Access Open-Assessment E-Journal, 12.*

Kritikos, A. S. (2014). Entrepreneurs and their impact on jobs and economic growth. *IZA World of Labor, 8.* Advance online publication. doi:10.15185/izawol.8

Küfeoğlu, S., & Özkuran, M. A. (2019). *Energy consumption of Bitcoin mining.* Cambridge Working Papers in Economics, University of Cambridge, Faculty of Economics.

Kumar, S., & Sharma, D. (2015). Factors Influencing ICT Development in BRICS Countries. *International Journal of Social and Organizational Dynamics in IT*, *4*(2), 30–40. doi:10.4018/IJSODIT.2015070103

Kuo, Y. K., Kuo, T. H., & Ho, L. A. (2014). Enabling innovative ability: Knowledge sharing as a mediator. *Industrial Management & Data Systems*, *114*(5), 696–710. doi:10.1108/IMDS-10-2013-0434

Kurjono, K., Mulyani, H., & Murtadlo, Y. (2020). The Effect of Psychological Factors on Entrepreneurial Intention. *Advances in Economics, Business and Management Research*, *117*, 150–152. doi:10.2991/aebmr.k.200131.032

Kurniawati, M. A. (2022). Analysis of the impact of information communication technology on economic growth: Empirical evidence from Asian countries. *Journal of Asian Business and Economic Studies*, *29*(1), 2–18. doi:10.1108/JABES-07-2020-0082

Laal, M. (2011). Knowledge management in higher education. *Procedia Computer Science*, *3*(1), 544–549. doi:10.1016/j.procs.2010.12.090

Lanzolla, G., & Anderson, J. (2008). Digital transformation. *Business Strategy Review*, *19*(2), 72–76. doi:10.1111/j.1467-8616.2008.00539.x

Latwal, G. S., Kommers, P., Mahajan, P., & Sharma, S. K. (2020). *Role of ICT in Higher Education: Trends, Problems and Prospects*. Apple Academic Press. doi:10.1201/9781003130864

Lazanyuk, I., & Revinova, S. (2019). Digital economy in the BRICS countries: myth or reality? In *International scientific and practical conference on digital economy (ISCDE 2019)* (pp. 227-231). Atlantis Press. 10.2991/iscde-19.2019.97

Lee, H., & Choi, B. (2003). Knowledge management enablers, processes, & organizational performance: An integrative view & empirical examination. *Journal of Management Information Systems*, *20*(1), 179–228. doi:10.1080/07421222.2003.11045756

Lee, N., & Rodríguez-Pose, A. (2021). Entrepreneurship and the fight against poverty in US cities. *Environment and Planning A. Economy and Space*, *53*(1), 31–52. doi:10.1177/0308518X20924422

Legalbank. (n.d.). https://legalbank.net/belge/kurumlar-vergisi-kanunu/1350080/kurumlar+vergisi

Leong, C. T., & Rasli, A. (2014). The Relationship between innovative work behavior on work role performance: An empirical study. *Procedia: Social and Behavioral Sciences*, *129*(1), 592–600. doi:10.1016/j.sbspro.2014.03.717

Levin, A., Lin, C. F., & Chu, C. (2002). Unit root tests in panel data: Asymptotic and finite-sample properties. *Journal of Econometrics*, *108*(1), 1–24. doi:10.1016/S0304-4076(01)00098-7

Lewin, C., Griffith, D., Assche, F. V., Anido, L., & McNicol, S. (2015). *Re-engineering the Uptake of ICT in Schools*. Springer International Publishing.

Liao, S.-H., Fei, W.-C., & Chen, C.-C. (2007). Knowledge Sharing, Absorptive Capacity, and Innovation Capability: An Empirical Study of Taiwan's Knowledge-Intensive Industries. *Journal of Information Science*, *33*(3), 340–359. doi:10.1177/0165551506070739

Li, F. (2020). The digital transformation of business models in the creative industries: A holistic framework and emerging trends. *Technovation*, *92*(1), 1–14. doi:10.1016/j.technovation.2017.12.004

Li, J., Li, N., Peng, J., Cui, H., & Wu, Z. (2019). Energy consumption of cryptocurrency mining: A study of electricity consumption in mining cryptocurrencies. *Energy*, *168*, 160–168. doi:10.1016/j.energy.2018.11.046

Lin, H. F. (2007). Knowledge sharing and firm innovation capability: An empirical study. *International Journal of Manpower*, *28*(3), 315–332. doi:10.1108/01437720710755272

Li, T., & Chen, Y. (2019). Will Virtual Reality be a Double-edged Sword? Exploring the Moderation Effects of the Expected Enjoyment of a Destination on Travel Intention. *Journal of Destination Marketing & Management*, *12*, 15–26. doi:10.1016/j.jdmm.2019.02.003

Litvin, S., Goldsmith, R., & Pan, B. (2008). Electronic word-of-mouth in hospitality and tourism management. *Tourism Management*, *29*(3), 458–468. doi:10.1016/j.tourman.2007.05.011

Liu, A. (2022). *Rumbaugh, Booch and Jacobson Methodologies*. Opengenus. https://iq.opengenus.org/rumbaugh-booch-and-jacobson-methodologies/

Liu, C., & Bhargava, R. (2022). *6 Experts Reveal the Technologies Set to Revolutionize Cancer Care*. World Economic Forum, Project on Health and Healthcare, Beijing China and The Fourth Industrial Revolution for Health Project, India. Retrieved from https://www.weforum.org/agenda/2022/02/cancer-care-future-technology-experts/

Liu, Y., & Phillips, J. S. (2011). Examining the antecedents of knowledge sharing in facilitating team innovativeness from a multilevel perspective. *International Journal of Information Management*, *31*(1), 44–52. doi:10.1016/j.ijinfomgt.2010.05.002

López, V. M. (2013). Leadership in organization knowledge to Mexico. *Procedia: Social and Behavioral Sciences*, *73*, 661–668. doi:10.1016/j.sbspro.2013.02.103

Lori, J. R., Munro, M. L., Boyd, C. J., & Andreatta, P. (2012). Cell Phones to Collect Pregnancy Data from Remote Areas in Liberia. *Journal of Nursing Scholarship*, *44*(3), 294–301. doi:10.1111/j.1547-5069.2012.01451.x PMID:22672157

Loukis, E., Kyriakou, N., Pazalos, K., & Popa, S. (2017). Inter-organizational innovation and cloud computing. *Electronic Commerce Research*, *17*(3), 379–401. doi:10.100710660-016-9239-2

Lucas, R. E. Jr. (1988). On the mechanics of economic development. *Journal of Monetary Economics*, *22*(1), 3–42. doi:10.1016/0304-3932(88)90168-7

Luck, D., & Lancester, G. (2003). E-CRM Customer Relationship Marketing in the Hotel Industry. *Managerial Auditing Journal*, *18*(3), 213–231. doi:10.1108/02686900310469961

Lv, Z. K., & Yang, R. D. (2018). Does women's participation in politics increase female labor participation? Evidence from panel data analysis. *Economics Letters*, *170*, 35–38. doi:10.1016/j.econlet.2018.05.013

Maccurtain, S., Flood, P. C., Ramamoorthy, N., West, M. A., & Dawson, J. F. (2008). *Top team trust, knowledge sharing and innovation*. Link Working Paper Series WP 05-08, 1-26.

Machado, J. A., & Silva, J. M. C. S. (2013). *Quantile regression and heteroskedasticity* [Unpublished manuscript]. Department of Economics, University of Essex.

Magruk, A. (2016). Uncertainty In The Sphere Of The Industry 4.0-Potential Areas To Research. Business, Management & Education / Verslas. *Vadyba Ir Studijos*, *14*(2), 275–291.

Majchrzak, A., Cooper, L. P., & Neece, O. (2004). Knowledge reuse for innovation. *Management Science*, *50*(2), 174–188. doi:10.1287/mnsc.1030.0116

Majdańska, A. (2021). *An Analysis of Cooperative Compliance Programmes*. IBFD.

Majeed, M. T., & Ayub, T. (2018). Information and communication technology (ICT) and economic growth nexus: A comparative global analysis. *Pakistan Journal of Commerce and Social Sciences*, *12*(2), 443–476.

Makun, K., & Jayaraman, T. K. (2022). *Tourism and economic growth in South Asian countries: Asymmetric analysis and lessons for mitigating the adverse effects of Covid-19 pandemic*. Academic Press.

Maneejuk, P., & Yamaka, W. (2020). An analysis of the impacts of telecommunications technology and innovation on economic growth. *Telecommunications Policy*, *44*(10), 102038. doi:10.1016/j.telpol.2020.102038

Mano, R. (2021). The institutionalization of ICT and civic participation: Evidence from eight European nations. *Technology in Society*, *64*, 101518. doi:10.1016/j.techsoc.2020.101518

Marasco, A., Buonincontri, P., Van Niekerk, M., Orlowski, M., & Okumus, F. (2018). Exploring the Role of Next-Generation Virtual Technologies in Destination Marketing. *Journal of Destination Marketing & Management*, *9*, 138–148. doi:10.1016/j.jdmm.2017.12.002

Mariscal, J., Mayne, G., Aneja, U., & Sorgner, A. (2019). Bridging the gender digital gap. *Economics, 13*(1).

Markham, A. N. (2004). Internet communication as a tool for qualitative research. In D. Silverman (Ed.), *Qualitative Research: Theory, Method and Practice*. SAGE. doi:10.4135/9781848608191.d28

Mark, W. (2006). *Laptops and Literacy*. Teachers' College Press.

Marsick, V. J., & Watkins, K. E. (2003). Demonstrating the Value of an Organization's Learning Culture: The Dimensions of the Learning Organization Questionnaire. *Advances in Developing Human Resources*, *5*(2), 132–151. doi:10.1177/1523422303005002002

Martindale, J. (2020). *What is Bitcoin Mining?* 15 Mayıs 2022, https://www.digitaltrends.com/computing/what-is-bitcoin-mining/

Martínez-Rodriguez, I., Callejas-Albiñana, F., & Callejas-Albiñana, A. (2020). Economic and socio-cultural drivers of necessity and opportunity entrepreneurship depending on the business cycle phase. *Journal of Business Economics and Management*, *21*(2), 373–394. doi:10.3846/jbem.2020.11848

Martini, M. (2020). Introduction: An Investigation of Co-operative Compliance Regimes and the ICAP. In Co-operative Compliance and the OECD's International Compliance Assurance Programme. Kluwer Law International BV.

Matejovsky, L., Mohapatra, S., & Steiner, B. (2014). The dynamic effects of entrepreneurship on regional economic growth: Evidence from Canada. *Growth and Change*, *45*(4), 611–639. doi:10.1111/grow.12055

Matt, C., Hess, T., & Benlian, A. (2015). Digital transformation strategies. *Business & Information Systems Engineering*, *57*(5), 339–343. doi:10.100712599-015-0401-5

Mayer, W., Madden, G., & Wu, C. (2020). Broadband and economic growth: A reassessment. *Information Technology for Development*, *26*(1), 128–145. doi:10.1080/02681102.2019.1586631

Meadows, J., & Leask, M. (2012). *Teaching and Learning Using ICT in the Primary School*. Taylor & Francis. doi:10.4324/9780203137048

Medical assessment. (n.d.). *Workplacetesting.com*. Retrieved July 24, 2022, from https://www.workplacetesting.com/definition/17/medical-assessment-occupational-health-and-safety

Medical Education. (n.d.). In *The Free Dictionary by Farlex*. Retrieved July 25, 2022, from https://medical-dictionary.thefreedictionary.com/medical+education

Mehrabani, E. S., & Mohamad, N. A. (2011). Identifying the important factors influencing the implementation of succession planning. *International Conference on Information and Finance, 21*, 37-41.

Mehta, G. B. (1998). Preference and Utility. In S. Barberà, P. J. Hammond, & H. Seidl (Eds.), *Handbook of Utility Theory* (pp. 1–47). Kluwer Academic Publishers.

Méndez-Picazo, M., Galindo-Martín, M., & Castaño-Martínez, M. (2021). Effects of sociocultural and economic factors on social entrepreneurship and sustainable development. *Journal of Innovation & Knowledge, 6*(2), 69–77. doi:10.1016/j.jik.2020.06.001

Menezes, E. A. C., Guimarães, T. A., & Bido, D. S. (2011). Dimensions of learning in organizations: Validation of the Dimensions of the Learning Organization Questionnaire (DLOQ) in the Brazilian context. *RAM. Revista de Administração Mackenzie, 12*(2), 4–29. doi:10.1590/S1678-69712011000200002

Michna, A. (2018). The mediating role of firm innovativeness in the relationship between knowledge sharing and customer satisfaction in SMEs. *The Engineering Economist, 29*(1), 93–103.

Mincer, J. (1962). Labour force participation of married women: A study of labour supply. Aspects of labour economics. In *A conference of the universities: National bureau committee for economic research* (pp. 63–105). Princeton University Press.

Minges, M. (2015). *Exploring the relationship between broadband and economic growth*. Academic Press.

Minghui, L. (2016). The Construction of Learning Oriented Enterprise Organization Based on Knowledge Management. *International Journal of Elementary Education., 5*(4), 40–46. doi:10.11648/j.ijeedu.20160504.11

Mirzaee, S., & Ghaffari, A. (2018). Investigating the impact of information systems on knowledge sharing. *Journal of Knowledge Management, 22*(3), 501–520. doi:10.1108/JKM-08-2017-0371

Mitchell, M., Hedt-Gauthier, B. L., Msellemu, D., Nkaka, M., & Lesh, N. (2013). Using Electronic Technology to Improve Clinical Care – Results from a before-after cluster trial to evaluate assessment and classification of sick children according to Integrated Management of Childhood Illness (IMCI) protocol in Tanzania. *BMC Medical Informatics and Decision Making, 13*(1), 95. doi:10.1186/1472-6947-13-95 PMID:23981292

Mohajan, H. (2017). The roles of knowledge management for the development of organizations. *Journal of Scientific Achievements, 2*(2), 1–27.

Molose, T., & Ezeuduji, I. O. (2015). Knowledge sharing, team culture and service innovation in the hospitality sector : The case of South Africa. *African Journal of Hospitality, Tourism and Leisure, 4*(1), 1–16.

Morabito, V. (2016). *The Future of Digital Business Innovation*. Springer International Publishing. doi:10.1007/978-3-319-26874-3

Mora, C., Rollins, R. L., Taladay, K., Kantar, M. B., Chock, M. K., Shimada, M., & Franklin, E. C. (2018). Bitcoin Emissions Alone Could Push Global Warming Above 2'C. *Nature Climate Change, 8*(11), 931–933. doi:10.103841558-018-0321-8

Morales-Gómez, D., & Melesse, M. (1998). Utilising information and communication technologies for development: The social dimensions. *Information Technology for Development, 8*(1), 3–13. doi:10.1080/02681102.1998.9525287

Morrar, R., Arman, H., & Mousa, S. (2017). The fourth industrial revolution (Industry 4.0): A social innovation perspective. *Technology Innovation Management Review, 7*(11), 12–20. doi:10.22215/timreview/1117

Morris, T. (2022). Re-examining female labor supply responses to the 1994 Australian pension reform. *Review of Economics of the Household, 20*(2), 419–445. doi:10.100711150-021-09561-y

Motyl, B., Baronio, G., Uberti, S., Speranza, D., & Filippi, S. (2017). How will change the future engineers' skills in the Industry 4.0 framework? A questionnaire survey. *Procedia Manufacturing, 11*, 1501–1509. doi:10.1016/j.promfg.2017.07.282

Mura, M., Lettieri, E., Radaelli, G., & Spiller, N. (2013). Promoting professionals' innovative behaviour through knowledge sharing: The moderating role of social capital. *Journal of Knowledge Management, 17*(4), 527–544. doi:10.1108/JKM-03-2013-0105

Murphy, K. (2004). The Role of Trust in Nurturing Compliance: A Study of Accused Tax Avoiders. *Law and Human Behavior, 28*(2), 187–209. doi:10.1023/B:LAHU.0000022322.94776.ca PMID:15141778

Murray, K., & Waller, R. (2007). Social networking goes abroad. *International Educator, 16*(3), 56–59.

Nabi, A. A., Tunio, F. H., Azhar, M., Syed, M. S., & Ullah, Z. (2022). Impact of Information and Communication Technology, Financial Development, and Trade on Economic Growth: Empirical Analysis on N11 Countries. *Journal of the Knowledge Economy*, 1–18. doi:10.100713132-022-00890-6

Nair, M., Pradhan, R. P., & Arvin, M. B. (2020). Endogenous dynamics between R&D, ICT and economic growth: Empirical evidence from the OECD countries. *Technology in Society, 62*, 101315. doi:10.1016/j.techsoc.2020.101315

Nambisan, S., Lyytinen, K., Majchrzak, A., & Song, M. (2017). Digital innovation management: Reinventing innovation management research in a digital world. *Management Information Systems Quarterly, 41*(1), 223–238. doi:10.25300/MISQ/2017/41:1.03

Nam, S. (1991). Determinants of Female Labor-Force Participation - a Study of Seoul, South-Korea, 1970-1980. *Sociological Forum, 6*(4), 641–659. doi:10.1007/BF01114405

Nasab, E. H., & Aghaei, M. (2009). The effect of ICT on economic growth: Further evidence. *International Bulletin of Business Administration, 5*(2), 46–56.

National Institute of Mental Health. (2019, September). *Technology and the Future of Mental Health Treatment*. Retrieved July 18, 2022, from https://www.nimh.nih.gov/health/topics/technology-and-the-future-of-mental-health-treatment

Ndung'u, N. (2019). Digital technology and state capacity in Kenya. Academic Press.

Neumann, T. (2021). The impact of entrepreneurship on economic, social and environmental welfare and its determinants: A systematic review. *Management Review Quarterly, 71*(3), 553–584. doi:10.100711301-020-00193-7

Ngoa, G. B. N., & Song, J. S. (2021). Female participation in African labor markets: The role of information and communication technologies. *Telecommunications Policy, 45*(9).

Nguyen, T. P., Dinh, T. T., Tran Ngoc, T., & Duong Thi Thuy, T. (2022). Impact of ICT diffusion on the interaction of growth and its volatility: Evidence from cross-country analysis. *Cogent Business & Management, 9*(1), 1–27. doi:10.1080/23311975.2022.2054530

Nguyen, T. T., Pham, T. A. T., & Tram, H. T. X. (2020). Role of information and communication technologies and innovation in driving carbon emissions and economic growth in selected G-20 countries. *Journal of Environmental Management, 261*, 110162. doi:10.1016/j.jenvman.2020.110162 PMID:32148259

Ngwenyama, O., Andoh-Baidoo, F. K., Bollou, F., & Morawczynski, O. (2006). Is There A Relationship Between ICT, Health, Education and Development? An Empirical Analysis of five West African Countries from 1997–2003. *Journal of Information Systems for Developing Country*, *23*(1), 1–11. doi:10.1002/j.1681-4835.2006.tb00150.x

Nicole. (2021, October 26). *The Growing Importance of SEO In Tourism Marketing*. Retrieved August 2022, from breaktheicemedia.com: https://breaktheicemedia.com/seo-in-tourism-marketing/

Niebel, T. (2018). ICT and economic growth – Comparing developing, emerging and developed countries. *World Development*, *104*(April), 197–211. doi:10.1016/j.worlddev.2017.11.024

Nikolov, H. S., & Krumova, M. Y. (2019). Hofstede's model in the context of e-government and open government in EU countries: Countries clustering based on similarities and differences *Smart Cities and Regional Development (SCRD). Journal*, *3*(1), 29–46.

Niles, S., & Hanson, S. (2003). A new era of accessibility. *URISA J.*, *15*, 35–41.

Nishikawa, S., & Higuchi, Y. (1981). Determinants of Female Labor-Force Participation. *Japanese Economic Studies*, *9*(2), 62–87. doi:10.2753/JES1097-203X090262 PMID:12313523

Nizami, U., Niazi, M. F. K., & Kadri, A. S. (2022). Impact of ICT Investment and Diffusion on Economic Growth: A Case Study of Pakistan. *Advances in World Economics*, *1*(1), 12–20.

Nodehi, H., Nehardani, H., & Keavanlou, F. (2013). The Relationship between Learning Organization and Knowledge Management. *International Research Journal of Applied and Basic Sciences*, *5*(10), 1303–1306.

Noguera, J. A., Quesada, F. J. M., Tapia, E., & Llàcer, T. (2014). Tax Compliance, Rational Choice, and Social Influence: An Agent-based Model. *Revue Française de Sociologie*, *55*(4), 765–804. doi:10.3917/rfs.554.0765

Nouri, B. A., Ghorbani, R., & Soltani, M. (2017). The Effect of Knowledge Management on Organizational Innovation with the Mediating Role of Organizational Learning (Case Study: Agricultural Bank in Iran). *Journal of Applied Economics and Business Research*, *7*, 194–211.

Nylén, D., & Holmström, J. (2015). Digital innovation strategy: A framework for diagnosing and improving digital product and service innovation. *Business Horizons*, *58*(1), 57–67. doi:10.1016/j.bushor.2014.09.001

O'Riordan, B. (2021). INNOVATION-Why Transformations Fail And How They Can Succeed With People Power. *Forbes*. https://www.forbes.com/sites/unit4/2021/10/11/why-transformations-fail-and-how-they-can-succeed-with-people-power/#:~:text=It%20is%20a%20bold%20decision,all%20major%20transformation%20projects%20fail

Obitko, M., & Jirkovský, V. (2015). Big data semantics in industry 4.0. In *International conference on industrial applications of holonic and multi-agent systems* (pp. 217-229). Springer.

Odhiambo, N. M. (2022). Information technology, income inequality and economic growth in sub-Saharan African countries. *Telecommunications Policy*, *46*(6), 102309. doi:10.1016/j.telpol.2022.102309

OECD. (2001). *Information and Communication Technologies and Rural Development*. doi:10.1787/9789264193079-en

OECD. (2013). *Co-operative Compliance: A Framework: From Enhanced Relationship to Co-operative Compliance*. OECD Publishing.

OECD. (2018). *Bridging the digital gender divide: Include, upskill, innovate*. Available at: www.oecd. org/digital/bridging-the-digital-gender-divide.pdf

OECD. (2019). *International Compliance Assurance Programme Pilot Handbook 2.0*. OECD.

OECD. (2021). *International Compliance Assurance Programme – Handbook for Tax Administrations and MNE Groups*. OECD.

OHCHR. (2017). *Report on ways to bridge the gender digital divide from a human rights perspective*. Available at: www.ohchr.org/en/issues/women/wrgs/pages/waystobridgethegenderdigital.aspx

OIC. (2022). *OIC Statistics Database (OICStat) - Query*. Retrieved from The Statistical, Economic and Social Research and Training Centre for Islamic Countries (SESRIC): https://www.sesric.org/query.php

Omarini, A. (2017). The Digital Transformation in Banking and The Role of FinTechs in the New Financial Intermediation Scenario. *Int J Financ Econ Trade.*, *1*(1), 1–6.

Öneren, M., Çiftçi, G. E., & Harman, A. (2016). Bilgi Paylaşiminin Yenilikçi Davranişa Ve Örgütsel Güvene Etkisi Üzerine Bir Araştirma. *Akademik Bakiş Dergisi*, *58*(1), 127–157.

Organisation for Economic Co-operation and Development. (2018). *Bridging the digital gender divide: include, upskill, innovate*. Retrieved August 21, 2022, from https://www.oecd.org/going-digital/bridging-the-digital-gender-divide.pdf

Orton, M., Agarwal, S., Muhoza, P., Vasudevan, L., & Vu, A. (2018). Strengthening delivery of health services using digital devices. *Global Health, Science and Practice*, *6*(suppl 1), S61–S71. doi:10.9745/GHSP-D-18-00229 PMID:30305340

Osipovskaya, A. V., & Mikhaylin, A. V. (2017). The Development of Financial Technologies in the field of banking services: The main directions. *Young Scientist*, *26*, 124–127.

Osundina, O. A. (2020). Sustainable development: Does improvement in education and health of women improve female labour force participation rate? *Sustainable Development*, *28*(1), 13–24. doi:10.1002d.1961

Ouakouak, M. L., & Ouedraogo, N. (2019). Fostering knowledge sharing and knowledge utilization: The impact of organizational commitment and trust. *Business Process Management Journal*, *25*(4), 757–779. doi:10.1108/BPMJ-05-2017-0107

Özkaya, M. H., & Alhuwesh, M. A. (2017). Determinants of the Republic of Yemen's Foreign Trade with the Gulf Cooperation Council Countries: Gravity Model. In *International Conference on Scientific Cooperation for the Future in the Economics and Administrative Sciences* (pp. 79-88). Thessaloniki: Usak University & Aristotle University of Thessaloniki.

Ozkoc, H. H., & Karalar, H. (2019). K12 ve lisans öğrencilerinin Endüstri 4.0 kavramına ilişkin algıları. *Eskişehir Osmangazi Üniversitesi Sosyal Bilimler Dergisi*, *20*, 243–258.

Padmaja, R., Rifana Meera, M., & Valli Mayil, M. (2017). Employee's Attitude towards Adoption of Electronic Banking Services: A Case of Public Sector Banks, International Journal of Banking. *Risk and Insurance*, *5*(2), 11–14.

Pan, W., Xie, T., Wang, Z., & Ma, L. (2022). Digital economy: An innovation driver for total factor productivity. *Journal of Business Research*, *139*, 303–311. doi:10.1016/j.jbusres.2021.09.061

Papagiannakis, G., Ponder, M., Molet, T., Kshirsagar, S., Cordier, F., Thalmann, N., & Thalmann, D. (2002). LIFEPLUS: Revival of life in ancient Pompeii, Virtual Systems and Multimedia. *Proceedings of VSMM*, 1-11.

Papavassiliou, G., Ntioudis, S., Mentzas, G., & Abecker, A. (2001). The DECOR approach to Business Process Oriented Knowledge Management (BPOKM). In *DEXA '01 Proceedings of the 12th International Workshop on Database and Expert Systems Applications*. IEEE Computer Society. https://astimen.wordpress.com/2009/10/28/the-decor-approach-to-business-process-oriented-knowledge-management-bpokm/

Parida, V., Sjödin, D. & Reim, W. (2019). *Reviewing literature on digitalization, business model innovation, and sustainable industry: Past achievements and future promises*. Academic Press.

Pavel, F. (2011). Grid Database—Management, OGSA and Integration. Academy of Economic Studies Romania, Bucharest. Database Systems Journal, 2(2).

Pérez-Castro, M. Á., Mohamed-Maslouhi, M., & Montero-Alonso, M. Á. (2021). The digital divide and its impact on the development of Mediterranean countries. *Technology in Society*, *64*, 101452. doi:10.1016/j.techsoc.2020.101452

Pérez-Morote, R., Pontones-Rosa, C., & Núñez-Chicharro, M. (2020). The effects of e-government evaluation, trust and the digital divide in the levels of e-government use in European countries. *Technological Forecasting and Social Change*, *154*, 119973. doi:10.1016/j.techfore.2020.119973

Pesaran, M. H. (2004). *General diagnostic tests for cross section dependence in panels*. CESifo Working Paper No. 1229.

Pesaran, M. H. (2004). *General Diagnostic Tests for Cross Section Dependence in Panels*. University of Cambridge, Discussion Paper No. 1240, 1-42.

Pesaran, M. H. (2007). A simple panel unit root test in the presence of cross-section dependence. *Journal of Applied Econometrics*, *22*(2), 265–312. doi:10.1002/jae.951

Pesaran, M. H., Shin, Y., & Smith, R. P. (1999). Pooled Mean Group Estimation of Dynamic Heterogeneous Panels. *Journal of the American Statistical Association*, *94*(446), 621–634. doi:10.1080/01621459.1999.10474156

Pesaran, M. H., Ullah, A., & Yamagata, T. (2008). A bias-adjusted LM test of error cross-section independence. *The Econometrics Journal*, *11*(1), 105–127. doi:10.1111/j.1368-423X.2007.00227.x

Pesaran, M. H., & Yamagata, T. (2008). Testing slope homogeneity in large panels. *Journal of Econometrics*, *142*(1), 50–93. doi:10.1016/j.jeconom.2007.05.010

Peterson, S. (2011). *Why it Worked: Critical Success Factors of a Financial Reform Project in Africa*. Faculty Research Working Paper Series. Harvard Kennedy School.

Pfeiffer, S. (2017). The vision of 'Industrie 4.0' in the making a case of future told, tamed, and traded. *NanoEthics*, *11*(1), 107–121. doi:10.100711569-016-0280-3 PMID:28435474

Pittino, D., Martinez, A. B., Chirico, F., & Galvan, R. S. (2018). Psychological ownership knowledge sharing and entrepreneurial orientation in family firms: The moderating role of governance heterogeneity. *Journal of Business Research*, *84*(1), 312–326. doi:10.1016/j.jbusres.2017.08.014

Polemis, M. L., & Tsionas, M. G. (2021). The environmental consequences of blockchain technology: A Bayesian quantile cointegration analysis for Bitcoin. *International Journal of Finance & Economics*, 1–20. doi:10.1002/ijfe.2496

Porter, M., Sachs, J., & McArthur, J. (2001). *Executive summary: Competitiveness and stages of economic development*. Academic Press.

Posada, J., Toro, C., Barandiaran, I., Oyarzun, D., Stricker, D., de Amicis, R., & Vallarino, I. (2015). Visual computing as a key enabling technology for industrie 4.0 and industrial internet. *IEEE Computer Graphics and Applications*, *35*(2), 26–40. doi:10.1109/MCG.2015.45 PMID:25807506

Posner, R. A. (1998). *Economic Analysis of Law* (5th ed.). Aspen.

Pottle, J. (2019, October). Virtual Reality and the Transformation of Medical Education. *Future Healthcare Journal*, *6*(3), 181-185. Retrieved from https://www.ncbi.nlm.nih.gov/pmc/articles/PMC6798020/?report=classic

Powell. (2016). *Quantile regression with nonadditive fixed effects*. Available at: https://works.bepress.com/david_powell/1/

Pradhan, R. P., Arvin, M. B., & Normancd, N. R. (2015). The dynamics of information and communications technologies infrastructure, economic growth, and financial development: Evidence from Asian countries. *Technology in Society*, *42*, 135–149. doi:10.1016/j.techsoc.2015.04.002

Pradhan, R. P., Mallik, G., & Bagchi, T. P. (2018). Information communication technology (ICT) infrastructure and economic growth: A causality evinced by cross-country panel data. *IIMB Management Review*, *30*(1), 91–103. doi:10.1016/j.iimb.2018.01.001

Pratap & Predovich. (2020). *The 2020 Gartner Magic Quadrant for IT Risk Management*. Gartner.

Pross, A., & Johnson, M. (2020). International Compliance Assurance Programme. In Co-operative Compliance and the OECD's International Compliance Assurance Programme. Kluwer Law International BV.

Pross, A., Knaepen, S., & Johnson, M. (2017). Embracing Tax Certainty Through Improved Dispute Resolution. *Int'l Tax Rev.*, *28*, 16–22.

Pryadarshini, K., & Bhargava, R. (2021). *These AI Projects are Improving Cancer Screening and Outcomes*. World Economic Forum, Fourth Industrial Revolution for Health, India and Worldwide Health Microsoft Asia. Retrieved from https://www.weforum.org/agenda/2021/07/ai-projects-improving-cancer-screening-outcomes

Pshenichnikov, V. V. (2018). Influence of Financial Technologies on Changing the Model of Customer Banking Services. Theory and Practice of Service: Economics, Social sphere. *Technology*, *1*(35), 48–52.

Psychiatric Disorder. (n.d.). In *The Free Dictionary by Farlex*. Retrieved July 24, 2022, from https://medical-dictionary.thefreedictionary.com/Psychiatric+disorders

Puncreobutr, V. (2016). Education 4.0: New challenge of learning. St. Theresa. *Journal of the Humanities and Social Sciences*, *2*(2), 92–97.

Putniņš, T. J., & Sauka, A. (2020). Why does entrepreneurial orientation affect company performance? *Strategic Entrepreneurship Journal*, *14*(4), 711–735. doi:10.1002ej.1325

Quality of Care. (n.d.). *World Health Organization*. Retrieved July 24, 2022, from https://www.who.int/health-topics/quality-of-care#tab=tab

Radaelli, G., Lettieri, E., Mura, M., & Spiller, N. (2014). Knowledge sharing and innovative work behaviorin healthcare: A micro-level investigation of direct and indirect effects. *Creativity and Innovation Management*, *23*(4), 400–414. doi:10.1111/caim.12084

Ragnedda, M., & Kreitem, H. (2018). The three levels of digital divide in East EU countries. *World of Media. Journal of Russian Media and Journalism Studies*, *1*(4), 5–26.

Rao Jada, U., Mukhopadhyay, S., & Titiyal, R. (2019). Empowering leadership and innovative work behavior: A moderated mediation examination. *Journal of Knowledge Management*, *23*(5), 915–930. doi:10.1108/JKM-08-2018-0533

Ravi, C. S. (2017). Digital payments system and rural India: A review of transaction to cashless economy. *International Journal of Commerce and Management Research*, *3*(5), 170–173.

RedHat. (2022). *Decision Model and Notation (DMN)*. RedHat. https://access.redhat.com/documentation/en-us/red_hat_process_automation_manager/7.1/html/designing_a_decision_service_using_dmn_models/dmn-elements-example-con

Regulating Power Sector Carbon Emissions. (2022). https://www.c2es.org/content/regulating-power-sector-carbon-emissions/

Revinova, S. (2016). Information technology development in the BRICS countries-comparative analysis. In *3rd International Multidisciplinary Scientific Conference on Social Sciences and Arts SGEM* (*Vol. 5*, p. 353). Academic Press.

Ricciardi, W., Barros, P. P., Bourek, A., Brouwer, W., Kelsey, T., & Lehtonen, L. (2019). Expert Panel on Effective Ways of Investing in Health (EXPH), How to govern the digital transformation of health services. *European Journal of Public Health, 29*(3), 7–12. doi:10.1093/eurpub/ckz165 PMID:31738442

Rijsbergen, H. (2020). The ICAP Experience: From A Tax Authority Perspective. In Co-operative Compliance and the OECD's International Compliance Assurance Programme. Kluwer Law International BV.

Roche, E. M. (2016). Information and Communication Technology Still a Force for Good? *Journal of Global Information Technology Management, 19*(2), 75–79. doi:10.1080/1097198X.2016.1172952

Rogers, E. M. (2001). The digital divide. *Convergence, 7*(4), 96–111. doi:10.1177/135485650100700406

Rogers, E. M. (2008). Diffusion of innovations. In D. Stacks & M. Salwen (Eds.), *An integrated approach to communication theory and research* (pp. 409–420). Routledge.

Romer, P. M. (1986). Increasing returns and long-run growth. *Journal of Political Economy, 94*(5), 1002–1037. doi:10.1086/261420

Romer, P. M. (1990). Endogenous technological change. *Journal of Political Economy, 98*(5, Part 2), S71–S102. doi:10.1086/261725

Rose, S., Feldman, J., & Jankowski, J. (2009). A Cognitive Approach to the Development of Early Language. *Child Development, 80*(1), 134–150. doi:10.1111/j.1467-8624.2008.01250.x PMID:19236397

Russo, V. (2020). Digital Economy and Society Index (DESI). European guidelines and empirical applications on the territory. In *Qualitative and Quantitative Models in Socio-Economic Systems and Social Work* (pp. 427–442). Springer. doi:10.1007/978-3-030-18593-0_31

Sabbagh, K., Friedrich, R., El-Darwiche, B., Singh, M., & Koster, A. (2013). Digitization for Economic Growth and Job Creation: Regional and Industry Perspectives. In *The Global Information Technology Report 2013 Growth and Jobs in a Hyperconnected World* (pp. 35-42). World Economic Forum.

Saidi, K., Ben Hassen, L., & Hammami, M. S. (2015). Econometric Analysis of the Relationship Between ICT and Economic Growth in Tunisia. *Journal of the Knowledge Economy, 6*(4), 1191–1206. doi:10.100713132-014-0204-9

Salleh, K. (2012). Knowledge Management in Public University: Empirical Relationships Between Learning Organization and Knowledge Management Process of Tacit Knowledge. *Knowledge Management International Conference (KMICe) 2012*, (July), 4-6.

Sall, M. A. (2022). Entrepreneurship and poverty alleviation in Africa. *Industrial Policy, 2*(1), 33–37.

Samargandi, N., Al Mamun, M., Sohag, K., & Alandejani, M. (2019). Women at work in Saudi Arabia: Impact of ICT diffusion and financial development. *Technology in Society, 59*, 59. doi:10.1016/j.techsoc.2019.101187

Sangeeta, M., & Librarian, N. D. (2015). The importance of Knowledge Management in the Higher Educational Institutes. *Turkish Online Journal of Distance Education*, (January), 162-183.

Sankowska, A. (2013). Relationships between organizational trust, knowledge transfer, knowledge creation, and firm's innovativeness. *The Learning Organization*, *20*(1), 85–100. doi:10.1108/09696471311288546

Sart, G., Bayar, Y., Corpădean, A. G., & Gavriletea, M. D. (2022). Impact of ICT and Globalization on Educational Attainment: Evidence from the New EU Member States. *Sustainability*, *14*(3039), 1–13. doi:10.3390u14053039

Sart, G., Sezgin, F. H., & Demir, N. (2018). Mobbingin Mesleki Tükenmişlik Algısı Üzerine Etkileri: Kadın Akademisyenler Örneği. *Beykoz Akademi Dergisi*, *6*(1), 118–135. doi:10.14514/BYK.m.21478082.2018.6/1.117-135

Sassi, S., & Goaied, M. (2013). Financial development, ICT diffusion and economic growth: Lessons from MENA region. *Telecommunications Policy*, *37*(4-5), 252–261. doi:10.1016/j.telpol.2012.12.004

Satoshi Nakamoto. (2008). *Bitcoin: A Peer-to Peer Electronic Cash System.* https://bitcoin.org/bitcoin.pdf

Saure, P., & Zoabi, H. (2014). International trade, the gender wage gap and female labor force participation. *Journal of Development Economics*, *111*, 17–33. doi:10.1016/j.jdeveco.2014.07.003

Schwarz, P. (2012). Tax disincentives and female employment in Organisation for Economic Co-operation and Development (OECD) countries. *Journal of European Social Policy*, *22*(1), 17–29. doi:10.1177/0958928711425267

Scott, S. G., & Bruce, R. A. (1994). Determinants of innovative behavior: A path model of individual innovation in the workplace. *Academy of Management Journal*, *37*(3), 580–607. doi:10.2307/256701

Scupola, A. (2018). Digital transformation of public administration services in Denmark: A process tracing case study. *Nordic and Baltic Journal of Information and Communications Technologies*, *2018*(1), 261–284. doi:10.13052/nbjict1902-097X.2018.014

Sen, A. (2010). The Mobile and the World. *Information Technologies and International Development*, *6*, 1–3.

Sepehrdoust, H., & Ghorbanseresht, M. (2019). Impact of information and communication technology and financial development on economic growth of OPEC developing economies. *Kasetsart Journal of Social Sciences*, *40*(3), 546–551.

Serrano-Cinca, C., Muñoz-Soro, J. F., & Brusca, I. (2018). A multivariate study of internet use and the digital divide. *Social Science Quarterly*, *99*(4), 1409–1425. doi:10.1111squ.12504

Sezgin, F. H. (2016). Bayesci Faktör Analizi ve Maslach Tükenmişlik Envanteri Uygulaması, *International Conference on Scientific Cooperation for the Future in the Social Sciences (USAK) Bildiriler Kitabı*, 1283-1296.

Sezgin, F. H. (2016). Bayesci Faktör Analizi ve Maslach Tükenmişlik Envanteri Uygulaması. *International Conference on Scientific Cooperation for the Future in the Social Sciences (USAK) Bildiriler Kitabı*, 1283-1296.

Sharma, S. (1995). *Applied Multivariate Techniques*. Wiley.

Sheninger, E.C. (2019). *Digital Leadership: Changing Paradigms for Changing Times.* Corwin Publication.

Shettar, R. (2019). Digital Banking An Indian Perspective. *Journal of Economics and Finance*, *10*(3), 1–5.

Shodiev, T., Turayey, B., & Shodiyev, K. (2021). ICT and Economic Growth Nexus: Case of Central Asian Countries. *Procedia of Social Sciences and Humanities*, *1*, 155–167. doi:10.21070/pssh.v1i.37

Shrivastava, U., Ofstein, L., & Golhar, D. (2021). Direct and indirect effects of ICT infrastructure, skills, and use on entrepreneurship: A cross-country empirical investigation. *Journal of Global Information Management*, *29*(6), 1–25. doi:10.4018/JGIM.20211101.oa48

Singh, A. (2003). *Information and Communication Technologies (ICT) and Sustainable Development.* Development Alternatives.

Singh, S. K. (2011). Leadership and organizational learning in knowledge management practices in global organizations. *Indian Journal of Industrial Relations, 47*(2), 353–365.

Skinner, C. (2014). *Digital Bank: Strategies to Launch Digital or Become a Bank.* Marshall Cavendish Business Pbc.

Solow, R. (1987). We'd better watch out. *New York Times Book Review, 36.*

Solow, R. M. (1956). A contribution to the theory of economic growth. *The Quarterly Journal of Economics, 70*(1), 65–94. doi:10.2307/1884513

Somekh, B. (2007). *Pedagogy and Learning with ICT: Researching the Art of Innovation.* Taylor & Francis. doi:10.4324/9780203947005

Sousa, M. J., & Rocha, A. (2019). Digital learning: Developing skills for digital transformation of organizations Futur. *Gener. Comput. Syst., 91*(1), 327–334. doi:10.1016/j.future.2018.08.048

Spanuth, T., & Wald, A. (2017). How to unleash the innovative work behavior of project staff? The role of affective and performance-based factors. *International Journal of Project Management, 35*(7), 1302–1311. doi:10.1016/j.ijproman.2017.07.002

Srinivasan, T. N. (1973). Tax Evasion: A Model. *Journal of Public Economics, 2*(4), 339–346.

Standish. (2011). *The Chaos Reports.* http://www.standish.com

Stanley, T. D., Doucouliagos, H., & Steel, P. (2018). Does ICT Generate Growth? A Meta-regression Analysis. *Journal of Economic Surveys, 32*(3), 705–726. doi:10.1111/joes.12211

Stitt, G., Stitt, F., Vahid, W., & Najjar, W. (2006). A Code Refinement Methodology for Performance-Improved Synthesis from C. *International Conference on Computer-Aided Design (ICCAD'06),* San Jose, CA.

Stoica, E. A., & Bogoslov, I. A. (2017, December). A comprehensive analysis regarding DESI country progress for Romania relative to the European average trend. In *Balkan Region Conference on Engineering and Business Education* (Vol. 2, No. 1, pp. 258-266). 10.1515/cplbu-2017-0034

Stoll, C., Klaaben, L., & Gallersdorfer, U. (2019). The carbon footprint of bitcoin. *Joule, 3*(7), 1647–1661. doi:10.1016/j.joule.2019.05.012

Stylianou, V., & Savva, A. (2016). Investigating the knowledge management culture. *Universal Journal of Educational Research, 4*(7), 1515–1521. doi:10.13189/ujer.2016.040703

Suazo, M. (2021). *The argument for mining Bitcoin from clean energy and waste energy streams.* https://www.emergingenergyinsights.com/2021/09/the-argument-for-mining-bitcoin-from-clean-energy-and-waste-energy-streams

Suchman, L. A. (1987). *Plans and situated actions: The problem of human-machine communication.* Cambridge University Press.

Suhaida, M. A., Nurulhuda, M. S., & Yap, S.-F. (2013). Access to ICT as Moderating Factor to Women's Participation in the Labor Force: A Conceptual Framework. *International Journal of Trade. Economics and Finance, 4*(4), 197–201. doi:10.7763/IJTEF.2013.V4.285

Suh, M. G. (2017). Determinants of Female Labor Force Participation in South Korea: Tracing out the U-shaped Curve by Economic Growth. *Social Indicators Research, 131*(1), 255–269. doi:10.100711205-016-1245-1

Taiba, S., Zeng, J., Muhammad, B., & Nausheen, S. (2021). Impact of ICT on entrepreneurial self-efficacy in emerging economy: Sustaining lock-down during COVID-19 pandemic. *Human Systems Management*, *40*(2), 299–314. doi:10.3233/HSM-201066

Tam, H. (2011). U-shaped female labor participation with economic development: Some panel data evidence. *Economics Letters*, *110*(2), 140–142. doi:10.1016/j.econlet.2010.11.003

Telemedicine. (n.d.). In *Merriam-Webster Dictionary*. Retrieved July 24, 2022, from https://www.merriam-webster.com/dictionary/telemedicine

Temel, K. (2019). Endüstri 4.0 farkındalığının belirlenmesi: Çanakkale Onsekiz Mart Üniversitesi örneği. *Girişimcilik ve Kalkınma Dergisi*, *14*(1), 31–44.

Thambi, M., & O'Toole, P. (2012). Applying a knowledge management taxonomy to secondary schools. *School Leadership & Management*, *32*(1), 91–102. doi:10.1080/13632434.2011.642350

The Open Group. (2011a). *Introduction to the Architecture Development Method (ADM)*. The Open Group.

The Open Group. (2021). *TOGAF Model Patterns*. The Open Group. https://sparxsystems.com/enterprise_architect_user_guide/15.2/model_domains/togaf_model_patterns.html

The United Nations Women. (2022). *Women and the Sustainable Development Goals (SDGs)*. Retrieved August 7, 2022, from https://www. unwomen.org/en/news/in-focus/women-and-the-sdgs

The World Bank, World Development Indicators. (2022a). *Individuals using the Internet (% of population)*. Retrieved August 7, 2022, from https://databank.worldbank.org/source/world-developmentindicators

The World Bank, World Development Indicators. (2022b). *Labor force participation rate, female (% of female population ages 15+) (modeled ILO estimate)*. Retrieved August 7, 2022, from https://databank.worldbank.org/source/world-developmentindicators

Tjoa, A. M., & Tjoa, S. (2016, September). The role of ICT to achieve the UN Sustainable Development Goals (SDG). In *IFIP world information technology forum* (pp. 3–13). Springer. doi:10.1007/978-3-319-44447-5_1

Toda Y., & Yamamoto T., (1995). Statistical inference in vector autoregressions with possibly integrated processes. *Hournal of Econometrics,* (66), 225-250.

Torgler, B. (2007). *Tax Compliance and Tax Morale: A Theoretical and Empirical Analysis*. Edward Elgar. doi:10.4337/9781847207203

Torun, N., & Cengiz, E. (2019). Endüstri 4.0 bakış açısının öğrenciler gözünden teknoloji kabul modeli ile ölçümü. *Uluslararası İktisadi ve İdari İncelemeler Dergisi*, *22*, 235–250.

Trad, A., & Kalpić, D. (2020a). *Using Applied Mathematical Models for Business Transformation*. IGI Global. doi:10.4018/978-1-7998-1009-4

Trunflo, M., & Campana, S. (2020). A visitors' experience model for mixed reality in the museum. *Current Issues in Tourism*, *23*(9), 1053–1058. doi:10.1080/13683500.2019.1586847

Tsetsi, E., & Rains, S. A. (2017). Smartphone Internet access and use: Extending the digital divide and usage gap. *Mobile Media & Communication*, *5*(3), 239–255. doi:10.1177/2050157917708329

Turkey Transfer Pricing Country Profile. (n.d.). https://www.oecd.org/tax/transfer-pricing/transfer-pricing-country-profile-turkey.pdf

Türkiye Cumhuriyeti Cumhurbaşkanlığı Strateji ve Bütçe Başkanlığı, On Birinci Kalkınma Planı (2019-2023). (n.d.). https://www.sbb.gov.tr/wp-content/uploads/2022/07/On_Birinci_Kalkinma_Plani-2019-2023.pdf

Tüzemen, S., Barış-Tüzemen, Ö., & Çelik, A. K. (2021). The relationship between information and communication technologies and female labour force participation in Turkey. *Economics and Business Review*, *7*(4), 121–145. doi:10.18559/ebr.2021.4.7

Uberti, L. J., & Douarin, E. (2022). The Feminisation U, cultural norms, and the plough. *Journal of Population Economics*.

Ukommi, A.S., Okeagu, B.N. (2015). Information and communication technology (ICT) education and youth entrepreneurship in Uyo, Akwa Ibom State of Nigeria. *IOSR Journal of Humanities and Social Science*, *20*(7), 01-14. doi:10.9790/0837-20740114

Umans, T., Kockum, M., Nilsson, E., & Lindberg, S. (2018). Digitalisation in the banking industry and workers subjective well-being: Contingency perspective. *International Journal of Workplace Health Management*, *11*(6), 411–423. doi:10.1108/IJWHM-05-2018-0069

UN. (2021). *Gender equality*. Available at: www.un.org/en/global-issues/gender-equality

Unal, O. O. (2022, February 20). *Metaverse'te Tatil Deneyimi*. Retrieved September 2022, from YeniŞafak Teknoloji: https://www.yenisafak.com/teknoloji/metaversete-tatil-deneyimi-3755740

UNDP. (2015). *Human Development Report*. available at: https://hdr.undp.org/content/human-development-report-2015

UNICEF. (2021). *International day of the girl 2021 digital generation. Our generation*. Available at: www.unicef.org/gender-equality/international-day-girl-2021

United Nations Conference on Trade and Development. (2022). *Data Center*. https://unctadstat.unctad.org/wds/ReportFolders/reportFolders.aspx?sCS_ChosenLang=en

Urbinati, A., Chiaroni, D., Chiesa, V., & Frattini, F. (2018). The role of digital technologies in open innovation processes: An exploratory multiple case study analysis. *R & D Management*, *50*(1), 136–160. doi:10.1111/radm.12313

Usai, A., Fiano, F., Petruzzelli, A. M., Paoloni, P., Briamonte, M. F., & Orlando, B. (2021). Unveiling the impact of the adoption of digital technologies on firms; innovation performance. *Journal of Business Research*, *133*, 327–336. doi:10.1016/j.jbusres.2021.04.035

Usman, A., Ozturk, I., Hassan, A., Zafar, S. M., & Ullah, S. (2021). The effect of ICT on energy consumption and economic growth in South Asian economies: An empirical analysis. *Telematics and Informatics*, *58*(101537), 1–9. doi:10.1016/j.tele.2020.101537

Valberg, S. (2020). ICT, Gender, and the Labor Market: A Cross-Country Analysis. In D. Maiti, F. Castellacci, & A. Melchior (Eds.), *Digitalisation and Development Issues for India and Beyond* (pp. 375–405). Springer Nature. doi:10.1007/978-981-13-9996-1_15

Van de Ven, A. H. (1986). Central problems in the management of innovation. *Management Science*, *32*(1), 590–607. doi:10.1287/mnsc.32.5.590

Vega, S. H., & Elhorst, J. P. (2017). Regional labour force participation across the European Union: A time-space recursive modelling approach with endogenous regressors. *Spatial Economic Analysis*, *12*(2-3), 138–160. doi:10.1080/17421772.2016.1224374

Verhoef, P. C., Broekhuizen, T., Bart, Y., Bhattacharya, A., Qi Dong, J., Fabian, N., & Haenlein, M. (2021). Digital transformation: A multidisciplinary reflection and research agenda. *Journal of Business Research, 122*(1), 889–901. doi:10.1016/j.jbusres.2019.09.022

Verkerk, V.-A. (2022). Virtual Reality:A Simple Substitute or New Niche? In *Information and Communication Technologies in Tourism 2022*. Springer Nature Switzerland AG. doi:10.1007/978-3-030-94751-4_3

Vial, G. (2019). Understanding digital transformation: A review and a research agenda. *The Journal of Strategic Information Systems*, *28*(2), 118–144. doi:10.1016/j.jsis.2019.01.003

Viju, M. (2010). Women entrepreneurship in Middle East: Understanding barriers and use of ICT for entrepreneurship development. *The International Entrepreneurship and Management Journal*, *6*(2), 163–181. doi:10.100711365-010-0144-1

Viollaz, M., & Winkler, H. (2022). Does the Internet Reduce Gender Gaps? The Case of Jordan. *The Journal of Development Studies*, *58*(3), 436–453. doi:10.1080/00220388.2021.1965127

Virtual reality. (n.d.). In *Merriam-Webster Dictionary*. Retrieved July 24, 2022, from https://www.merriam-webster.com/dictionary/virtual%20reality

Vlahakis, V., Karigiannis, J., Tsotros, M., Gounaris, M., Almeida, L., Stricker, D., ... Ioannidis, N. (2001). Archeoguide: first results of an augmented reality, mobile computing system in cultural heritage sites. In *VAST November 28 - 30: Proceedings of the 2001 conference on Virtual reality, archeology, and cultural heritage* (pp. 131-140). Glyfada Greece: Association for Computing Machinery. 10.1145/584993.585015

Vogel-Heuser, B., & Hess, D. (2016). Guest editorial Industry 4.0-Prerequisites and visions. *IEEE Transactions on Automation Science and Engineering*, *13*(2), 411–413. doi:10.1109/TASE.2016.2523639

Vrasidas, C., & Glass, G. V. (2005). *Preparing Teachers to Teach with Technology*. Information Age Publishing.

Vu, K. M. (2011). ICT as a source of economic growth in the information age: Empirical evidence from the 1996–2005 period. *Telecommunications Policy*, *35*(4), 357–372. doi:10.1016/j.telpol.2011.02.008

Wagler, A., & Hanus, M. (2018). Comparing Virtual Reality Tourism to Real-Life Experience: Effects of Presence and Engagement on Attitude and Enjoyment. *Communication Research Reports*, *35*(5), 456–464. doi:10.1080/08824096.2018.1525350

Wallner, T., & Wagner, G. (2016). Academic education 4.0. *Proceedings of the international conference on education and new developments END 2016 icinde*, 155-159.

Walsh, K. (2012). Understanding Taxpayer Behaviour–New Opportunities for Tax Administration. *The Economic and Social Review*, *43*(3), 451–475.

Wamuyu, P. K. (2015). The impact of information and communication technology adoption and diffusion on technology entrepreneurship in developing countries: The case of Kenya. *Information Technology for Development*, *21*(2), 253–280. doi:10.1080/02681102.2014.948372

Wang, E. H. H. (1999). ICT and economic development in Taiwan: Analysis of the evidence. *Telecommunications Policy*, *23*(3-4), 235–243. doi:10.1016/S0308-5961(99)00005-1

Wangpipatwong, S. (2009). Factors influencing knowledge sharing among university students. *Proceedings of the 17th International Conference on Computers in Education, ICCE 2009*, 800-807.

Ward, K. B., & Pampel, F. C. (1985). Structural Determinants of Female Labor-Force Participation in Developed Nations, 1955-75. *Social Science Quarterly*, *66*(3), 654–667.

Watanabe, R. M., & Senoo, D. (2011). The role of a strong corporate mission for knowledge management. *Journal of Information & Knowledge Management, 10*(2), 109–122. doi:10.1142/S0219649211002869

Watson, T., Corliss, M., & Le, M. (2018). Digitalisation and Women's Workforce Participation in the Indo-Pacific. *Australian Journal of Labour Economics, 21*(1), 45–74.

WB. (2002). *Information and Communication Technologies: A World Bank Group Strategy.* The World Bank.

WB. (2012). *ICT for Greater Development Impact: World Bank Group Strategy for Information and Communication Technology, 2012-2015.* World Bank. Retrieved from https://openknowledge.worldbank.org/bitstream/handle/10986/27411/715400WP0WBG0I0sclosed0July02502012.pdf?sequence=1&isAllowed=y

WB. (2022). *Indicators.* Retrieved from World Bank Open Data: https://data.worldbank.org/indicator

Weinberg, B. D., & Pehlivan, E. (2011). Social spending: Managing the social media mix. *Business Horizons, 54*(3), 275–282. doi:10.1016/j.bushor.2011.01.008

Wei, S., Ren, G., & O'Neill, E. (2014). Haptic and audio displays for augmented reality tourism applications. *IEEE Haptics Symposium*, 485-488.

Wenzel, M. (2003). Tax Compliance and the Psychology of Justice: Mapping the Field. In Taxing Democracy: Understanding Tax Avoidance and Tax Evasion. Routledge.

Wenzel, M. (2002). The Impact of Outcome Orientation and Justice Concerns on Tax Compliance: The Role of Taxpayers' Identity. *The Journal of Applied Psychology, 87*(4), 629–645. doi:10.1037/0021-9010.87.4.629 PMID:12184568

Westerman, G., Bonnet, D., & McAfee, A. (2014). The nine elements of digital transformation. *MIT Sloan Management Review, 55*(3), 1–6.

White, M. (2012). Digital workplaces: Vision and reality. *Business Information Review, 29*(4), 205–214. doi:10.1177/0266382112470412

Woolfolk, A. (2010). *Educational Psychology* (9th ed.). Pearson Education Inc.

World Bank. (2002). *Information and Communication Technologies: A World Bank Group Strategy.* World Bank. https://openknowledge.worldbank.org/handle/10986/15243

World Bank. (2022). *World Bank Country and Lending Groups.* https://datahelpdesk.worldbank.org/knowledgebase/articles/906519-world-bank-country-and-lending-groups

World gross electricity production by source. (2022). https://www.iea.org/data-and-statistics/charts/world-gross-electricity-production-by-source-2019

Wu, W.-L., Lee, Y.-C., & Shu, H.-S. (2013). Knowledge management in educational organizations: A perspective of knowledge spiral. *The International Journal of Organizational Innovation, 5*(4), 7–13.

Xiang, Z. A., & Pan, B. (2011). Travel queries on cities in the United States: Implications for search engine marketing for tourist destinations. *Tourism Management, 32*(1), 88–97. doi:10.1016/j.tourman.2009.12.004

Xie, H. J., Miao, L., Kuo, P. J., & Lee, B. Y. (2011). Consumers' Responses to Ambivalent Online Hotel Reviews: The Role of Perceived Source Credibility and Predecisional Disposition. *International Journal of Hospitality Management, 30*(1), 178–183. doi:10.1016/j.ijhm.2010.04.008

Yang, D. (2011). The Effect of Knowledge Management on Product Innovation-Evidence from the Chinese Software Outsourcing Vendors. *iBusiness, 3*, 16-22.

Yang, L., Huang, C., & Hsu, T. (2014). Knowledge leadership to improve project and organizational performance. *International Journal of Project Management, 32*(1), 40–53. doi:10.1016/j.ijproman.2013.01.011

Yasmeen, R., & Tufail, S. (2015). Impact of Internet Technology on Economic Growth in South Asia with Special Reference to Pakistan. *Pakistan Journal of Social Sciences, 35*(2), 777–784.

Yazdanifard, R., & Yee, L. T. (2014). Impact of Social Networking Sites on Hospitality and Tourism Industries. *Global Journal of Human-Social Science: E Economics, 14*(8), 1-5.

Yazıcıoğlu, Y., & Erdoğan, S. (2004). *SPSS Uygulamalı Bilimsel Araştırma Yöntemleri*. Detay Yayıncılık.

Yıldız, M., Çakan, C., & Altınışık, E. (2022). The Impact of Covid-19 on Financial Markets and Investment Preferences. *Çankırı Karatekin University Journal of Institute of Social Sciences, 13*(1), 121-140.

Yirka, B. (2022). *Bitcoin carbon emissions rise as mining moves to US and other countries*. 20 Mayıs 2022, https://phys.org/news/2022-02-bitcoin-carbon-emissions-countries.html

Ylimäki, T. (2008). *Potential Critical Success Factors for Enterprise Architecture*. University of Jyväskylä, Information Technology Research Institute.

Yoo, Y., Boland, R. J. Jr, Lyytinen, K., & Majchrzak, A. (2012). Organizing for innovation in the digitized world. *Organization Science, 23*(5), 1398–1408. doi:10.1287/orsc.1120.0771

Yousefi, A. (2011). The impact of information and communication technology on economic growth: Evidence from developed and developing countries. *Economics of Innovation and New Technology, 20*(6), 581–596. doi:10.1080/10438599.2010.544470

Yuan, F., & Woodman, R. W. (2010). Innovative behaviorin the workplace: The role of performance and image outcome expectations. *Academy of Management Journal, 53*(2), 323–342. doi:10.5465/amj.2010.49388995

Yung, R., & Khoo-Lattimore, C. (2019). New realities: A systematic literature review on virtual reality and augmented reality in tourism research. *Current Issues in Tourism, 22*(17), 2056–2081. doi:10.1080/13683500.2017.1417359

Zhang, H., Fu, X., Cai, L. A., & Lu, L. (2014). Destination Image and Tourist Loyalty: A Meta-Analysis. *Tourism Management, 40*, 213–223.

Zhang, W., Sun, S. L., Jiang, Y., & Zhang, W. (2019). Openness to experience and team creativity: Effects of knowledge sharing and transformational leadership. *Creativity Research Journal, 31*(1), 62–73. doi:10.1080/10400419.2019.1577649

Zhao, S., Jiang, Y., Peng, X., & Hong, J. (2021). Knowledge sharing direction and innovation performance in organizations: Do absorptive capacity and individual creativity matter? *European Journal of Innovation Management, 24*(2), 371–394. doi:10.1108/EJIM-09-2019-0244

Zhou, J., & Shalley, C. E. (2003). Research on employee creativity: A critical review and directions for future research. *Research in Personnel and Human Resources Management, 22*, 165–218. doi:10.1016/S0742-7301(03)22004-1

Zurovac, D., Sudoi, R. K., Akhwale, W. S., Ndirtu, M., Hamer, D. H., Rowe, A. K., & Snow, R. W. (2011). The Effect of Mobile Phone Text-message reminders on Kenyan Health Workers' Adherence to Malaria Treatment Guidelines: A Cluster-randomized trial. *Lancet, 378*(9793), 795–803. doi:10.1016/S0140-6736(11)60783-6 PMID:21820166

About the Contributors

Yilmaz Bayar was awarded the PhD degree in the field of Economics in Istanbul University Institute of Social Sciences in 2012. He worked as an Assistant Professor in the Faculty of Business Administration, Karabuk University during the period September 2012-March 2015 and as an Associate Professor in the Faculty of Economics and Administrative Sciences, Usak University during the period March 2015-October 2020 and now working as a Professor in the Faculty of Economics and Administrative Sciences, Bandirma Onyedi Eylul University, since October 2020. His main research areas: Macroeconomics, financial economics, entrepreneurship, international economics, environmental economics, energy economics.

Lina Karabetyan completed her undergraduate education in Health Management Department of Faculty of Health Sciences in Marmara University. She did master of business administration degree (MBA) in Institute of Social Sciences in Istanbul University, and PhD at business administration in Istanbul Arel University. Her main research areas are information management, strategic management, technology and innovation management, management and strategy, human resources management, corporate social responsibility, organizational behavior, health institutions management, and international business. (646b09fe-b0b7-492d-a182-b3c93b397ea2)

* * *

Gülsüm Akarsu is working in Ondokuz Mayıs University as a research assistant. She completed her PhD in 2013 at Middle East Technical University, Ankara, Turkey. Her research interests cover Macroeconomics, Applied Econometrics, Financial Economics and Energy Economics.(4b9c5627-f8ea-4de4-a901-3d03a4dda0e8)

İmran Arıtı Erdem graduated from Ankara University, Faculty of Law in 2013. She awarded her master's degree in Public Finance in Ankara Yıldırım Beyazıt University, Institute of Social Sciences in 2018. She is currently a PhD candidate in the field of Fiscal Law in Ankara Yıldırım Beyazıt University, Institute of Social Sciences. She is working as a research assistant in Ankara Yıldırım Beyazıt University, Faculty of Political Sciences, Department of Public Finance since 2015.

Betül Garda awarded the PhD degree in the field of Production Management and Marketing from Selcuk University Social Sciences Institute in 2010. She worked as a lecturer at Selcuk University Social Sciences Vocational School between 1999-2015. She has been working as an Assistant Professor at Selcuk University, Social Sciences Vocational School, in the marketing and advertising department since 2015.(bb2d7eeb-dfbf-48cc-b181-0cbdda133afb)

Mitchell Alan Kaplan is a certified professional sociological practitioner in New York City. He received his doctorate in Sociology from the City University of New York Graduate Center in 1987. He also received a Postdoctoral Research Fellowship from the National Institute on Drug Abuse. Dr. Kaplan has spent the last 34 years of his professional career as a research scientist and consultant for several non-profit and city government agencies across the social service and public health spectrum. His articles and reviews have been featured in professional journals on substance abuse treatment, vocational rehabilitation, opioid abuse, and chronic pain, healthcare policy, criminal justice reform, global aging, and regional history, as well as academic textbooks, magazines on higher education and cultural healing in the Hispanic community, and in New York newspapers. You can contact Dr. Kaplan at his email address drmkaplan@aol.com.(a75c9ef6-4c8b-4004-a4db-2638c93c97d6)

Hakkı Odabaş graduated from Middle East Technical University, Faculty of Economic and Administrative Sciences, Department of Political Science and Public Administration in 1993. He completed the master's degree in Public Finance in Afyon Kocatepe University, Institute of Social Sciences in 1995 and awarded the PhD degree in the field of Public Finance in Dokuz Eylül University, Institute of Social Sciences in 2001. He worked as a research assistant in Afyon Kocatepe University, Faculty of Economics and Administrative Sciences, Department of Public Finance between 1993-2002, as an Assistant Professor during the period 2002-2006 and also worked as an Associate Professor in the Faculty of Economics and Administrative Sciences, Usak University, during the period 2006-2011. He is now working as a Professor in the Ankara Yıldırım Beyazıt University, Faculty of Political Sciences, Department of Public Finance since 2017. He also currently holds the position of dean of the same faculty. (9b5600c8-26d6-47a0-94df-f36a7618ef6b)

Gamze Sart was awarded the PhD degree in the field of Educational Sciences in Bogazici University in 2013. She worked as an Assistant Professor in the Faculty of Education, Yeditepe University in 2013 and Hasan Ali Yucel Faculty of Education, Istanbul University during the period 2014-2018 and as an Associate Professor in the Hasan Ali Yucel Faculty of Education, Istanbul University-Cerrahpaşa, since 2018. Her main research areas: Educational Studies, Higher Education Policies, Higher Education Management, Lifelong Learning, Sustainable Development. (15188ead-fdc7-45ee-9ad7-9d1aa1816b24)

Funda Hatice Sezgin completed her undergraduate education in Faculty of economics in University of Istanbul. She did master of business administration degree (MBA) in Econometrics of University of Istanbul, and doctor of econometrics in Marmara University. Her research interests are time series analysis, panel data analysis and statistical methods. She works in the department of industrial engineering at Istanbul University-Cerrahpasa.(5dac132a-39fe-4317-bb85-164e99d24a76)

Gokce Tekin Turhan completed her undergraduate education at Anadolu University, Faculty of Economics and Administrative Scıence, Department of Finance. She completed her master's degree in the Istanbul Ticaret University, Institute of Social Sciences, Departman of International Banking and Finance. She received her PhD from Kadir Has University, Institute of Social Sciences, Departman of International Banking and Finance. She works as a Assistant Professor Dr. at Nisantasi University. Her main research areas are international finance, behavioral finance and banking. (cc1e33fe-58a9-452e-a0f3-e237687687df)

Pınar Tokal, PhD completed her undergraduate education at Yildiz Technical University, Faculty of Arts & Science, Department of Mathematics. She completed her master's degree in the Department of Mathemetics at the same university and Vienna Technical University, Department Technical Mathematics. She received her doctorate from Istanbul University, Faculty of Economics, Department of Econometrics. She works as a Assistant Professor Dr. at Netkent Mediterranean Research and Science University, Cyprus. Her main research areas are artificial neural network, data mining, panel data analysis. (5f5e7b11-37a5-42e2-8448-77e1a9687fb3)

Aleyna Uca was born on January 4, 1999 in Ankara. She completed his primary and high school education here. She graduated from Gazi University Banking and Insurance Department in 2016 as the second in the department and the third in the higher school. She holds her master's degree from Marmara University, Institute of Banking and Insurance, on Capital Markets and the Stock Exchange.(2bf65300-4180-41fb-bbe7-447d9e5e0f34)

Oguzhan Yelkesen is a Research Assistant at the department of Public Finance, Bandirma Onyedi Eylul University. He has a bachelor's degree (Public Finance) from Ankara Yildirim Beyazit University and holds a Master's degree (Public Finance) from the same university. He has been doing his Ph.D. at the department of Public Finance, Hacettepe University. He is interested in public economics, fiscal policy, and sovereign debt.(5c3eb67c-eb9a-445e-8152-1f1b64413457)

Index

A

Advanced Ruling 268, 275
Arabian Peninsula Region Countries 59
Assessment 20-21, 24-26, 29, 32, 35, 37, 45, 54, 79, 91, 108, 195, 209, 228, 237, 253, 259-262, 264-269, 271
Augmented Reality (AR) 134-135, 140, 143-144, 147
Avoidance 237, 254, 270, 272-275

B

Banking Sector 171-173, 175, 178-179
Blockchain 139-140, 151-157, 160, 167-169, 178
BRICS 73, 75, 77-80, 83-90
Business Transformation Projects 220

C

Cancer 26-28, 34-36
Causality Analysis 92, 97-99
Clinical Trial 26, 32, 34, 36
Cluster Analysis 61, 88, 104, 110, 113, 116
Connectivity 110, 121, 172
Cooperative Compliance 252-260, 265, 267-275
Creativity and Innovation 207
Cryptocurrencies 142, 151-156, 159-160, 162-167, 169

D

DeFi 151
DESI 104, 106, 108-111, 115-120
Developed Country 121
Developing Country 119, 121
Digital Divide 4, 7, 13-14, 19, 39, 75, 85-87, 104-109, 111, 116-121
Digital Inclusion 13-14, 18, 118
Digital Marketing 134-140, 147, 149, 171
Digital Technology 19-20, 23-27, 29, 33-34, 36, 90, 105-106, 110, 125, 146
Digital Transformation 36, 108, 119, 123-124, 129-132, 134, 171-173, 178-180, 209-210, 212, 218-219
Digitalization 109, 122-126, 129-130, 132-133, 153, 171-173, 175, 178-180, 212
Distance Learning 20-21, 34, 37, 45, 55

E

Economic Growth 2-5, 7, 17, 19, 38, 59-77, 79, 81, 83-91, 93-94, 99, 101-102, 106, 108, 116, 129, 269
Economics of Marriage 4, 18
Educational Technology (Edtech) 58
Effective Learning 184
Enterprise Architecture 220, 226, 251
Entrepreneurial Activity 92-97, 99-100
Evasion 254-255, 270, 272-275
Extended Reality (ER) 146
Extrinsic Motivation 50-51, 58

F

Female Labor Force Participation 1-2, 14, 16-17
Female Labour Force Participation 1-2, 16-19
Feminisation U Hypothesis 3, 7, 16, 19

G

Gender Digital Divide 7, 13, 19, 105-107, 111, 116, 118-120
GII 104, 106, 109-111, 113, 115-116
Glass Ceiling 14, 19
Green Energy 151, 167

H

HDI 104, 106, 108-116
High Income Countries 92, 96
Human Capital 4, 19, 75, 84, 86-87, 107, 110, 121, 178

I

ICAP 252-254, 258-274
Inclusion 4-5, 13-15, 18, 24, 32, 41-42, 49, 58, 88, 105, 118
Industry 4.0 123-124, 129, 132, 204, 209-219
Industry 4.0 Conceptual Awareness 209-210, 213-216
Information and Communication Technologies 1, 17-19, 23, 34, 57, 69, 71, 73-74, 89-91, 103-104, 107, 116, 119-120, 148, 150
Innovation 2, 7, 26, 33, 40, 42, 53, 57-58, 70, 72, 74-75, 84-85, 87, 90, 92-94, 99-103, 122-126, 129-133, 147, 161, 179-180, 184, 192-199, 204-208, 218-219, 222, 241, 249
Innovative Behaviour 196, 200, 206-207
Instructional Design 43-44, 58
Intrinsic Motivation 51, 58

K

Knowledge Management 88, 120, 130, 181-188, 190-196, 205-207, 222, 234, 243, 250
Knowledge Sharing 191, 196-201, 203-208

L

Leapfrogging 13, 19, 87

M

Macro Level 107, 121
Medical Assessment 35, 37
Medical Diagnostics 20
Medical Education 20-23, 26, 33-37
Metaverse 134-135, 139-143, 146-149, 163
Micro Level 107, 121, 125
Mixed Reality (MR) 134-135, 140, 143, 145, 147
Multinational Enterprises (MNE's) 275

O

OECD 2-3, 13, 17, 78, 88-89, 91, 96, 101, 103, 105-106, 108, 119, 252-271, 273-275
Open-Ended Learning 47-48, 58
Organizational Learning 101, 181-186, 188, 190-195

P

Panel ARDL 73, 75, 82, 86
Panel Data 1, 3, 5-8, 10, 12-13, 15-19, 61-63, 69, 71, 77, 80-82, 88-89, 111, 122, 126, 132-133, 153
Panel Data Analysis 1, 17, 19, 69, 80, 88, 122, 133
Parent Company 276
Permanent Establishment 276
Problem Solving 39, 48, 51, 191, 240
Psychiatric Disorder 35, 37

Q

Quality of Care 36-37
Quantile Regression 104, 106, 109-111, 113, 115-116, 118-119, 171-172, 175-177, 179

R

Refine Process 220
Reinforcement Schedule 52, 58

S

Science and Technology 38-40, 47, 58, 122, 129, 134
Self-Regulation 41, 58
Social Advancement 38
Solow Residual 76, 91
Source Code Components 220
Structural Equation Model 145, 181, 187, 190, 196, 201
Sustainability 54, 68, 71, 95, 100-101, 116, 118, 135, 143, 148, 151-154, 156, 159-160, 162, 164-167, 169, 179-180, 191, 193, 206, 244, 249

T

Tax Authority 255, 260, 263, 267, 274, 276
Tax Compliance 252-257, 259, 264, 267, 269-276
TCDM 226, 229-233, 235, 240-244, 246, 248-249
Teachers' Attitudes and Beliefs 38
Technology Integration 38, 50, 53
Telemedicine 20-21, 34, 36-37
Theory of Investment in Human Capital 4, 19
Theory of Marriage 4, 15, 18
Theory of Time Allocation 4, 19
TOGAF 221-222, 224, 228-230, 232-233, 235, 242-243, 245-246, 251
Transcendent Software Engineering 220
Transfer Pricing 259, 262-265, 268-269, 271, 274-276
Treatment Services 20, 22, 24, 30, 33
Turkey 1, 3, 6, 14, 18, 59, 73, 92, 104, 107, 111, 122, 134, 151, 159, 171, 181, 209, 211-212, 216-217, 252-253, 267-272, 274

U

Universities 17, 49, 181, 184-186, 190-191, 197, 199-200, 204, 210-211, 216
University Students 208-209, 211-212, 214-217

V

Variation Rate 109, 111-112, 121
Virtual Reality (VR) 134, 140, 143, 147

W

WID 104, 106, 109-111, 113-116

www.ingramcontent.com/pod-product-compliance
Lightning Source LLC
LaVergne TN
LVHW062143300926
849161LV00010B/2611